# Human Biology

**Exclusively endorsed by OCR for GCE Human Biology**

YORK
COLLEGE

Heinemann is an imprint of Pearson Education Limited, a company incorporated in England and Wales, having its registered office at Edinburgh Gate, Harlow, Essex, CM20 2JE. Registered company number: 872828

www.heinemann.co.uk

Heinemann is a registered trademark of Pearson Education Limited

Text © Pearson Education Limited 2008

First published 2008

12 11 10 09
10 9 8 7 6 5 4 3

British Library Cataloguing in Publication Data is available from the British Library on request.

ISBN 978 0 435692 10 0

Edited by Jill Bailey
Designed by Wearset Ltd, Boldon, Tyne and Wear
Project managed and typeset by Wearset Ltd, Boldon, Tyne and Wear
Original illustrations © Pearson Education Limited 2008
Illustrated by Wearset Ltd, Boldon, Tyne and Wear
Cover photo © Steve Allen/Brand X/Corbis
Printed in China (SWTC/03)

### Acknowledgements

We would like to thank Professor Mick Fuller, Colin Geatrell, Dr Jennifer Marwick, Carol Pemberton and Anne Wilson for their invaluable help in the development of this course.

The authors and publisher would like to thank the following for permission to reproduce photographs:

**p2** JS Abbott; **p3** Dr. E. Walker/Science Photo Library; **p6 TL** Juergen Berger/Science Photo Library; **p6 ML** Carolina Biological Supply Company/Photolibrary; **p6 BL** PHOTOTAKE Inc./Alamy; **p6 BR** PHOTOTAKE Inc./Alamy; **p7** Dennis Kunkel/Photolibrary; **p15** Eye of Science/Science Photo Library; **p22** FreezeFrameStudio; **p26** Pearson Education Ltd. Jules Selmes. 2004/Cumulus; **p27** Microsan/Photolibrary; **p30 T** PhotoCreate/Shutterstock; **p30 B** The Stock Asylum, LLC/Alamy; **p35** Michael Ross/Science Photo Library; **p38** AJ Photo/Science Photo Library; **p40** Ian Hooton/Science Photo Library; **p43** CNRI/Science Photo Library; **p47** Richard Smith. 2004/Cumulus; **p48** Carolina Biological Supply Company/Photolibrary; **p49 TL** Susumu Nishinaga/Science Photo Library; **p49 BR** Biophoto Associates/Science Photo Library; **p53** Ian Hooton/Science Photo Library; **p54** Pearson Education Ltd. Gareth Boden. 2005/Cumulus; **p62 TL** Dr Gopal Murti/Science Photo Library; **p62 TR** Dr Gopal Murti/Science Photo Library; **p62 BL** Dr Gopal Murti/Science Photo Library; **p62 BR** Dr Gopal Murti/Science Photo Library; **p65** Scott Camazine/Alamy; **p70** ISM/Photolibrary; **p71** David Wall/Alamy; **p72 TL** Horizon International Images Limited/Alamy; **p72 ML** BSIP/Laurent/Muriel; **p72 BL** ImageState/Alamy; **p72 BR** Science Photo Library; **p73 L** Du Cane Medical Imaging Ltd/Science Photo Library; **p73 R** Corbis/Cumulus; **p76** Alamy. Carlos Davila/Cumulus; **p77** Getty Images. PhotoDisc/Cumulus; **p83** Adrian T Sumner/Science Photo Library; **p84 T** PhotoDisc/Cumulus; **p84 M** PhotoDisc/Cumulus; **p84 B** Medical-on-Line/Alamy; **p85** Dreamstime; **p86** Pauline Breijer/bigstock; **p89** ISM/Photolibrary; **p90** Saturn Stills/Science Photo Library; **p96** Zephyr/Science Photo Library; **p97** The Illustrated London News Picture Library/Cumulus; **p98** National Cancer Institute; **p100** Hank Morgan/Photolibrary; **p102** Geoffrey Kidd/Alamy; **p103** Robert Bird/Alamy; **p117** Pearson Education Ltd. Gareth Boden. 2005/Cumulus; **p118** Adam Hart-Davis/Science Photo Library; **p140** Homas Deerinck, NCMIR/Science Photo Library; **p141** CNRI/Science Photo Library; **p149** Matt Dunham/Associated Press; **p152** Dr Tim Evans/Science Photo Library; **p156** Eye of Science/Science Photo Library; **p164** Professor P.M. Motta et al/Science Photo Library; **p172** Hank Morgan/Photolibrary; **p174** Ed Reschke/Photolibrary; **p182** Karel Prinsloo/Associated Press; **p185** Pascal Goetgheluck/Science Photo Library; **p187** Tim Scoones/Photolibrary; **p195** Conor Caffrey/Science

Photo Library; **p197 T** Leonard Lessin/Photolibrary; **p197 B** Peter Arnold, Inc./Alamy; **p198** Andrew Syred/Science Photo Library; **p199** Kath Lovatt/NPS UK; **p201** L. Willatt, East Anglian Regional Genetics Service/Science Photo Library; **p208** Chris Priest/Science Photo Library; **p211** Siberia/Shutterstock; **p217** Paul Parker/Science Photo Library; **p220 T** Simon Fraser/Royal Victoria Infirmary, Newcastle upon Tyne/Science Photo Library; **p220 B** Zephyr/Science Photo Library; **p227** PHOTOTAKE Inc./Alamy; **p228 T** David Levinson/Alamy; **p228 M** Simon Fraser/Royal Victoria Infirmary, Newcastle upon Tyne/Science Photo Library; **p228 B** Du Cane Medical Imaging Ltd/Science Photo Library; **p231** PhotoDisc/Cumulus; **p244** Conge, ISM/Science Photo Library; **p250** Peter Arnold, Inc./Alamy; **p259** Ronnie McMillan/Alamy; **p264** Getty Images. PhotoDisc/Cumulus; **p267 T** Cordelia Molloy/Science Photo Library; **p267 B** Cordelia Molloy/Science Photo Library; **p270** Will & Deni McIntyre/Science Photo Library

The authors and publisher would like to thank the following for permission to use copyright material:

**p25 Fig 5:** Tom Linder, Ph.D, University of Washington; **p40 Fig 1:** © State of New South Wales through the Department of Education and Training and Charles Stuart University; **p66 Fig 1:** G. Toole and S. Toole, *Essential AS Biology*, 2004 Nelson Thornes; **p69 Fig 1:** From http://www.training.seer.cancer.gov; funded by the U.S. National Cancer Institute's Surveillance, Epidemiology and End Results (SEER) Program, via contract number N01-CN-67006, with Emory University, Atlanta SEER Cancer Registry, Atlanta, Georgia, U.S.A.; **p91 article:** Babies overfed to meet flawed ideal', issue 2601 of *New Scientist* magazine, 26 April 2007, page 6-7; **p98 Fig 1:** http://www.dkimages.com/; **p99 Fig 3:** Avert; **p113 Fig 1 (HIV):** WHO/UNAIDS http://www.who.int/; **p113 Fig 1 (AIDS):** Johns Hopkins Public Health magazine/Paul Mirocha; **p118 Fig 2:** The National Heart, Lung, and Blood Institute, a part of the National Institutes of Health and the U.S. Department of Health and Human Services; **p120 Fig 1:** Dr M Laker, *Understanding Cholesterol*, Family Doctor Books; **p125 case study:** C. Bunting, 'Pets 'cut chance of children having asthma'', *Independent Online* 28th August 2002; **p154 Fig 1:** M. Cox and D. Nelson, *Lehninger Principles of Biochemistry*, 4th Edition, 2004 Worth; **p154 Fig 2:** http://hemoglobin.navajo.cz/hemoglobin-4.png; **p159 Fig 2:** J. Hubbard and D. Mechan, *The Physiology of Health and Illness - With Related Anatomy*, 1997 Nelson Thornes; **p163 Fig 4:** Greg Smith; **p164 Fig 1:** J. Hubbard and D. Mechan, *The Physiology of Health and Illness - With Related Anatomy*, 1997 Nelson Thornes; **p169 Fig 1b:** SHine SA; **p177 Fig 2:** M. Jones and G. Jones, *Human Biology for A2 Level*, 2005 Cambridge University Press; **p183 Fig 2b:** © 2006 Compare Infobase Limited; **p187 Fig 2:** from Kimball's Biology Pages (http://biology-pages.info); **p197 Fig 2a:** http://en.wikipedia.org/wiki/Malaria; **p200 Fig 1:** J. Gregory, *Applications of Genetics (Cambridge Advanced Sciences)*, 2000 Cambridge University Press; **p203 Fig 2:** BBC Education Scotland http://www.bbc.co.uk/scotland/education/bitesize/standard/img/biology; **p205 Fig 2:** Stanford University; **p206 case study:** B. Hirschler, 'Doctors test gene therapy to treat blindness', *Reuters*, 1st May 2007 http://www.reuters.com; **p207 case study:** T. Castle, 'UK approves human-animal embryo research', *Reuters*, 5th September 2007 http://www.reuters.com; **p212 Fig 1:** Richard Conan-Davies, ClearlyExplained.com; **p218 Fig 1:** M. Jones and G. Jones, *Mammalian Physiology and Behaviour (Cambridge Advanced Sciences)*, 2002 Cambridge University Press; **p219 case study:** Laura Spinney; **p225 Fig 1–1:** taken from D.R. Siegfried, *Anatomy and Physiology for Dummies*, Copyright © 2002 John Wiley & Sons, reproduced with permission from John Wiley & Sons, Inc.; **p225 Fig 1–2:** http://www.CoolSchool.ca; **p225 Fig 1–6:** http://www.cidpusa.org/details.html; **p230 case study:** I. Sample, 'Human stem cells allow paralysed mice to walk again', *The Guardian*, 20th September 2005; **p232 case study:** Stanford University School of Medicine; **p247 Fig 3:** http://www.edren.org; **p249 Fig 2a:** © McKesson; **p251 Fig 5:** M.K. Davies, C.R. Gibbs and G.Y.H. Lip, 'ABC of heart failure: Management: Diuretics, ACE inhibitors , and nitrates', *British Medical Journal*, 12th February 2000 BMJ Publishing Group Ltd.; **p253 Fig 3b:** Peter Agre, Chem. Br., 2003, 39(11), 11-20 (http://www.rsc.org/chemistryworld/Issues/2003/November/theflow.asp) – Reproduced by permission of The Royal Society of Chemistry

Every effort has been made to contact copyright holders of material reproduced in this book. Any omissions will be rectified in subsequent printings if notice is given to the publisher.

### Websites

There are links to relevant websites in this book. In order to ensure that the links are up-to-date, that the links work, and that the sites are not inadvertently linked to sites that could be considered offensive, we have made the links available on the Heinemann website at www.heinemann.co.uk/hotlinks. When you access the site, the express code is 1806P.

# OCR

## Human Biology

AS / A2

Exclusively endorsed by OCR for GCE Human Biology

Barbara Geatrell, Pauline Lowrie and Alan Tilley
Series editor: Fran Fuller

**www.heinemann.co.uk**
✓ Free online support
✓ Useful weblinks
✓ 24 hour online ordering

**01865 888080**

**Heinemann**

# Contents

# OCR Human Biology

## Introduction

This book has two main aims. It is intended to offer clear and concise coverage of all the material you require for OCR Human Biology and it aims to help you to use your knowledge to succeed in the AS and A2 Human Biology exams.

## The OCR Course

OCR Human Biology is an exciting and stimulating course that covers how the human body functions from a cellular level up to the physiology that keeps us functioning as living organisms. Furthermore, it also looks at how scientific advances are used to improve health care – this course is a 'real life' approach to Human Biology. It explores the implications of recent scientific developments such as DNA technology for both the individual and for society. The roles of various health professionals, from radiographers to genetic counsellors, are examined. Practical procedures such as CPR and renal dialysis appear as learning outcomes. In other words, you will learn how people apply their biological knowledge and skills as part of their day-to-day lives.

The course delivers all the areas you would expect in Biology at A level but does this in the context of the human body. For example, you will be introduced to enzymes in the context of the blood-clotting cascade. Cell structure is covered – but not until you have looked at how a blood sample is taken and a slide prepared so you can see the cells. The course encourages a practical approach so you may even find yourself looking at a sample of your own cells! Enzymes are taught through exploring the conditions for storing blood and blood products and linked to the effect of temperature, co-enzymes and pH on enzyme activity. In A2, athletic performance is used as the context to explore cellular respiration and protein synthesis and appears in the same unit as muscle contraction and the effects of training on athletic performance.

You will spend time looking at the some of the issues which recent advances in science have raised. Stem cell technology, DNA technology, new emerging diseases such the H5N1 virus ('bird flu'), the implications of a population which is becoming more obese, carbon footprints and global warming – these topics appear on an almost daily basis in the media. You will cover the underpinning science as part of your course and, hopefully, you will be equipped to discuss these and other issues in an informed way both in lessons and as part of your exams. What is the role of NICE? How are clinical trials set up and evaluated? How can infertility be treated? Should it be treated? These give you a flavour of the topics covered by the specification but there is so much more!

## The Heinemann OCR Human Biology AS and A2 Textbook

The AS qualification consists of two taught units, Molecules, Blood and Gas Exchange (F221) and Growth, Development and Disease (F222). The A2 qualification also consists of two units, Energy, Reproduction and Populations (F224) and Genetics, Control and Ageing (F225). The two A2 unit exams can also ask synoptic questions. The aim of this book is to support you through the whole course. You will see that each spread is linked to a 'learning outcome'. Each unit on the course is broken down into learning outcomes and exam questions are written to test these learning outcomes. So, page by page, the book follows through each of the units you need to cover for the AS and A2 exam.

Why have both the AS and A2 in one book? As an AS student you can see what the course has 'in store', and as an A2 student, you have access to the AS material you will need to refer back to for your 'synoptic questions' on A2.

The book has been written by experienced examiners and each spread contains several features to support your learning and to help you maximise your performance in the exam. Look for examiner tips.

You will also see examples of case studies. In the second AS unit exam, you will have two questions based on pre-release material or case studies. The questions are based on the material in the case studies and you will see how some of the 'principles' or key ideas you have covered can apply in a range of different situations. This gives you the opportunity to practise looking at the application of your knowledge to different situations – and questions based on different situations often prove to be the challenging ones at both AS and A2.

Another key feature of this book is the signposting of How Science Works.

## How Science Works

This could be a margin feature or signposted with a coloured background. This feature guides you through some of the implications of science for society such as the clinical trialling of drugs or the possible consequences of an ageing population.

Questions on 'How Science Works' can appear in all the unit exams – this book gives you ample opportunity to consider this across a range of topics and again provides excellent opportunities to polish up your exam technique.

Each double-page spread contains illustrations and diagrams to support the explanations in the text and there are questions at the end of each set of 'learning outcomes' to help you to reinforce your understanding.

You will also find worked examples of calculations you could be asked to carry out in an exam.

There are also exam-style questions covering topics on each of the units. The format of each of the exams is explained and, again, there are examiner tips to help you to identify what a 'How Science Works' question may look like or how synoptic questions may be structured. The questions also illustrate how 'QWC' marks – written communication marks – might be awarded and what a 'stretch and challenge' question might ask you to do.

Finally, running throughout the book, you will find 'key definitions'. Why 'key definitions'? Because they are fundamental to the course, you are expected to know them and … they are asked for in exam questions! Remember the two aims of this book – to cover the specification content and to help you get the exam grade you deserve.

**Examiner tip**

Watch out for these in the margins! They will point out common mistakes that students make on a topic, or guide you to links with other topics that could appear in synoptic questions.

**Case study**

Look out for these on the spreads! They give you 'real life' examples of relevant applications of the learning outcomes covered on that page. They show you how to look at the ideas you have covered in different settings – and this is exactly what exam questions will also do, particularly in F222 and on the two A2 units.

**Key definition**

These pick out the key terms and ideas that come up so often as exam questions – a real 'you must know this' signpost!

## NewScientist

**Reinforce your learning and keep up to date with recent developments in science by taking advantage of Heinemann's unique partnership with New Scientist. Visit www.heinemann. co.uk/newscientistmagazine for guidance and discounts on subscriptions.**

# ① Blood tests

It was just the worst time for Emily to feel ill, as she was revising for her January AS modules. She told her GP that she was feeling very tired, and had a sore throat. When the GP examined her, she noticed that Emily had swollen lymph nodes in her neck. The GP thought it would be best to take a sample of blood to get it tested. She explained to Emily that the blood sample would be taken by a phlebotomist.

A few days later, the GP told Emily that she had glandular fever. Unfortunately, Emily would feel like this for a few weeks or even months, but she would eventually get better without any treatment.

## Taking a blood sample

The study of the cells in the blood can provide valuable information about health, and help to diagnose a number of conditions. Sometimes a blood sample is taken that is large enough for several different tests to be carried out.

Most blood samples are taken from a vein, commonly from those around the elbow. You can see this in Figure 1.

First you tie a band (tourniquet) around the arm to make the vein stand out. This makes it much easier to target the needle. Then you clean the area around the vein with an alcohol-based solution. Then you push a sterile needle into the vein. The needle is attached to a sterile syringe. As you pull back the plunger of the syringe, blood is sucked into the syringe. When the necessary volume of blood has been extracted, you remove the needle and hold a little ball of cotton wool over the wound. You press this for one to two minutes until bleeding has more or less stopped, then apply a dressing.

Some of the tests that may be done on a blood sample involve looking at the stained cells under a microscope.

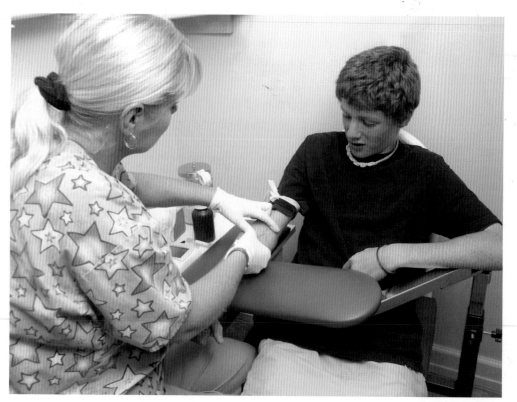

**Figure 1** Taking a blood sample

# Making and staining a blood film

In a pathology laboratory, blood films are made and stained by a machine. Usually, they are labelled with a barcode so that the information can be stored on a computer system.

(a)

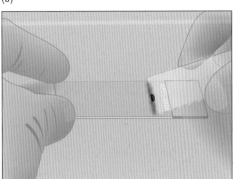

Place a very small drop of blood near the end of a clean microscope slide

(b)

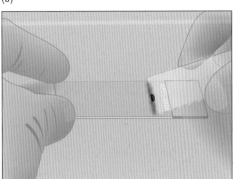

Place the end of another slide (the 'spreader') on the sample slide

(c)

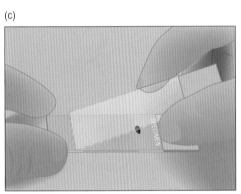

Hold the spreader at an angle of about 30° and push it along the slide, spreading the drop of blood as a smear

(d)

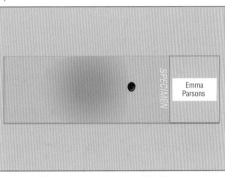

Label the slide with the patient's details. Allow the slide to dry in the air, so the cells stick to the slide

(e)

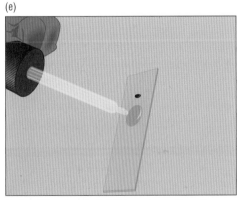

Fix the slide using alcohol. This preserves the cells

**Figure 2** Making a blood film

(f)

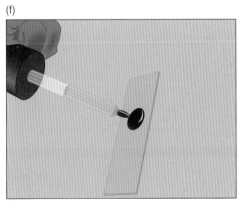

Stain the slide using a Romanowsky stain, such as Wright's or Leishman's stain. The stain is poured over the slide and left for about 2 minutes before it is washed off with water

**Key definition**

A **differential stain**, such as Leishman's stain, makes some structures appear darker or different in colour from other structures. In a normal blood sample the nuclei of the leucocytes will be stained purple, allowing neutrophils, lymphocytes and monocytes to be identified by the shape of the nucleus.

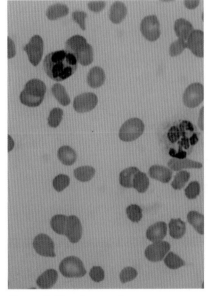

**Figure 3** Blood smear from a person with megaloblastic anaemia. In megaloblastic anaemia, the red blood cells are larger than normal and they can have an abnormal shape. The cause is usually a deficiency of Vitamin B12 and this type of anaemia is referred to as 'pernicious' – increasing levels of iron will have no effect. Compare this with the blood smears on page 6

# Questions

1 Suggest why
   **(a)** the equipment used to take a blood sample is sterile
   **(b)** the skin is cleaned with an alcohol-based solution before taking a blood sample
   **(c)** blood is taken from a vein rather than an artery. (Hint: think about the structure and functions of veins and arteries.)
2 When making a blood film, suggest why
   **(a)** the blood film must be very thin
   **(b)** a stain is used
   **(c)** the slide must be very clean and grease-free.

The number of different kinds of cell in a blood sample is usually counted by a machine in a **pathology** laboratory. But for some kinds of test a device called a **haemocytometer** may be used.

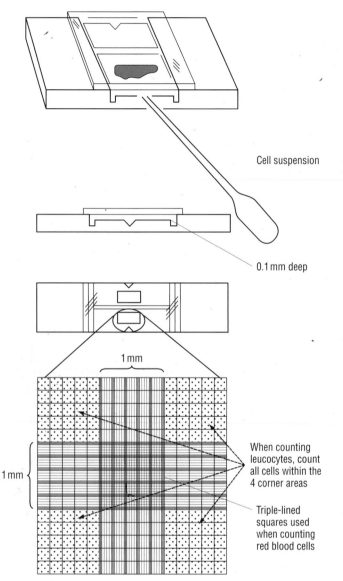

Cell suspension

0.1 mm deep

1 mm

1 mm

When counting leucocytes, count all cells within the 4 corner areas

Triple-lined squares used when counting red blood cells

**Figure 1** A haemocytometer

## A haemocytometer

A haemocytometer is a special counting chamber designed for counting blood cells. It is rather like a microscope slide. Look at Figure 1. You will see that the haemocytometer has a central platform with grooves each side of it. There is a tiny grid etched on this platform, like microscopic graph paper. In the centre of this grid, you will see that there are some triple-lined squares. These squares measure exactly 0.2 × 0.2 mm. The haemocytometer is designed so that when you put a cover slip on top, the platform is exactly 0.1 mm below the cover slip. This means that when you view one of the triple-lined squares under the microscope, you are looking at a volume of exactly 0.1 x 0.2 x 0.2 mm, i.e. 0.004 mm³.

### Counting red blood cells

The haemocytometer can be used to count the number of **erythrocytes** (red blood cells) in a sample of blood. Before the red blood cells can be counted, they have to be diluted very carefully. A special blood pipette is used for this. You can see this in Figure 2. You will see that the pipette has markings at 0.5, 1 and 101. First you fill the pipette with blood to the 0.5 mark.

Then you draw a fluid called Dacie's fluid into the pipette up to the 101 mark. There is a bulge in the pipette, and inside this section is a little bead. This bead means that the blood and the Dacie's fluid get mixed together. This gives a 1 in 200 dilution.

Let the first few drops of mixture run out of the pipette – these will not be used. Then place the diluted blood on the haemocytometer. Count the number of red blood cells in five triple-lined squares.

Sometimes cells lie on top of the triple lines around the edge of the square. If we count all these cells, the cell count will be too high. If we miss them out, the cell count will be too low. For this reason, there is a rule called the **northwest rule**. If a cell lies on the middle of the triple lines on the north or west of the grid, we count it in. If a cell is lying on the middle of the triple lines on the east or south of the grid, we miss it out.

Remember that each triple-lined square has a volume of 0.004 mm³ and that the blood was diluted 200 times. The total volume of the five triple-lined squares is 0.02 mm³. If the number of cells counted in the five triple-lined squares is E, then the number of red blood cells in 1 mm³ of the original blood sample = $1/0.02 \times E \times 200 = E \times 10\,000$.

## Calculating the number of red blood cells in a sample

There are 10 cells in the triple-lined square shown in Figure 3. We will assume that this square is typical of all the squares on the haemocytometer grid although, normally, an average is taken from at least five cells.

This means there are 10 cells in $0.1 \times 0.2 \times 0.2 \, mm^3 = 0.004 \, mm^3$

Therefore the number of cells in $1 \, mm^3$ of the diluted blood $= 1/0.004 \times 10$

$= 2500$

But the blood sample was diluted 200 times.

Therefore $1 \, mm^3$ of the original blood contained $2500 \times 200$ cells

$= 500\,000$ cells.

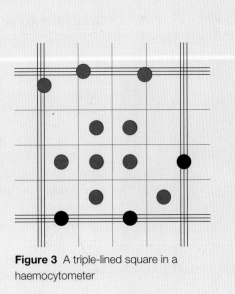

**Figure 3** A triple-lined square in a haemocytometer

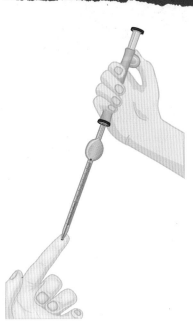

**Figure 2** Red blood cell pipette

### Counting leucocytes

This is done using a white blood cell pipette. This is like the red blood cell pipette, but it is graduated 0.5, 1 and 11. First you fill the pipette with blood to the 0.5 mark. Then you add a diluting fluid to the 11 mark and mix together the blood and diluting fluid. This means you have diluted the blood by a factor of 20. The diluting fluid causes the red blood cells to burst, but the white blood cells remain intact. (You will find out more about how this happens when you study osmosis in spread 1.1.2.4.)

Allow the first one-third of the mixture from the pipette to run out; this will not be used. Add the diluted blood to the haemocytometer. This time, use the larger squares on the corners of the haemocytometer grid. These larger squares measure $1 \times 1 \, mm$. Remember that the depth of the haemocytometer chamber is 0.1 mm. This means that each square has a volume of $1 \times 1 \times 0.1 = 0.1 \, mm^3$.

Count the number of white blood cells in the four corner squares. These four squares together have a volume of $0.4 \, mm^3$. If the number of cells counted in these four squares = L, then the number of leucocytes per $mm^3$ of the original blood sample $= 1/0.4 \times L \times 20$

$= 50 \times L$.

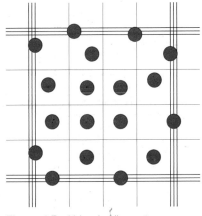

**Figure 4** Red blood cell count

## Questions

1 Suggest why the first sample of diluted blood from the blood pipette is not used. (Hint: look carefully at the shape of the pipette.)
2 Explain why several squares on the haemocytometer are counted, rather than just one.
3 When white blood cells are counted, we use a diluting fluid that will burst the red blood cells. Explain why it is important to do this.
4 The diluting fluid used to count white blood cells contains a stain. Explain why this is necessary.
5 Figure 4 shows one triple-lined square on a haemocytometer that is being used to count red blood cells. Assume that this square is typical of all the triple-lined squares on the haemocytometer. The blood has been diluted 200 times. Calculate the number of red blood cells in $1 \, mm^3$ of the original blood sample. Show your working.

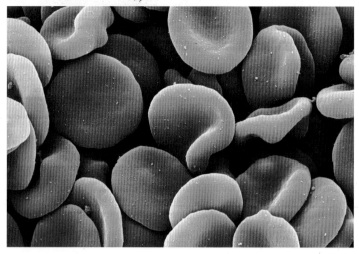

**Figure 1** Erythrocytes

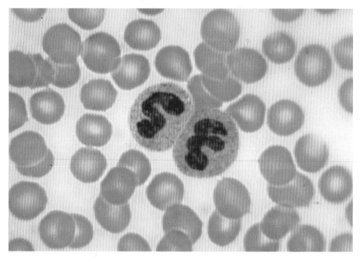

**Figure 2** Neutrophils

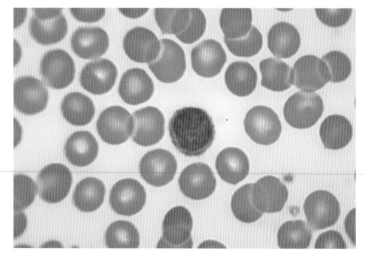

**Figure 3** A lymphocyte

## Identifying cells

Studying the relative numbers of different kinds of cell in the blood can give valuable information about a person's health. It can also help to diagnose a number of medical conditions. For example, someone with too few red blood cells may have anaemia, and someone with a high white cell count may have an infection.

Red blood cells, or **erythrocytes**, are biconcave discs. They transport oxygen and carbon dioxide. Their shape means they have a relatively large surface area-to-volume ratio to speed up gas exchange. Their cytoplasm is packed with a pigment, a protein called **haemoglobin** that associates reversibly with oxygen. Mature red blood cells have no nucleus, which gives more room inside them for haemoglobin. Also, they are very small and flexible, so they can be flattened against capillary walls. This reduces the distance that gases have to diffuse across, and again speeds up gas exchange.

**Leucocytes** are white blood cells. Some of the different kinds of leucocytes are described below.

**Neutrophils** have small granules in the cytoplasm. These cells engulf microorganisms by phagocytosis (see spread 1.1.2.5).

**Lymphocytes** have a large, darkly-stained nucleus surrounded by a thin layer of clear cytoplasm. There are two kinds of lymphocyte, B and T lymphocytes, which look the same but have different functions. Both are cells of the immune system. B lymphocytes produce antibodies. T lymphocytes have several functions, including cell destruction.

**Monocytes** are the largest kind of leucocyte. They have a large, bean-shaped nucleus and clear cytoplasm. They spend 2–3 days in the circulatory system, then move into the tissues. Here they become **macrophages**, engulfing microorganisms and other foreign material.

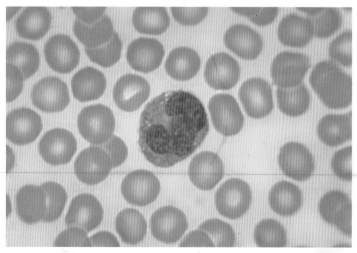

**Figure 4** Monocyte (macrophage)

In addition to red and white blood cells, there are also **platelets**. These are fragments of giant cells called megakaryocytes. They are involved in blood clotting.

## Calculating magnification

Many biological structures are very small and can only be seen using a microscope. Many of these structures are so small we need to measure them in **micrometres**. A micrometre (μm) is one-millionth of a metre, or one-thousandth of a millimetre.

Many illustrations in textbooks like this show the magnification of the structure. This tells you how many times bigger the illustration is than the real structure.

$$\text{Magnification} = \frac{\text{size of structure in the picture}}{\text{real size of the structure}}$$

You can also rearrange the formula to find the real size of the structure, if you know the magnification.

$$\text{Real size} = \frac{\text{size of structure in the picture}}{\text{magnification}}$$

**Key definition**

A **micrometre** (μm) is one millionth of a metre, i.e. $10^{-6}$ m or one $\frac{1}{1000}$ of a mm $10^{-3}$ mm.

**Examiner tip**

A micrometre is one-thousandth of a millimetre. In other words, 1 μm = $10^{-3}$ mm. Always measure the structure in millimetres. This means you can convert this to micrometres by multiplying by 1000, i.e. by adding three zeros.

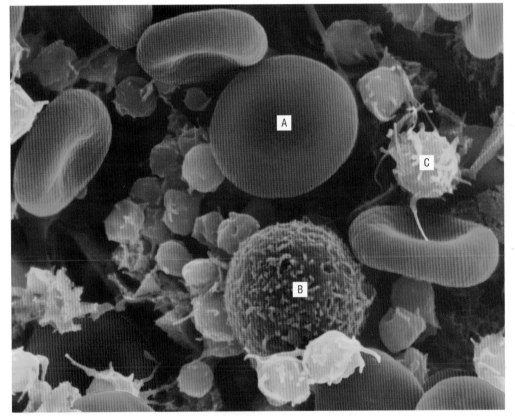

**Figure 5** SEM (magnification ×2500) and labels A, B and C for different cell types

## Questions

1 Name the type of cell represented by A, B and C in Figure 5.
2 Calculate the actual diameter of cell A and cell B.

## Membranes

Cell membranes are made up mainly of two kinds of molecule:
- **phospholipids**, which form the bulk of the membrane
- proteins, which are scattered around in the membrane.

Some molecules of carbohydrate and **cholesterol** may also be present.

### Phospholipids

A phospholipid is a special kind of **lipid** molecule. It is made of a **glycerol** molecule with a phosphate group and two **fatty acid** chains attached. The phosphate group is **hydrophilic** (hydro = water, philic = loving) because it has a charge. It is soluble in water. The fatty acid chains are made of hydrocarbons. They are **hydrophobic** (hydro = water, phobic = hating). They do not have a charge and they are insoluble in water.

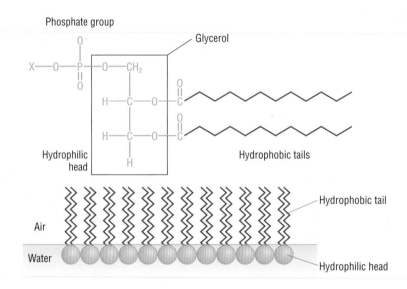

**Figure 1** Structure of a phospholipid

Phospholipids pack together in a membrane as shown in Figure 2. They form a double layer called a **bilayer**. Remember that there is water inside and outside a cell.
- The fatty acid 'tails', which are hydrophobic, pack together away from water.
- The hydrophilic 'heads' arrange themselves on the outside of the membrane, facing the water.

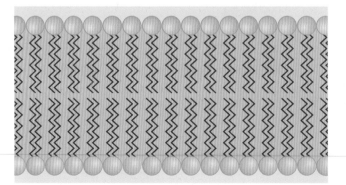

**Figure 2** A phospholipid bilayer

## Proteins

In a membrane there are many different sorts of protein with complex shapes. The proteins are scattered around in the membrane and are always changing their position.

- There are large proteins that span the whole bilayer. These are called **intrinsic proteins**.
- There are smaller proteins on just one side of the bilayer. These are called **extrinsic proteins**.

Some of the proteins have carbohydrate chains attached to them. They are called **glycoproteins**. In the plasma membrane these proteins are involved in *cell recognition*. This means that other molecules may attach to them. Some of the carbohydrate chains join to other carbohydrate chains in the glycoproteins of adjacent cells. This causes the cells to join together – *cell adhesion*. Sometimes carbohydrate chains attach to phospholipids to form **glycolipids**.

In many membranes, **cholesterol** is present. This helps to keep the membrane stable. Cholesterol is a lipid like substance called a **steroid**.

### The fluid-mosaic model of membrane structure

The way the molecules in a membrane are arranged is called the **fluid-mosaic model** of membrane structure (fluid = moving, mosaic = made of pieces). You can see this in Figure 3.

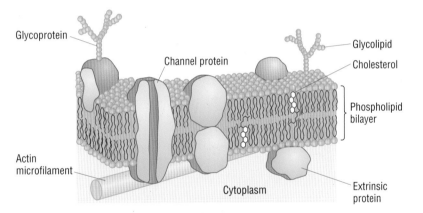

**Figure 3** The fluid-mosaic model of membrane structure

# Membranes around and within cells

Figure 3 is a diagram of the plasma membrane – the membrane at the cell's surface. You can see that the glycoproteins are on the outside. This membrane separates the cell from its environment (the plasma or tissue fluid) and from other cells. Glycoproteins are also important in cell-to-cell signalling. The binding of ions or other chemicals to these receptors can trigger reactions inside the cell including the switching on or off of genes in the nucleus. Blocking these receptors is one way that drugs can have an effect.

But membranes are also found inside cells to separate the cell into 'compartments'. A list of membrane bound structures or 'organelles' is given in the examiner tip. The membranes allow complex processes within the cell to be separated. For example, the enzymes involved in respiration are found in the mitochondria.

# Questions

1. Where in the cell would a protein be formed?
2. Where would it be modified to form a glycoprotein?
3. Suggest why the carbohydrate chains on a membrane glycoprotein are on the outside of the cell membrane.
4. Outline the role of membranes both inside cells and at the cell surface.

**Examiner tip**

The composition of membranes is not always the same. For example, Myelin, which insulates neurones, contains only 18% protein and 76% lipid. Mitochondrial inner membranes contain 76% protein and only 24% lipid.

Plasma membranes of human red blood cells contain nearly equal amounts of proteins (44%) and lipids (43%). The structure of lipids and proteins is part of this unit. Which component of the membrane will contain nitrogen? Which will contain phosphorus?

| Membrane system | Function |
|---|---|
| Plasma membrane | Partially permeable. Retains cell contents |
| Rough endoplasmic reticulum | Ribosomes synthesise proteins. Membranes package them for distribution around the cell |
| Smooth endoplasmic reticulum | Synthesis of lipids including steroids |
| Golgi apparatus | Synthesis of glycoproteins, polysaccharides and hormones, production of lysosomes |
| Nuclear envelope | Regulates exchange between cytoplasm and nucleus. |
| **Organelles** | |
| Lysosomes | Contain enzymes for intracellular digestion |
| Nucleus | Contains DNA and regulates cell activity |
| Mitochondrion | Aerobic respiration and production of ATP |
| Chloroplast | Absorbance of light energy and production of carbohydrates in photosynthesis. |

## Eukaryotic cells

Cells with a nucleus are **eukaryotic cells**. This means 'cells with a true nucleus'. Animal cells, such as a human **leucocyte**, and plant cells, such as **palisade mesophyll** cells, are eukaryotic.

In spread 1.1.3 you looked at the appearance of some leucocytes under a light microscope. But this kind of microscope does not show the detailed structure of the cells. To see all the structures inside the cell in detail, we need to look at its **ultrastructure**, using an electron microscope.

### A leucocyte

A leucocyte is one example of a kind of animal cell. Figure 1 shows the ultrastructure of a **lymphocyte** that is capable of making **antibodies**. You will remember that this is a kind of leucocyte. The figure shows the organelles that can be seen using an electron microscope.

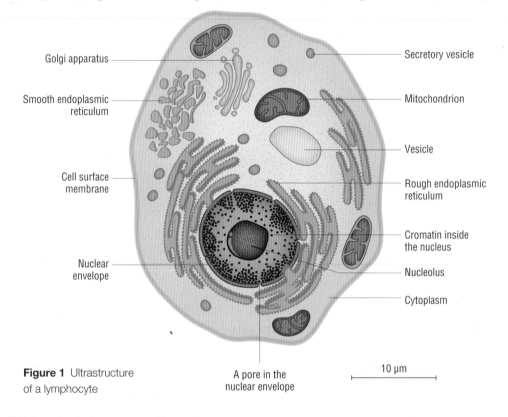

**Figure 1** Ultrastructure of a lymphocyte

### Collaboration between organelles

You will learn about the functions of these organelles over the next few pages. However, it is also important that you appreciate that a function like the secretion of antibodies requires several organelles to work together. Antibodies are modified proteins. The genes which code for these proteins are held on chromosomes in the **nucleus**. A messenger molecule takes the code from these genes to the **ribosomes** on the **rough endoplasmic reticulum** and here the proteins are made. They are then transported in small membrane sacks called **vesicles** to the **Golgi apparatus**. Here they are modified and re-packaged into more vesicles. These make their way to the **cell surface membrane**. They fuse with this to release antibodies from the cells. The energy to make all this possible comes in the form of ATP made in the **mitochondria**. In another type of leucocyte, a neutrophil, enzymes are made on the ribosomes and packaged by the Golgi apparatus, but this time the vesicles with the enzymes stay in the cytoplasm as **lysosomes**. This type of cell engulfs bacteria and these enzymes will then be released to destroy the bacteria in the cell.

**Plasmamembrane**

**Exocytosis** releases contents of vesicle into the extracellular environment.

**Lysosome** contains hydrolytic enzymes which may digest ingested materials, or damaged organelles.

**Cisternae stack of Golgi** 'processes' molecules often by adding or modifying carbohydrate 'signals'. This directs the molecules to the correct cellular compartment.

**Product molecules** are moved through the stack in a precisely defined sequence.

**Protein synthesis** at ribosomes on E.R. Newly-synthesized protein carries a 'signal' which means that the protein will enter the cisterna ready to be packaged and delivered to the Golgi apparatus.

**Endoplasmic reticulum** 'buds off' membranous sacs containing proteins, and may be for export (◎) or for use within the cell (●).

**Nuclear pore** can control the entry (e.g. ribosomal proteins) and exit (e.g. ribosomal subunits, messenger RNA) of molecules in and out of the nucleus.

**Messenger RNA** carries code from genes to ribosomes for protein synthesis.

**Nuclear envelope** is a double membrane. The outer membrane is continuous with the E.R.

**Nucleus** contains information for protein synthesis as DNA in series of genes on the chromosomes.

**Nucleolus** where ribosomal subunits are made.

**Figure 2** Organelles collaborate to carry out a cell's function

### A palisade mesophyll cell

Palisade mesophyll cells are present in plant leaves. You may remember from GCSE that they carry out photosynthesis. They are one example of a kind of plant cell. Figure 3 shows the ultrastructure of a palisade mesophyll cell. You will notice that some of the organelles found in the lymphocyte are also found in a palisade mesophyll cell. But there are some organelles found in the palisade mesophyll cell that are not found in a lymphocyte.

### Comparing animal and plant cells

| Organelles | Animal cell | Plant cell |
|---|---|---|
| Nucleus | √ | √ |
| Nucleolus | √ | √ |
| Ribosomes | √ | √ |
| Cell wall | ✕ | √ |
| Plasma membrane | √ | √ |
| Golgi apparatus | √ | √ |
| Rough endoplasmic reticulum | √ | √ |
| Smooth endoplasmic reticulum | √ | √ |
| Mitochondria | √ | √ |
| Chloroplasts | ✕ | √ only in leaf/green parts |
| Permanent vacuole | ✕ | √ |
| Cytoskeleton | √ | √ |

**Table 1** Summary of the similarities and differences between animal and plant cells

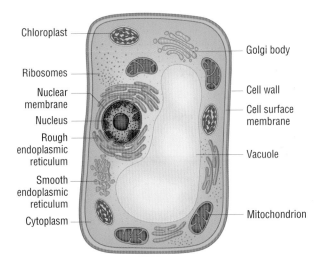

**Figure 3** Ultrastructure of a palisade mesophyll cell

Chloroplast
Golgi body
Ribosomes
Cell wall
Nuclear membrane
Cell surface membrane
Nucleus
Rough endoplasmic reticulum
Vacuole
Smooth endoplasmic reticulum
Cytoplasm
Mitochondrion

## Questions

1 Which of the structures in Figure 3 would not be seen in an animal cell?
2 Name a type of plant cell in which you would be unlikely to find chloroplasts.

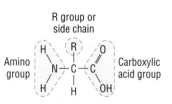

**Figure 1** The structure of an amino acid

## Key definition

A **condensation reaction** is a reaction in which two molecules are joined together by removing a hydrogen atom from one molecule and OH from the other molecule. H and OH combine to form a molecule of water. Condensation reactions are a key type of **anabolic** or 'building up' reaction.

The **primary structure** of a protein is the number and sequence of amino acids in the polypeptide chain.

**Secondary structure** – the amino acid chain is coiled into alpha-helix or folded into beta-pleated sheets. The shape is held by hydrogen bonds between the oxygen atoms and the hydrogen atoms around peptide bonds. R groups are not involved.

## Blood chemistry

Investigating the chemistry of the blood is a relatively simple way of assessing the health of an individual. Red blood cells are packed with a protein called **haemoglobin**, which is important in carrying oxygen around the body to where it is needed. We are now going to look at the structure of haemoglobin in more detail.

### Amino acid

Haemoglobin is a protein. This means that, like all proteins, haemoglobin is a polymer and the monomers are amino acids. You can see the structure of an amino acid in Figure 1. Every amino acid has a **carboxylic acid group** and an **amino group**. There are 20 different amino acids which appear commonly in proteins and each of them has a different R group.

Two amino acids can join together to form a **dipeptide**. You can see this in Figure 2. You will see that a hydrogen atom is removed from the amino group of one amino acid, and an oxygen atom and a hydrogen atom are removed from the carboxylic acid group of the other amino acid. The H and OH are removed and a molecule of water is produced. This kind of reaction is called a **condensation** reaction. The bond formed is called a **peptide bond**. Many amino acids can join together to form a **polypeptide**.

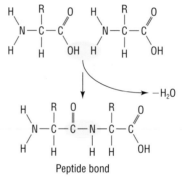

**Figure 2** The formation of a dipeptide

### Primary structure

The primary structure of a protein is the number and sequence of amino acids in the polypeptide chain. Haemoglobin is a very complex protein. It is made up of four polypeptide chains. Two chains are identical and are called alpha chains. The other two are also the same as each other and are called beta chains. Each chain has over 100 amino acids in it, so the molecule is very big. You can see this in Figure 2 and on spread 4.1.2.3.

### Secondary structure

The polypeptide chains are so long that they twist or coil to form different shapes. Weak bonds called **hydrogen bonds** form between different peptide bonds in different parts of the chain. You can see some different shapes that the polypeptide chain can fold into in Figure 4. The shape that the polypeptide chain folds into is called the **secondary structure** of the protein. Two common kinds of secondary structure are the *alpha-helix* and the *beta-pleated sheet*.

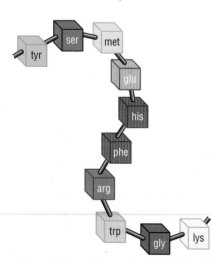

**Figure 3** The primary structure of a protein. Each coloured block represents a different amino acid

## Tertiary structure

The polypeptide chains can fold up again, to produce a complex three-dimensional shape. You can see this in Figure 5. This overall shape is called the **tertiary structure**. This is held in place by bonds that form between the R groups of amino acids in the chain. Several different kinds of bonds hold the tertiary structure together. These include weak hydrogen bonds and much stronger bonds called disulfide bonds. Cysteine is an amino acid with sulfur in its R group. Two cysteine molecules in different parts of a chain can join together by forming a disulfide bond between two sulfur atoms. You can see the tertiary structure of haemoglobin in Figure 5.

## Quaternary structure

Proteins that contain more than one polypeptide chain are said to have a **quaternary structure**. This is the way the polypeptide chains fit together.

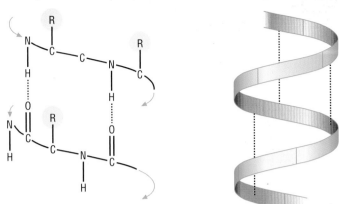

·········· Hydrogen bonds

An α-helix has 36 amino acids per 10 turns of the coil. H-bonds form between one amino acid and the one 'four places' along the chain

**Figure 4** The secondary structure of a protein

---

Tertiary structure in proteins is stabilised by a number of bonds.

1 Disulfide bonds
The amino acid cysteine contains sulfur. Where two cysteines are found close to each other a covalent bond can form

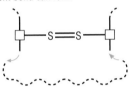

2 Ionic bonds
R groups sometimes carry a charge, either +ve or −ve. Where oppositely charged amino acids are found close to each other an ionic bond forms

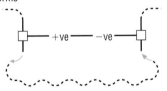

3 Hydrogen bonds
As in secondary structure. Wherever slightly positively charged groups are found close to slightly negatively charged groups hydrogen bonds form

**Figure 5** The tertiary structure of a protein

4 Hydrophobic and hydrophilic interactions
In a water-based environment, hydrophobic amino acids will be most stable if they are held together with water excluded. Hydrophilic amino acids tend to be found on the outside in globular proteins, with hydrophobic amino acids in the centre

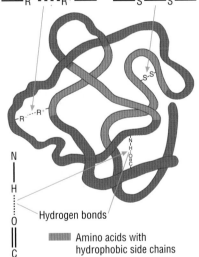

Ionic bond between ionised R groups —R⁻ ---- R⁺—

Disulfide bond between R groups containing +SH groups —S—S—

Hydrogen bonds

▓▓▓ Amino acids with hydrophobic side chains

▓▓▓ Amino acids with hydrophilic side chains

## Questions

1 Complete the table (left) with ticks or crosses to show which bonds are involved in holding together the primary, secondary and tertiary structures of a protein.
2 The amino acid cysteine contains sulfur. In which part of the amino acid will this be present?

| Bond | Primary structure | Secondary structure | Tertiary structure |
|------|-------------------|---------------------|--------------------|
| Peptide bond | ✓ | ✓ | ✓ |
| Hydrogen bond | ✗ | ✓ | ✓ |
| Disulfide bond | ✗ | ✗ | ✓ |

## Globular and fibrous proteins

The final three-dimensional shape of the protein results in two different classes of protein. Some proteins are **fibrous proteins**. In a fibrous protein, the polypeptides join together to form long fibres or sheets. These proteins are strong and insoluble in water. They tend to have structural functions, for example, keratin, which makes up human hair. **Haemoglobin**, on the other hand, is a **globular protein**. These proteins are roughly spherical, or globular in shape. They are usually soluble in water and tend to have biochemical functions. For example, **enzymes** are globular proteins.

Globular proteins are folded so that hydrophilic R groups are on the outside. This enables them to be soluble in water-based liquids such as cytoplasm and blood plasma.

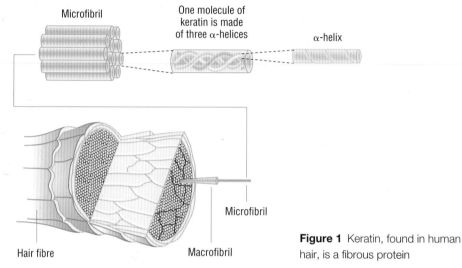

**Figure 1** Keratin, found in human hair, is a fibrous protein

### Denaturation

The tertiary structure of a globular protein is held together mostly by fairly weak bonds, such as hydrogen bonds. This means that when the temperature increases, the molecule vibrates more. If the protein vibrates too much, this will break some of these weak bonds and the shape of the molecule will change. When this happens, we say that the protein is **denatured**. Different proteins denature at different temperatures, but most proteins denature at temperatures around 45 °C.

**Hydrogen bonds** depend on a very weak attraction between a slightly positively charged hydrogen and a slightly negatively charged oxygen in different parts of the protein molecule. Some other kinds of bonds, called **ionic bonds**, also depend on tiny charges like this. As a result, these bonds can also be broken if the pH changes.

**pH** measures the concentration of hydrogen ions. Hydrogen ions are positively charged. As the concentration of hydrogen ions surrounding the protein increases or decreases, this affects the charges holding together the folds of the globular protein. As these weak bonds break, the protein becomes denatured.

### Carrying oxygen

Haemoglobin molecule is made of four **polypeptide** chains, each with a **haem** group attached to it. You can see this in Figure 2. Haem is a **prosthetic group**.

In the middle of each haem group there is an iron ion, which can associate with one oxygen molecule. Since haemoglobin has four haem groups, each haemoglobin molecule can carry four oxygen molecules. As the first haem group combines with an oxygen molecule, the shape of the haemoglobin molecule changes slightly. This exposes the next haem group, making it easier for it to pick up more oxygen.

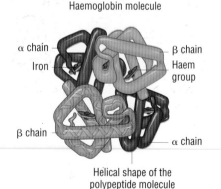

**Figure 2** Quaternary structure of haemoglobin. Each colour represents a different polypeptide chain

## When things go wrong

There are different kinds of haemoglobin. These result from the haemoglobin molecule having a different shape, or **tertiary structure**.

There is a disorder, called beta thalassaemia, which is common in Greek and Italian families. People with this condition have haemoglobin in which the beta chains are shorter than normal. This means that their haemoglobin does not carry as much oxygen as normal haemoglobin.

Diabetes is a condition in which a person's blood glucose levels are not well controlled. You will learn more about this in spread 2.4.3.1. People with diabetes are likely to have high blood glucose levels unless the condition is well controlled. When blood glucose levels are high, glucose attaches to haemoglobin in the red blood cells. This forms 'glycosylated haemoglobin'. Glycosylated haemoglobin picks up oxygen very readily, but it is not as good as normal haemoglobin at releasing oxygen to respiring tissues. If a person has too much glycosylated haemoglobin in their blood, this can lead to damage in certain parts of the body. For example, diabetics with high levels of glycosylated haemoglobin are more at risk of developing diabetic retinopathy. This is a condition in which the blood vessels in the eye become damaged, and this can eventually lead to blindness.

**Examiner tip**

You will meet sickle cell anaemia again in A2. Diabetes is also covered again at A2 – look for 'synoptic links'.

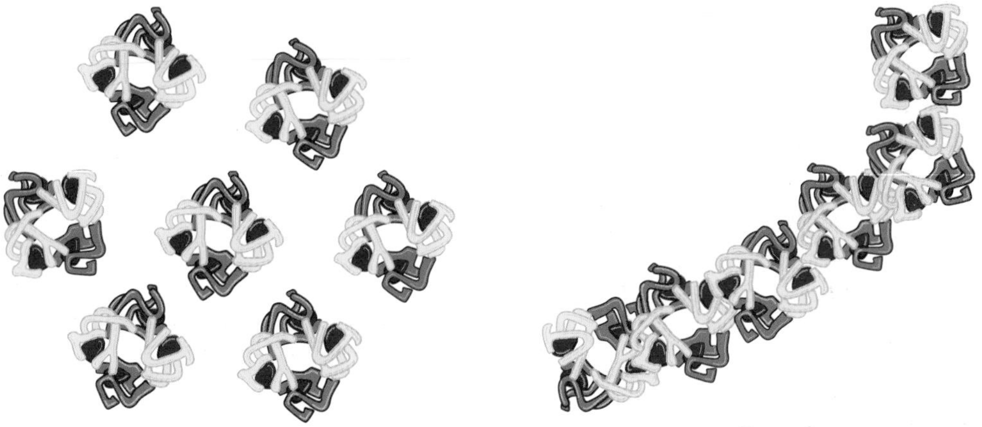

Normal haemoglobin          Clumped haemoglobin

**Figure 3** Sickle cell haemoglobin

Sickle cell anaemia is another condition in which the haemoglobin is different. It is most common among people of African ancestry. People with sickle cell anaemia have haemoglobin in which the alpha chains are normal, but the beta chains have the amino acid valine present instead of the amino acid glutamic acid. You might think that having only one amino acid different in a very long chain should make no difference. Sickle cell haemoglobin is quite like normal haemoglobin when there is a high concentration of oxygen, for example, in the lungs and in the arteries. However, when it loses its oxygen in the capillaries in the tissues, the molecules tend to stick together. You can see this in Figure 3. These chains of haemoglobin are long and stiff. They cause the red blood cells to change shape. Most of the red blood cells simply become misshapen, but some become crescent or sickle-shaped. It is this that gives the disorder its name. You can see this in Figure 4. The sickle cells cannot pass down the capillaries properly. This causes a 'sickle cell crisis' and means that tissues become short of oxygen. It is a very painful condition.

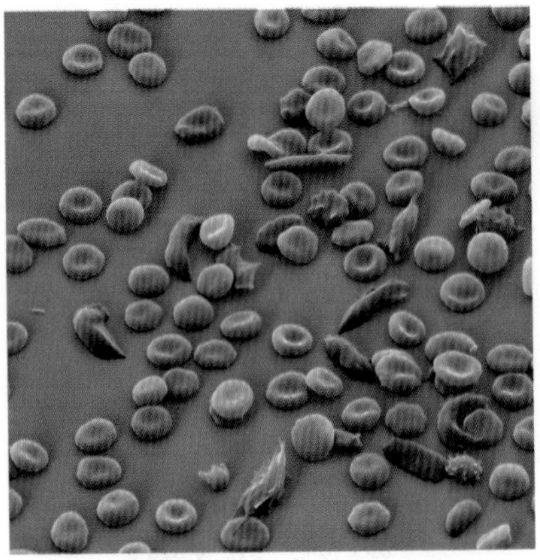

**Figure 4** Red blood cells of a person with sickle cell anaemia

## Questions

1 Egg white is a protein. In a raw egg, the egg white is clear and runny, but when it is cooked it becomes white and hard. Use your knowledge of protein structure to explain what happens to the egg white protein when it is cooked.

2 Suggest why it is not possible to turn cooked egg white back to being clear and runny.

3 Explain why high levels of glycosylated haemoglobin can lead to blindness.

## Water – a special molecule

Our bodies contain about 60% water. There can be no life as we know it without water, because most biochemical reactions occur in water. Water has some unique properties that make it very important as a biological molecule.

Look at Figure 1. You can see that a molecule of water is made of an oxygen atom joined to two hydrogen atoms. The oxygen atom has a slightly negative charge and the hydrogen atoms have a slightly positive charge. A molecule like this, which has different areas of positive and negative charges, is said to be **polar**.

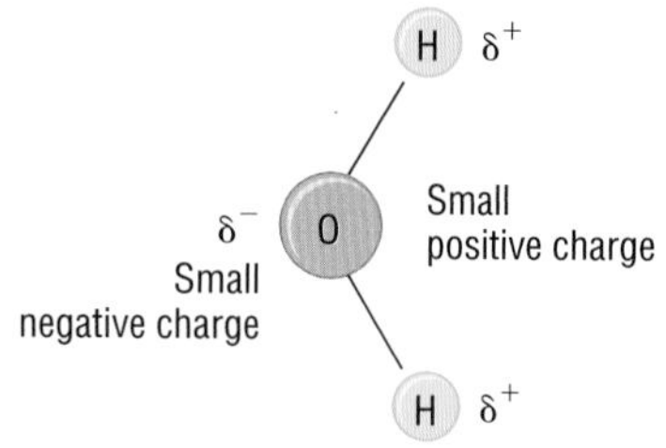

Water molecules are attracted to each other. The area of positive charge on the hydrogen atom of one water molecule attracts the negative charge of the oxygen atom of another water molecule. These forces of attraction are **hydrogen bonds**. Water molecules therefore 'stick' to each other, as you can see in Figure 2.

Water molecules are also attracted to other polar particles. Water molecules form a 'shell' around ions and many other molecules that have slight charges on their surfaces, causing them to dissolve. This makes it a very good solvent for ions and other polar molecules such as sugars, amino acids, etc. It is sometimes called the 'universal solvent' because so many different chemicals will dissolve in it. This makes water an excellent transport medium.

### Blood plasma

**Blood plasma** is the liquid part of the blood. Its exact composition varies, depending on which part of the body it is taken from. However, its main components are water plus:

- Proteins, such as **fibrinogen**, used in blood clotting; **antibodies**, produced by the immune system to destroy pathogens; and **albumin**, used to maintain the osmotic balance of the blood.
- Ions, such as sodium, potassium, calcium and chloride ions. These maintain the osmotic balance of the blood. Some ions have other uses as well. For example, calcium ions are involved in blood clotting.
- Hormones, such as **insulin** and **oestrogen**. These travel in the blood plasma to their target organ, where they change the way it works.
- Dissolved food substances, such as amino acids, glucose and glycerol. These travel from the gut to the cells where they are needed.
- Oxygen. Although most of the oxygen is carried by the haemoglobin in the red blood cells, some oxygen is dissolved in the blood plasma.
- Waste products, such as **urea** and carbon dioxide (some in solution and some in the form of the hydrogen carbonate ion). These are taken from the cells where they are produced to the part of the body where they can be excreted.
- Heat. Blood plasma transfers heat around the body, helping to keep the body temperature constant.

### Serum, tissue fluid and lymph

When the blood-clotting protein, fibrinogen, has been removed from the blood plasma, the solution which remains is called **serum**. In other ways, it is the same as blood plasma. Serum can be useful to treat patients in hospital. You will learn more about this in spread 1.1.4.2.

**Tissue fluid** is formed when blood passes through the **capillaries**. Capillary walls are permeable to everything in the blood except most blood cells and the large plasma proteins. At the arterial end of the capillaries, the blood is under enough pressure for the

H δ⁺

δ⁻  O  Small positive charge

Small negative charge

H δ⁺

**Figure 1** A water molecule

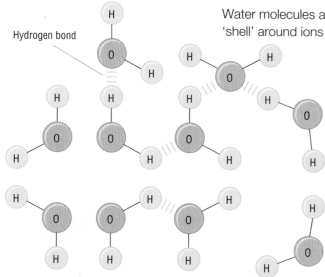

Hydrogen bond

**Figure 2** Weak hydrogen bonds exist between the hydrogen atom of one water molecule and the oxygen atom of another water molecule

### Examiner tip

Non-polar molecules cannot be carried on their own in plasma. They have to be combined with a protein – see LDLs in spread 1.1.2.7

**Solvent properties:**
Allow water to act as a transport medium for polar molecules, for example, transport via blood and lymph and removal of metabolic wastes such as urea in urine

**Lubricant properties:**
Water's cohesive and adhesive properties mean that it is viscous, making it a useful lubricant, for example,

**Pleural fluid**
Minimises friction between lungs and thoracic cage (ribs) during breathing;

**Mucus**
Permits easy passage of faeces down the colon, and lubricates the penis and vagina during intercourse.

**Thermoregulation:**
The high specific heat capacity of water means that bodies composed largely of water (cells are typically 70–80% water) are very thermostable. This means they are less prone to heat damage by changes in environmental temperatures.
The high latent heat of vaporisation of water means that a body can be cooled by evaporative water. This is used in sweating – a part of thermoregulation

**Supporting role:**
The cohesive forces between water molecules mean that it is not easily compressed, thus it is an excellent medium for support, e.g. *amniotic fluid* (which supports and protects the mammalian fetus)

**Transparency:**
Water allows light to penetrate so plants can photosynthesise

**Figure 3** The properties of water are fundamental to human life

smaller components of the blood plasma to be squeezed out. This allows the exchange of materials between the blood and the tissues. You can see this in Figure 4.

Some of this tissue fluid returns to the blood capillaries at the **venule** end. The rest of the tissue fluid drains into blind-ended **lymphatic** capillaries. Once the fluid is inside these vessels, it is called **lymph**. The lymph capillaries drain into larger lymph vessels. The fluid moves through these lymph vessels by contractions of the body's muscles that surround them. The lymph vessels also contain valves to prevent backflow. The fluid moves very slowly through the lymph system, but eventually the lymph is returned to the bloodstream at a vein in the neck region.

Plasma filtered under pressure – tissue fluid leaves capillary

Tissue fluid enters capillary

High hydrostatic pressure

Low hydrostatic pressure

Arteriole

Venule

Red blood cell

Tissue cells

Capillary

Tissue fluid

Lymph capillary

Tissue fluid

**Figure 4** The relationship between plasma, tissue fluid and lymph

# Questions

1 Hydrogen bonds are found between molecules of water. Name **one** other example of where hydrogen bonds are found.

2 Complete the table with 'higher' or 'lower' to compare the composition of blood plasma in two different blood vessels.

| Component | Vein leaving a leg muscle | Aorta |
|---|---|---|
| Oxygen | | |
| Glucose | | |
| Carbon dioxide | | |

3 Complete the table to compare serum, blood plasma, tissue fluid and lymph.

| Component | Blood plasma | Serum | Tissue fluid | Lymph |
|---|---|---|---|---|
| Fibrinogen | Present | | | |
| Albumin | Present | | | |

## Diffusion

Diffusion is the net movement of a substance from a region where it is in higher concentration to a region where it is in lower concentration. This continues until the molecules are evenly distributed. It results from the random movement of molecules, and does not require additional energy. Therefore, diffusion is said to be a **passive** process.

Small, lipid-soluble molecules can diffuse across the plasma membrane through the phospholipid bilayer. The structure of cell membranes can be seen in spread 1.1.1.4.

### Facilitated diffusion

Molecules that are soluble in water, or charged particles such as ions, cannot diffuse through the phospholipid bilayer. These molecules have to diffuse through the plasma membrane with the help of proteins. This kind of diffusion is called **facilitated diffusion** (facilitated means 'helped').

Some of these are **protein channels** that are permanently open. The protein channel is lined with **hydrophilic** amino acids and water.

Molecules can also diffuse through the membrane by binding to **carrier proteins**. You can see how these work in Figure 1. The molecule binds to the carrier protein. This causes the protein to change shape and release the molecule on the other side of the membrane. No additional energy is used. This means that facilitated diffusion, like simple diffusion, is a passive process. One example of facilitated diffusion is the diffusion of glucose into red blood cells through carrier proteins.

Specific molecule fits into carrier, e.g. glucose

Carrier 'flip-flop' can carry specific molecule in either direction – movement depends on diffusion gradient

**Figure 1** Diffusion and facilitated diffusion

### Key definition

**Osmosis** is the movement of water molecules from a region of high water potential to a region of lower (more negative) water potential across a selectively permeable membrane. The weak nature of hydrogen bonds means water molecules commute easily relative to each other

### Examiner tip

- Pure water has the highest possible water potential, which is 0 kPa.
- Solutions always have a lower water potential than pure water, so they have negative water potentials.
- Water always moves from a region of higher water potential to a region of lower (more negative) water potential.
- If one cell has a water potential of −6 kPa and the next one has a water potential of −4 kPa, water will move from the cell at −4 kPa to the cell at −6 kPa, i.e. from a higher water potential to a lower water potential.

## Osmosis

Some solutes cannot pass freely through cell membranes, and this allows osmosis. **Osmosis** is really a special kind of diffusion. To explain osmosis, we use the term **water potential**. Water potential is the tendency of a solution to gain or lose water. Pure water has the highest possible water potential of zero. Adding solutes to water decreases the water potential, i.e. it makes the water potential more negative. Water molecules move by osmosis from a region of higher water potential to a region of lower water potential across a selectively permeable membrane. This occurs until the water potential is the same on both sides of the membrane. At this point, **equilibrium** has been reached. You can see this in Figure 2.

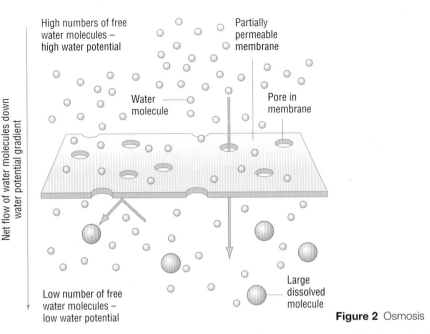

Figure 2 Osmosis

**Labels in Figure 2:**
High numbers of free water molecules – high water potential
Partially permeable membrane
Water molecule
Pore in membrane
Net flow of water molecules down water potential gradient
Low number of free water molecules – low water potential
Large dissolved molecule

## Measuring electrolytes

Electrolytes are measured by a process known as **potentiometry**. This method measures the voltage that develops between the inner and outer surfaces of an ion selective electrode. The electrode (membrane) is made of a material that is selectively permeable to the ion being measured. This potential is measured by comparing it to the potential of a reference electrode. Since the potential of the reference electrode is held constant, the difference in voltage between the two electrodes must be equivalent to the concentration of the ion. The process involved placing a small sample of body fluid (blood, plasma or urine) in a machine. The rest of the process is automated, with readings being produced quickly. In the case of potassium, getting the result back to the clinician quickly can make the difference between life and death!

Water can cross a membrane in two ways. All membranes are permeable to water to a small degree. Although it is a polar molecule, it is very small, so water can slip between the phospholipids in the bilayer. However, some membranes can be up to 1000 times more permeable to water due to the presence of membrane channels. A solution with the same water potential as a cell is said to be **isotonic** with the cell.

A solution with a lower water potential than the cell is said to be **hypertonic**.

A solution with a higher water potential than the cell is said to be **hypotonic**.

# Keeping the osmotic balance

Glucose and other solutes will dissolve in blood plasma and lower the water potential but it is mainly the concentration of **electrolytes** in plasma and in cells that is responsible for maintaining a water potential balance. Electrolytes are **ions** with a positive or negative charge. Positively charged ions are called **cations** and those with a negative charge are called **anions**. A test for electrolytes includes the measurement of sodium, potassium, chloride, and bicarbonate ions but other plasma ions can also be tested for in plasma such as calcium, magnesium and phosphate.

Levels of electrolytes in body fluids such as plasma and urine are kept within a narrow range. Monitoring these levels is necessary for the diagnosis and management of many conditions such as diabetes and kidney disease which you will cover in Unit 5. But just to illustrate how important the levels are, consider just potassium ions. Potassium plasma levels should be between 3.5 and 5.0 mmol dm$^{-3}$. In an accident and emergency department, a reading for potassium is often asked for immediately since levels which fall below 3.0 mmol dm$^{-3}$ can lead to tachycardia and cardiac arrest (see page 39) while values above 6.0 mmol dm$^{-3}$ are associated with bradycardia and heart failure.

# Questions

1 Give one similarity and one difference between diffusion and osmosis.
2 Explain why water is described as a 'polar' molecule.
3 What will happen to a red blood cell if it is put into
   (a) a hypotonic solution?
   (b) a hypertonic solution?
4 Figure 3 shows three red blood cells. They have been placed in pure water. Copy the diagram and add arrows, showing the direction in which water will move between the cells and the water.

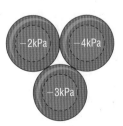

Figure 3

## Using energy

So far, we have looked at how substances enter a cell down their concentration gradient. However, it is a characteristic of living things that they can have a composition very different from their environment. The composition of their cytoplasm can also be very different from the composition of the solution that surrounds it. This can happen because substances sometimes enter or leave a cell against their concentration gradient. These substances are moved by a process called **active transport**.

You can see how active transport occurs in Figure 1.

Molecule being actively transported

Shape change of active transport protein requires energy from ATP. The shape change does not allow the molecule to go the 'wrong way'

Active transport pump is shaped so that the molecule it carries fits on one side of the membrane only

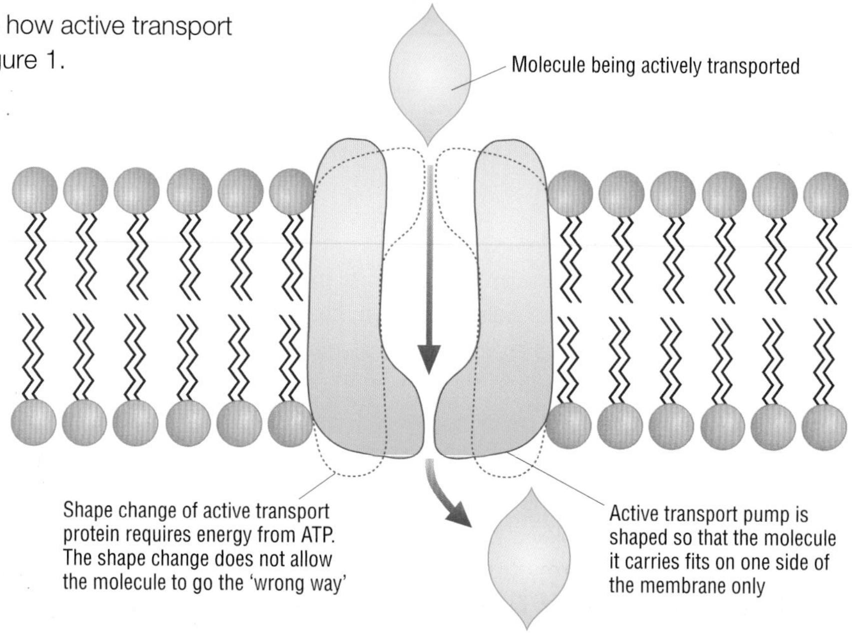

**Figure 1** Active transport

Most cells, including red blood cells, contain a protein called a **sodium–potassium pump**. This is a carrier protein that uses **ATP** energy to transport sodium ions out of the cell and potassium ions into the cell. ATP is dealt with in more detail in spread 1.1.4.1.

## Three kinds of membrane transport

In the plasma membrane of cells such as those lining the kidney tubule, you can see how the transport mechanisms we have seen so far operate together. At 1, active transport is being used to pump sodium ions out of the cell and potassium ions in. At 2, a process called co-transport is happening. The low concentration of sodium ions inside the cell allows them to diffuse in 'bringing' other ions and molecules in at the same time using the same carrier proteins. You can also see water moving in by osmosis. On the side of the cell next to the capillary, you can see how these substances will then move by diffusion into the capillary (from 3 to 4).

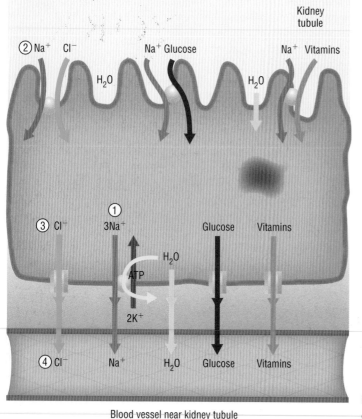

**Figure 2** Plasma membrane of cells lining kidney tubule

Blood vessel near kidney tubule

### Endocytosis and exocytosis

Sometimes substances that are larger than single molecules need to enter or leave a cell. Some kinds of white blood cells, such as neutrophils, engulf **pathogens** (for example, bacteria) that have entered the body. You can see this in Figure 3. The bacterium is engulfed by a kind of **endocytosis** called **phagocytosis**. After the bacterium has been digested, the waste products left behind are released back out of the cell by **exocytosis**. This is also an energy-requiring process that uses ATP.

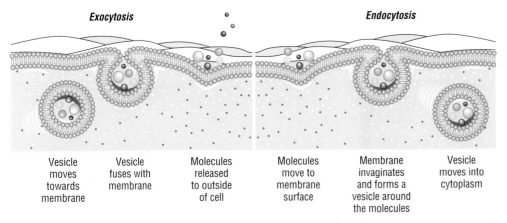

| Exocytosis | | | Endocytosis | | |

| Vesicle moves towards membrane | Vesicle fuses with membrane | Molecules released to outside of cell | Molecules move to membrane surface | Membrane invaginates and forms a vesicle around the molecules | Vesicle moves into cytoplasm |

**Figure 3** Endocytosis and exocytosis

Cholesterol is taken into cells by endocytosis. Substances such as enzymes and antibodies are secreted from cells by exocytosis.

Cholesterol is a non-polar molecule. As you already know, this means it must be transported through the blood attached to proteins. Cholesterol is transported through the bloodstream in the form of lipoprotein particles, the most common of which is called **low-density lipoprotein**, or **LDL**. LDLs have to bind to a specific protein in the cell membrane. Then the LDL is taken into the cell by endocytosis. The cholesterol is then released for use by the cell, and the receptor protein returns to the cell surface membrane to be used again.

Some people have an inherited condition known as familial hypercholesterolemia. Patients with this condition have very high levels of cholesterol in their blood and suffer heart attacks early in life. Scientists have found that many people with this condition have cell membranes that do not have the specific receptors for LDLs. This means that the cells cannot take up LDLs from the blood.

YORK
COLLEGE
Sim Balk Lane
York
YO23 2BB

## Questions

1  Give one similarity and two differences between active transport and facilitated diffusion.
2  The table shows the concentrations of two ions in a cell and in the solution that surrounds the cell. Explain how these ions enter the cell.

| | Ion concentration/mg dm$^{-3}$ | |
|---|---|---|
| | Sodium | Chloride |
| Solution surrounding cell | 0.55 | 0.61 |
| Cell | 0.04 | 0.60 |

## Measuring blood glucose concentration

Blood glucose levels can be measured by a doctor, or by an individual, using a blood glucose meter. This is particularly important for people who have diabetes mellitus. This is a condition in which the person's blood glucose level is not well controlled. If a person with diabetes knows that their blood glucose level is too high or too low, they can take action to bring about a change.

Most blood glucose meters use a test strip. This contains an enzyme, glucose dehydrogenase or glucose oxidase. When a blood sample comes into contact with the enzyme, glucose in the blood is converted to gluconolactone, and a small electric current is produced. This is detected by an electrode on the test strip. The meter gives a digital reading for the blood glucose level within 15–30 seconds.

To obtain a reading from a blood glucose meter, a person washes their hands thoroughly and swabs the skin with alcohol. They use a sterile lancet to prick the skin on the top of their finger. A small drop of blood is placed on a clean glucose test strip that has been placed in the glucose test meter. After 15–30 seconds, the blood glucose concentration will appear on the screen of the meter in the form of a digital display. You can see this in Figure 1.

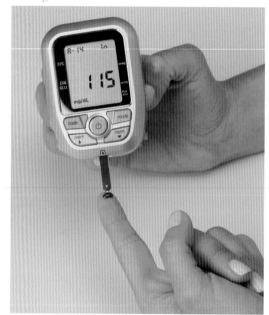

**Figure 1** A blood glucose meter

Doctors advise patients on how often to test their blood glucose. The frequency of testing depends on the kind of diabetes and how well controlled it is.

Blood glucose is measured in millimoles per litre ($mmol\,dm^{-3}$). Normal blood glucose levels stay within $4\text{–}8\,mmol\,dm^{-3}$ throughout the day, although it can vary. Ideally, blood glucose should be $4\text{–}7\,mmol\,dm^{-3}$ before meals, less than $10\,mmol\,dm^{-3}$ 90 minutes after a meal, and around $8\,mmol\,dm^{-3}$ at bedtime.

### Monosaccharides

Glucose is an example of a monosaccharide. Monosaccharides are simple sugars. You can see the structure of glucose in Figure 2. The kind of glucose found in human blood is alpha glucose.

Glucose is used as a respiratory substrate. This means that it is easily broken down by cells during cellular respiration. The energy released from the glucose is used to make ATP. Glucose is also very soluble, so it is easy to transport in the blood plasma.

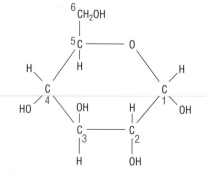

**Figure 2** The structure of alpha glucose

## Disaccharides

Two monosaccharides can be joined together to make a disaccharide. When two glucose molecules join together, the disaccharide maltose is formed. The two molecules join together by a condensation reaction. The bond produced is called a glycosidic link. You have already learned about condensation reactions on spread 1.1.2.1.

Figure 3 shows how two glucose molecules join together to form maltose. Sucrose, the sugar you use in tea or coffee, is made from a molecule of glucose and a molecule of fructose.

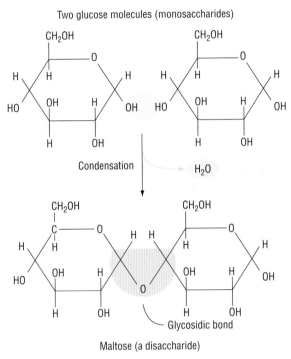

**Figure 3** The formation of maltose

## Polysaccharides

Polysaccharides are formed when many monosaccharides join together by condensation reactions. When many alpha glucose molecules join together in human cells, a polysaccharide called **glycogen** is formed. One glucose residue in a chain can form a glycosidic link with three different glucose molecules. This produces a branched molecule. You can see this in Figure 4. Glycogen is stored in liver and muscle cells.

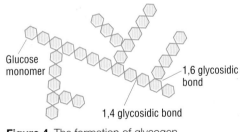

**Figure 4** The formation of glycogen

Glycogen is an ideal storage molecule because:
- It is insoluble, so it does not affect the water potential of cells.
- It is compact, so a lot of glucose can be stored in a small space.
- It is a branched molecule, so there are lots of 'ends' where glucose can be released quickly when needed.

## Questions

1 Use your knowledge of water potential to explain what will happen to a red blood cell if a person's blood glucose level is too high.
2 Explain why glycogen is stored in liver and muscle cells, rather than glucose.
3 Nerve cells require large amounts of ATP for active transport. They do not store glycogen. Why might low blood glucose levels lead to loss of concentration?

Lipids are a group of molecules that include fats, oils and cholesterol. Lipids are non-polar (that is, they are not charged) and hydrophobic, so they are insoluble in water.

## Fats and oils

Fats and oils are also known as triglycerides. They are made up of a glycerol molecule joined to three fatty acids. A fatty acid is made up of a hydrocarbon chain with a carboxylic acid group attached.

Three fatty acids join on to a glycerol molecule by a condensation reaction. The bonds formed are called ester bonds. The molecule formed is called a triglyceride.

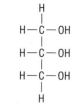

**Figure 1** Glycerol

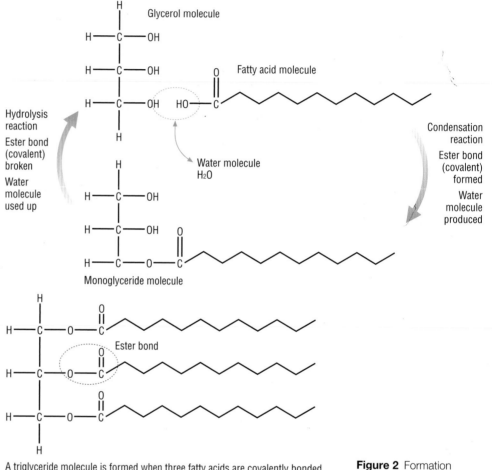

A triglyceride molecule is formed when three fatty acids are covalently bonded to a glycerol molecule

**Figure 2** Formation of a triglyceride

### Fatty acids

Look at Figure 3. The fatty acid in 3a has a hydrocarbon chain in which all the carbon–carbon bonds are single. This means that it has the maximum number of hydrogen atoms in the chain. It is **saturated** with hydrogen atoms. We call this kind of fatty acid a **saturated fatty acid**.

The fatty acid in 3b has some double bonds between the carbon atoms. This means that it does not have room for so many hydrogen atoms in the chain. This kind of fatty acid is called an **unsaturated** fatty acid.

Saturated fatty acids tend to have higher melting points than unsaturated fatty acids. Oils, which are liquid at room temperature, tend to contain more unsaturated fatty acids. Fats, which are solid at room temperature, tend to contain more saturated fatty acids.

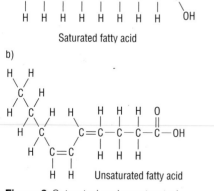

**Figure 3** Saturated and unsaturated fatty acids

## Uses of lipids in the body

- Triglycerides are the main energy storage molecules in the human body. Triglycerides are insoluble in water, and contain more energy per gram than carbohydrates. This makes them ideal to be stored in adipose tissue. Adipose tissue is the correct name for what we usually call body fat.
- Adipose tissue acts as heat insulation. It lies under the skin and around delicate organs, helping to reduce heat loss from the body. It also protects delicate organs, such as the kidney, from mechanical damage.
- Fat soluble vitamins, such as vitamins A and D, are stored in lipid globules inside liver cells.

## Cholesterol

Cholesterol is a kind of steroid. Steroids are usually classed as lipids, but they do not have the same kind of structure at all. You will remember from spread 1.1.2.5 that cholesterol is important in keeping cell membranes fluid. Other steroids are important as hormones, for example testosterone and oestrogen.

## Phospholipids

You will remember from spread 1.1.1.4 that phospholipids are polar molecules. They form a bilayer that is important in cell membranes. Phospholipids contain glycerol and two fatty acids. However, instead of the third fatty acid, a phosphate group is attached. You can see this in Figure 4.

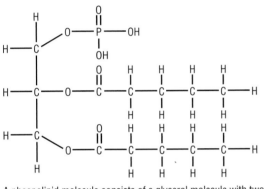

A phospolipid molecule consists of a glycerol molecule with two fatty acids and a phosphate group.

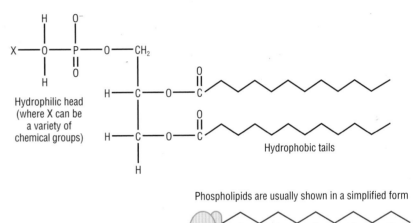

Hydrophilic head (where X can be a variety of chemical groups)

Hydrophobic tails

Phospholipids are usually shown in a simplified form

Phosphate 'head'    Fatty acid 'tails'

**Figure 4** Formation of a phospholipid

## Transport of lipids

Lipids are not soluble in water, so they cannot be transported round the body dissolved in the blood plasma. When the triglycerides in the adipose tissue are broken down, they form glycerol and fatty acids. Glycerol is able to dissolve in blood plasma. However, the fatty acids combine with plasma proteins and are carried in the blood as small globules.

# Questions

1 Give two other examples of condensation reactions.
2 Look at Figure 5. Explain why this structure is able to carry lipids in the blood.

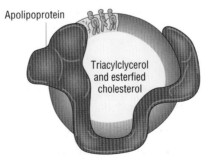

Apolipoprotein

Triacylclycerol and esterfied cholesterol

**Figure 5** Lipoproteins

## Preventing excessive blood loss

When a blood vessel is damaged and bleeding occurs, a cascade of reactions occurs in the blood. These will result in the formation of a blood clot to stop continued blood loss. But if the injury is severe, there will be too much blood loss for a clot to form.

If a person is bleeding badly, you should seek medical help immediately. There are some first aid procedures that you can carry out to reduce blood loss until medical help arrives.

- If you can, put on a pair of disposable gloves.
- Reassure the person, and get them to sit or lie down.
- Look carefully at the wound. You may need to cut the clothing away to see it clearly. Check that there is nothing in the wound, such as a piece of glass.
- If there is nothing in the wound, place a large pad of clean cloth onto the wound and press it down firmly using your hand.
- Use a bandage to hold the pad firmly in place.
- If there is an object in the wound, such as glass, do not remove it. Instead, make a pad in the shape of a ring and place this on the wound so that it surrounds the object.
- Now use a bandage to apply pressure on the ring around the sides of the wound. The pressure should push the edges of the wound together.
- If the wound is in an arm or a leg, raise the limb higher, for example, by holding the person's arm in the air, or raising the leg up on cushions.
- If blood soaks through the first pad, do not remove the first pad but apply another on top.

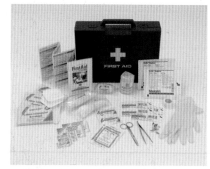

**Figure 1** First aid medical kit

### How blood clots

When tissues are damaged, they are exposed to the air. **Collagen** fibres in the connective tissue are exposed, and **platelets** stick to them. The platelets release a chemical that makes other platelets sticky, so the platelets clump together to form a plug, forming an initial barrier to further blood loss. Calcium is also needed for this process.

The white blood cells (leucocytes) that collect at the site, and the exposed tissues just below the endothelium, release an enzyme called **thromboplastin**. Platelets also break down and release thromboplastin. Thromboplastin then catalyses the conversion of an inactive **plasma protein**, **prothrombin**, into **thrombin**. This reaction also requires calcium ions. Thrombin is an active enzyme. Thrombin then hydrolyses a large, soluble plasma protein called **fibrinogen** into smaller units. These smaller units then join together (polymerise) to form long, insoluble fibres of the protein **fibrin**. This process also requires the presence of calcium ions. The fibrin fibres pile up in tangles that form a mesh over the wound. Blood cells become trapped in this mesh, forming a blood clot. The clot dries to form a scab that prevents further blood loss. It also stops pathogens getting into the wound. This process is shown in Figure 2.

### A closer look at thrombin

Thrombin is an enzyme. An enzyme is a protein that catalyses a specific reaction in the body. It does this by reducing the **activation energy** needed for a reaction to occur.

For molecules to react, they need to have a certain amount of energy. We call this energy **activation energy**. Enzymes lower the amount of energy needed for a reaction to occur. This means that reactions can take place at body temperature and atmospheric pressure, which might otherwise need much greater temperatures or pressures. You can see how an enzyme lowers the activation energy needed in Figure 4.

(a) Blood vessels cut

Platelets
Blood cells
Blood clotting factors

(b) Platelets form a plug at cut, interspersed with clotting factors

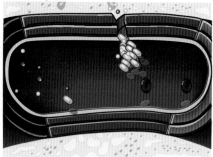

(c) Fibrin mesh forms scab, sealing cut from outside and in

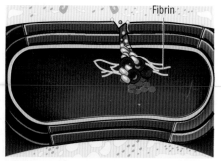

Fibrin

**Figure 2** The process of blood clotting

Enzymes lower the activation energy needed because they have a very specific three-dimensional shape. You will remember from spread 1.1.2.1. that enzymes are globular proteins with a highly specific **tertiary structure**. Enzymes have a particular place within them called the **active site**. This is exactly the right shape for one specific substrate to fit in. It works rather like a key fits into a lock – in a way, the key has the opposite shape to the lock. When the substrate is in the active site, it forms an **enzyme-substrate complex**. Because they fit together so closely, this enables the enzyme to exert forces on the substrate. This lowers the activation energy required. You can see this in Figure 5. After the reaction has taken place, the enzyme is unchanged and can be used over and over again.

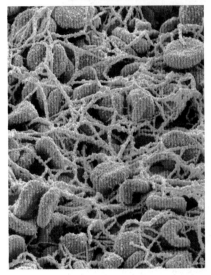

**Figure 3** Electronmicrograph of a blood clot showing erythrocytes and fibrin

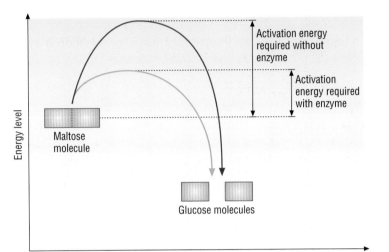

Adding the enzyme maltose reduces the amount of activation energy required for the reaction to take place

**Figure 4** How enzymes lower activation energy

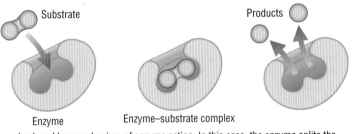

Lock and key mechanism of enzyme action. In this case, the enzyme splits the substrate molecule into two smaller products. it is worth noting that given the right circumstances, the enzyme can work in reverse.

**Figure 5** Mechanism of enzyme action

**Examiner tip**

Remember that the enzyme has the active site and that the substrate has a complementary shape to the active site. Never say that the substrate has an active site, or that it has the *same* shape as the enzyme's active site!

# Questions

1 Explain why it is a good idea to wear disposable gloves when giving first aid for bleeding.
2 Suggest why you should not remove an object, such as a piece of glass, that is in the wound.
3 A bandage placed on a wound should be firm but not too tight. Suggest why.
4 Suggest why it is a good idea to raise a limb if it is wounded.
5 Explain why you should apply a second pad over the first if blood soaks through, rather than removing the first pad.
6 Thrombin converts soluble fibrinogen into fibrin by a hydrolysis reaction. What is a hydrolysis reaction? Give two other examples of a hydrolysis reaction.
7 Explain why thrombin catalyses the reaction of fibrinogen into fibrin, but not any other reaction.

## The effect of enzyme concentration

The concentration of enzymes affects the rate at which an enzyme-controlled reaction occurs.

You will see that when the enzyme concentration is low, the rate of reaction is also low. However, as more enzyme is added, the rate of reaction becomes faster. When the enzyme concentration is low, only a few of the substrate molecules can form an enzyme-substrate complex at any one time. The rest of the substrate molecules are 'waiting' for an active site to become available. However, as you add more enzyme molecules, there are more active sites to be filled. More substrate molecules can form enzyme-substrate complexes, and so the product is produced more quickly.

This is true only if there is an excess of substrate available. In other words, there is a constant high concentration of substrate molecules available.

### Investigating a 'clotting' reaction

The enzyme rennin is used to 'clot' the proteins in milk during the making of cheese. This is a good model for investigating the effect of enzyme concentration on the rate of an enzyme controlled reaction.

**MATERIALS**
- 50 cm³ 1% rennin (available from Health Food shops, supermarkets and chemists as Vegeren®) or from suppliers such as NCBE.
- 50 cm³ pasteurised milk
- distilled water
- test tubes and rack
- graduated pipettes or the equivalent
- water bath and thermometer
- stop watch and black card

**Safety note**
The hazard of a 1% enzyme solution is very low but avoid contact with skin or rubbing eyes with contaminated hands. Wash off any splashes. If the enzyme solution is made using powdered enzyme, avoid inhaling the powder as this can cause an allergic reaction.

**BASIC METHOD**
Make up a series of dilutions of the rennet by varying the ratio of rennet to distilled water. A range of 5 concentrations (0.2, 0.4, 0.6, 0.8 and 1.0%) and one tube with just water could be used as a 'control'.

Set up a water bath and keep the temperature around 35 °C. Monitor this with a thermometer and adjust it. Temperature is another **variable** which affects the rate of any reaction so this must be controlled.

Place 5 cm³ of milk and 5 cm³ of enzyme solution in separate test tubes in the water bath and leave for 5 minutes so they are both at the correct temperature – this is **equilibration**.

Then mix the enzyme solution and milk together and start the timer. Check the contents of the tube every minute by holding it against the black card and tilting the tube. When the milk proteins begin to 'clot' you will see small granules appearing. You need to repeat this at least three times for each concentration to get a **reliable** result.

**KEY TERMS**
*Independent variable*
This is the factor whose effect you are investigating – 'What you will change'. In this case, it is the enzyme concentration. The independent variable always goes in the first column of the table and on the *x*-axis of a graph.

*Dependent variable*

This is the factor you are measuring – in this case the time taken for the milk to clot. This always goes on the *y*-axis of the graph. All other variables such as temperature and volume and, in this case, the substrate concentration (milk) need to be kept constant. These are sometimes called the control or CONFOUNDING variables.

*Control*

In this experiment, the water acts as a control. This will confirm that any 'clotting' you see is due to the enzyme activity. In this case, if the milk clotted on its own, this would not be a **valid** investigation.

**QUESTION**

What would be the effect on the clotting time if the experiment was carried out at a) a lower temperature and b) a temperature above 50 °C?

**ANSWER**

a) time would be longer (as rate would be slower). b) at high temperatures, the enzyme could be denatured so no clotting may occur.

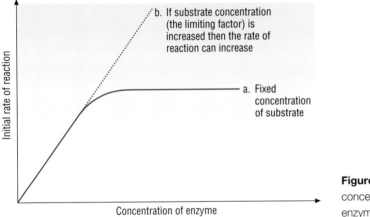

**Figure 1** Effect of enzyme concentration on the rate of an enzyme-controlled reaction

## The effect of substrate concentration

If the amount of enzyme is kept the same, but only a low concentration of substrate is available, there will be a low rate of reaction. You can see this in Figure 2. This is because there are some active sites still available. As more substrate is added, the rate of reaction increases. However, when the substrate concentration is very high, there is a constantly high rate of reaction. This is because all the active sites are being filled, and now some of the substrate molecules are 'waiting' for an active site.

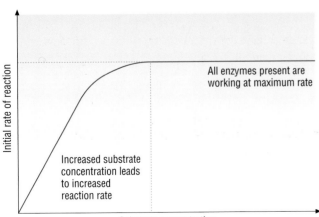

**Figure 2** Effect of substrate concentration on the rate of an enzyme-controlled reaction

## Question

1 Afibrogenaemia is a rare condition in which there is a low level of fibrinogen in the blood. Suggest the symptoms of this condition.

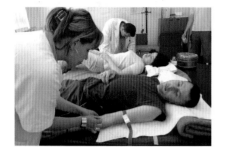

**Figure 1** A blood donating session

## Blood transfusion

### Storing blood

You can see from Figure 1 that blood is collected from a patient and stored in a plastic bag as shown in Figure 3. A small sample is also collected for testing. Blood taken in an evening donor session needs to be stored overnight, and then the blood needs to be stored before it is used. It is very important that the blood is stored in the right conditions, so that it is in perfect condition when given to a person who needs it. It is also important that the blood does not clot.

### The effect of temperature on enzyme activity

Blood clotting results from the activities of enzymes. You will also remember that enzymes are globular proteins. Temperature affects enzyme activity, as you can see in Figure 2. At low temperatures, an enzyme-controlled reaction occurs only very slowly, or not at all. This is because the enzyme and substrate do not have enough kinetic energy, and so they do not collide with each other very often. As temperature increases, the enzyme and substrate have more kinetic energy. They collide more often and the rate of reaction is faster. However, once the temperature passes a certain point, called the **optimum temperature**, the rate of reaction starts to slow down again. This is because the higher temperature is making the enzyme vibrate more. You will remember that a protein is held in its tertiary shape by some weak bonds, such as hydrogen bonds. At higher temperatures, these weak bonds break, changing the shape of the protein. This means that the active site of the enzyme is no longer the right shape for the substrate to fit into. We say that the enzyme is **denatured**.

You can see that blood must be stored at a temperature that is low enough to prevent enzyme activity taking place. It is also important that other blood proteins, such as haemoglobin, are not allowed to denature. However, if we were to freeze blood, ice crystals would form inside the blood cells. These would damage the cell membranes so that the cells would be destroyed when the blood was thawed out. Therefore, blood is stored at 4 °C.

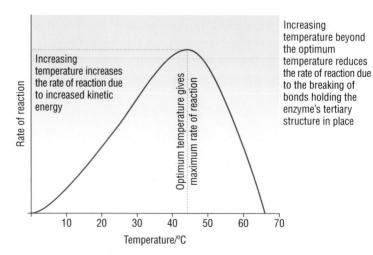

**Figure 2** The effect of temperature on an enzyme-controlled reaction

**Figure 3** A blood bank. The blood is kept at a constant temperature of 4 °C. If the temperature fluctuates by more than 2 °C, an alarm sounds to warn the staff

### The effect of pH on enzyme activity

pH is another factor that can affect enzyme activity. pH is a measurement of how acid or alkaline a solution is.

The weak bonds that hold a protein in its globular, tertiary shape rely on weak positive and negative charges. The more acid a solution is, the more hydrogen ions ($H^+$) are present. These can affect the charges on the molecule, causing weak bonds such as hydrogen bonds to break. This means that the enzyme's tertiary structure is altered, and the active site is no longer the right shape for the substrate to fit into it. In other words, the enzyme is denatured.

If the pH of stored blood changes, not only would enzymes denature, but blood proteins would also become denatured. To stop this happening, a **buffer solution** is added to the stored blood. A buffer solution keeps the pH of the blood the same.

Hydrogen bonds hold structures like an α-helix in place in protein molecules

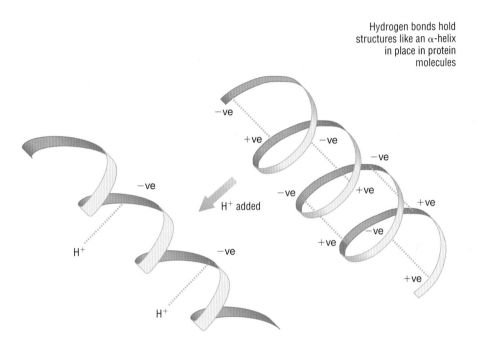

H⁺ added

As $H^+$ is increased in concentration, the positive charges are attracted to the negative charges on the α-helix and so 'replace' the hydrogen bonds

**Figure 4** The effect of pH on enzyme activity

### Enzyme cofactors

Cofactors are substances that are needed for an enzyme-controlled reaction to occur. You will remember from spread 1.1.3.1 that calcium ions are needed for blood clotting enzymes to work. Calcium ions are normally present in blood plasma. They are also released by damaged platelets. When blood is stored for transfusion, it is important that calcium ions are removed, to stop the blood clotting. A substance such as sodium citrate may be added to the blood. This removes the calcium ions.

## Questions

1 Once a protein has denatured, it cannot return to its original shape. Suggest why.
2 Haemoglobin is a protein found in the blood. Name two other proteins found in blood.

### Case study

When Emily was 15, she was diagnosed with aplastic anaemia. This is a rare disorder in which the bone marrow does not produce enough blood cells. The normal blood-forming cells in the bone marrow become replaced with fat cells. However, Emily is able to lead a normal life because she has regular blood transfusions. She is very grateful to the many people who have donated blood.

Emily's older brother, David, decided to become a blood donor. Any healthy person between the ages of 17 and 59 can donate blood. David was rather worried about giving blood before he attended his first session. He went along one evening after work to a session held in his local community centre. It only took about an hour and didn't hurt at all. Now he gives blood regularly and is encouraging his friends at work to do the same.

## Types of blood products

When a blood donation is taken from a donor, it is not always used directly for transfusion. Most of the blood is processed into different forms. These are needed for different purposes.

Some types of stored blood product are listed below.

| Type of stored blood product | Uses |
|---|---|
| Whole blood | This contains everything that is in blood, i.e. **erythrocytes** (red blood cells), **leucocytes** (white blood cells), **platelets** and **plasma**. It was more commonly used in the past. It is rarely used now, except for cases like severe blood loss |
| Leuco-depleted blood | This is blood that has had as many of the leucocytes removed as possible. This is particularly important for patients who receive repeated transfusions, e.g. Emily in the case study on page 31. White blood cells are more likely to provoke the patient's immune system into making **antibodies**. This can cause problems with future transfusions |
| Packed red cells | The red blood cells are separated from the rest of the blood and stored. When needed, the cells are diluted with a salt and sugar solution. There will be no white blood cells present in the final solution. This is used in many kinds of transfusion, especially for people with anaemia, or to replace lost red blood cells after surgery or childbirth |
| Platelets | Platelets are particularly useful for patients with bone marrow failure. They are also used following transplant and chemotherapy treatments, and for patients with leukaemia |
| Clotting factors | Plasma from blood donations can be processed to provide **clotting factors**. There are many soluble proteins in plasma that help the blood to clot. Some people may have a condition in which one of these blood clotting factors is missing. This means that their blood is slow to clot. For example, a person with classical haemophilia does not have factor VIII (Factor 8) in their blood. These people can be treated with transfusions of factor VIII |
| Plasma | Plasma is blood from which all the blood cells have been removed. Fresh frozen plasma is used during cardiac surgery, to reverse any anti-coagulant treatment and when a woman has lost blood during childbirth. It is also used to replace clotting factors after major transfusions or when clotting factors are not being produced in sufficient quantities, such as liver disease |

### How science works

In addition to blood being screened for antibodies, it is now possible to detect the nucleic acid from the virus even before viral antibodies are present. However, HIV can also be transmitted in BLOOD PRODUCTS and as plasma is pooled to produce a single batch, one HIV infected donor could potentially infect huge numbers of people. For this reason, although blood cannot be treated to destroy the virus, blood products can be. Treatment involves the use of cold ethanol in increasing concentrations. Heat can be used, but only if the blood product is stable. Improved safety and purity of blood products has resulted in an increase in their cost. One way around the problem has been to use recombinant DNA technology to produce factor 8 (see spreads 5.1.2.1 and 5.1.2.2).

### Screening blood

When a person volunteers to donate blood, it is important that their health is checked. They will be asked a number of questions about their health. Some of the guidelines are shown in Figure 1.

You will remember from page 30 that a small sample of the blood is kept aside for testing. Donated blood is screened for a number of infections, including HIV and hepatitis C. These are both viral infections. If a person has been exposed to these viruses, their blood will contain antibodies to the virus. The blood is tested by adding a small drop of blood to the antigens from these viruses. If the blood contains the right antibody, it will attach to the antigen. This screening test is carried out automatically, using a machine.

The blood is also tested to find out the **blood group**. This is because patients must receive blood of the correct blood group when they need a transfusion.

## Blood group distribution

All of us will belong to one of the ABO blood groups – groups A, B, AB and O. We inherit the genes which determine our blood group from our parents.

However, the distribution of the different blood groups around the world varies considerably. For example, blood group A is present in about 21% of the world's population but in some groups of people, such as the Lapps in Northern Scandinavia, over 50% of people are blood group A. Blood group B is rarer, with only 16% of people having this group but, in Central Asia for example, some populations have over 25% with this blood group, while in South America the figure rarely rises above 5%. Blood group O is the most common but even this varies. About 63% of humans are blood group O, but in Central and South America this rises to almost 100%!

In the United Kingdom, the National Blood Service has being developing a strategy to attract more people from BME communities (black and minority ethnic communities) to give blood. A glance at Table 1 shows that, while this has been successful in that there has been a growth in donations from BME communities, there is still a long way to go.

| Ethnicity | 2001/2002 | 2002/2003 | 2003/2004 | 2004/2005 | 2005/2006 |
|---|---|---|---|---|---|
| White | 98.9% | 98.6% | 98.2% | 97.9% | 97.7% |
| Mixed | 0.37% | 0.45% | 0.53% | 0.59% | 0.63% |
| Asian | 0.32% | 0.50% | 0.70% | 0.86% | 0.94% |
| Black | 0.26% | 0.30% | 0.36% | 0.37% | 0.41% |
| Chinese | 0.07% | 0.10% | 0.14% | 0.16% | 0.17% |
| Other | 0.07% | 0.09% | 0.11% | 0.13% | 0.17% |

**Table 1** Blood donations in the UK according to ethnicity

**You should not give blood if:**

1 You've already given blood in the last 12 weeks (normally, you must wait 16 weeks).
2 You have a chesty cough, sore throat or active cold sore.
3 You're currently taking antibiotics or you have just finished a course within the last seven days.
4 You've had hepatitis or jaundice in the last 12 months.
5 You've had ear or body piercing or tattoos in the last 6 months.
6 You've had acupuncture in the last 6 months outside the NHS (unless you can produce the approved certificate from your acupuncturist or physiotherapist).
7 A member of your family (parent, brother, sister or child) has suffered with CJD (Creutzfeld-Jakob Disease).
8 You've ever received human pituitary extract (which was used in some growth hormone or fertility treatments before 1985).
9 You have received blood or think you may have received blood during the course of any medical treatment or procedure anywhere in the world since 1st January 1980.

**You may not be able to give blood if:**

1 You've had a serious illness or major surgery in the past or are currently on medication. Please discuss this with the clinical staff. The reason you're taking medicines may prevent you from donating.
2 You've had complicated dental work. Simple fillings are OK after 24 hours, as are simple extractions after 7 days.
3 You've been in contact with an infectious disease or have been given certain immunisations in the last four weeks.
4 You're presently on a hospital waiting list or undergoing medical tests.
5 You do not weigh over 50kgs (7st 12).

**Pregnancy**

You should not give blood if you are pregnant or you are a woman who has had a baby in the last 9 months.

**Travel abroad**

Please wait 6 months after returning from a malarial area before giving blood. Please also tell us if you have visited Central/South America at any time. (Those who've had malaria, or an undiagnosed illness associated with travel, may not however be able to give blood.)

If you are unsure please call our 24 hour donor helpline on 0845 7 711 711.

**West Nile Virus**

Have you been to or plan to go to CANADA or the UNITED STATES this Summer? If yes, please visit http://www.blood.co.uk/pages/west_nile.html as it might affect you giving blood.

**Figure 1** Advice from the National Blood Transfusion Service

## Questions

1 Why can blood plasma be stored frozen, but other blood products cannot?
2 Explain why people should not give blood if they:
   (a) are pregnant
   (b) have recently had a tattoo or body piercing
   (c) have worked as a prostitute.

# 1.1 Molecules and blood summary

## Self-check questions

### Fill the blanks

1  Red blood cells or .................... are biconcave discs containing a protein called .................... . They carry oxygen and carbon dioxide. .................... are white blood cells. There are several different kinds:

- .................... have small granules in the cytoplasm and engulf bacteria by .................... .
- .............. have a large darkly-stained .............. . B lymphocytes produce .................... and T lymphocytes destroy cells.
- .................... have a large, bean-shaped .................... and clear cytoplasm. They mature and move into the tissues, where they become .................... that engulf .................... .

2  Cells containing a nucleus are .................... .................... make proteins. The .................... endoplasmic reticulum makes and transports proteins in the cell. The .................... endoplasmic reticulum is concerned with making steroids. The .................... modifies proteins and packages them in vesicles for secretion from the cell. .................... contain digestive enzymes, and .................... produce ATP in aerobic respiration. A palisade mesophyll cell is found in the .................... of a plant. It contains all the same organelles as a lymphocyte, but also contains .................... that carry out photosynthesis, a cell wall made of ...................., and a permanent cell .................... .

3  A protein such as haemoglobin is made up of many .................... joined together by .................... reactions. The .................... structure of a protein is the sequence of amino acids it contains. The .................... structure of a protein is the way the polypeptide chain folds. The .................... structure of a protein is its overall three-dimensional shape. Proteins that have more than one .................... chain, such as haemoglobin, have a .................... structure.

4  Diffusion is the movement of molecules from a region of .................... concentration to a region of .................... concentration. Lipid-soluble substances can diffuse into cells through the ....................bilayer. Other molecules can diffuse through the membrane with the help of protein carrier molecules or .................... proteins. This is called .................... diffusion. .................... is the movement of water from a region of .................... water potential to region of .................... water potential. .................... transport is when a protein carrier molecule uses .................... to transport a molecule into or out of a cell .................... its concentration gradient.

5  Glucose is an example of a .................... . Two monosaccharides join together by a .................... reaction to form a .................... . Many monosaccharides can join together to form a .................... such as glycogen .................... . Glycogen acts as an energy store in liver and .................... cells. Triglycerides are formed when a .................... molecule combines with three .................... molecules by a .................... reaction. .................... fatty acids have single bonds between all the carbon atoms, while .................... fatty acids have some double bonds between carbon atoms.

6  .................... are globular proteins that catalyse .................... metabolic reactions. Each enzyme has an .................... which is specific to the shape of the enzyme's .................... . By forming an enzyme-substrate complex, enzymes lower the .................... energy needed for the reaction to occur. The rate of an enzyme-catalysed reaction is affected by factors such as enzyme or .................... concentration, pH and .................... .

# Summary questions

**1** **(a)** Draw the structure of an amino acid. [2]

**(b)** **(i)** Name the bond formed when two amino acids join together. [1]

**(ii)** Name the type of reaction that occurs when two amino acids join together. [1]

**(c)** Give **two** differences between the secondary and tertiary structure of a protein. [2]

**2** The photograph shows a type of leucocyte.

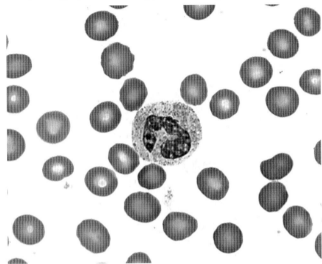

**(a)** Name the type of leucocyte shown in the photograph. [1]

**(b)** The diameter of this cell is 14 µm across. Calculate the magnification of the photograph. Show your working. [2]

**(c)** Describe how you would make a blood film like the one shown in the photograph. [4]

**3** **(a)** Complete the following table giving information about three different macromolecules.

| Macromolecule | Example | Sub-unit molecule(s) | Chemical element(s) |
|---|---|---|---|
| | Glycogen | | Carbon, hydrogen, oxygen |
| | Haemoglobin | | Carbon, hydrogen, oxygen, nitrogen, sulfur |
| | Triglycerides | Glycerol and fatty acids | |

[6]

**(b)** Glycogen is stored in liver and muscle cells. Explain why glycogen is a useful storage molecule. [3]

**4** The diagram shows a palisade mesophyll cell from a leaf.

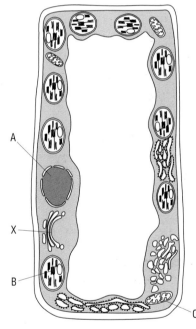

**(a)** Name organelles **A**, **B** and **C**. [3]

**(b)** Describe the function of organelle **X**. [2]

**(c)** Give **two** differences between this cell and a lymphocyte. [2]

# (1) The heart

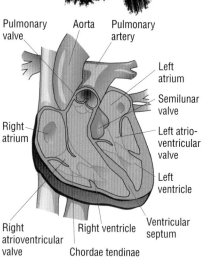

Pulmonary valve
Aorta
Pulmonary artery
Left atrium
Semilunar valve
Left atrio-ventricular valve
Left ventricle
Ventricular septum
Chordae tendinae
Right ventricle
Right atrioventricular valve
Right atrium

**Figure 1** Internal structure of the heart

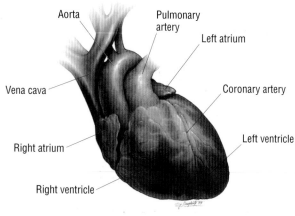

Aorta
Pulmonary artery
Left atrium
Coronary artery
Left ventricle
Vena cava
Right atrium
Right ventricle

**Figure 2** External features of the heart ×0.3

## Monitoring the heart

### How the heart works

The heart pumps blood all around your body. In fact, the heart is really two pumps, side by side. Look at Figure 1. The right side of the heart receives deoxygenated blood that has been round the body, and pumps it to the lungs. The left side of your heart receives oxygenated blood from the lungs and pumps it all around the body.

Blood enters the heart into the atria. It is then pushed into the ventricles. The ventricles pump the blood out of the heart. The heart muscle contracts and relaxes continuously throughout life. The two atria contract together, then the two ventricles contract together.

### The cardiac cycle

The first stage of the cardiac cycle is **diastole**. This is when the heart muscle is relaxed, and the atria of the heart, and then the ventricles, are filling with blood.

There are valves between the atria and the ventricles, called the **atrio-ventricular valves**. As the atria fill with blood, the blood pushes the atrio-ventricular valves open and enters the ventricles (Figure 4(a)).

The next stage of the cardiac cycle is atrial systole.

The muscular wall of the atria contracts. This pushes the blood from the atria into the ventricles, so that the atria are now emptied. This pushes the atrio-ventricular valves fully open.

The next stage of the cardiac cycle is ventricular systole.

The muscular walls of the ventricles contract. The atrio-ventricular valves are forced shut (Figure 4(b)). You will see that they can only open one way. When there is a greater pressure in the ventricles than in the atria, the atrio-ventricular valves close. They are prevented from going 'inside-out' by the tendons that attach them to the wall of the ventricles. These tendons are called **chordae tendinae**. The pressure in the ventricles forces the blood out of the ventricles.

The blood in the right ventricle is pumped through semi-lunar valves into the pulmonary artery. The blood in the left ventricle is pumped through semi-lunar valves into the aorta. The pressure of the blood in the ventricles pushes these valves open. These valves prevent any backflow into the heart.

Following ventricular systole, the heart muscle relaxes. This is diastole. The chambers of the heart refill. The cycle repeats itself about 75 times every minute.

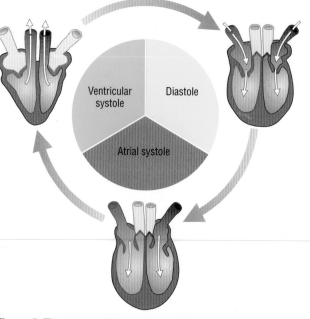

Ventricular systole
Diastole
Atrial systole

**Figure 3** The stages of the cardiac cycle

### Case study

Martin visited his GP because he had a bad, chesty cough. His doctor examined his chest. He told Martin that he would need some antibiotics to clear up the infection, and then added, 'By the way, did you know that you have a very slight heart murmur?' Martin was not aware of this, and was a little worried at first. His GP reassured him. 'It's only a very slight heart murmur. This means there is a very tiny hole in one of your heart valves. Clearly it's caused you no problems up until now. I'll refer you to a specialist to get it checked out. However, I'd like to assure you that most minor heart murmurs like this cause no problems at all, and don't need any treatment.'

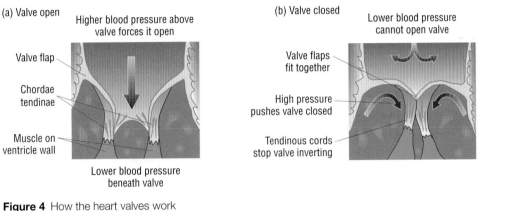

**Figure 4** How the heart valves work

### Pressure and volume changes in the heart

Figure 5 shows the pressure and volume changes in the heart during one cardiac cycle.

You will see that the pressure in the ventricle increases a lot when the muscular wall of the ventricle contracts. As the ventricle contracts, the pressure of blood inside it increases. When the pressure of blood in the ventricle exceeds the pressure in the artery, blood is forced out and the volume of the ventricle decreases.

The pressure in each atrium increases a little as it fills with blood, and then as it contracts, but it never gets very high. This is because the atrium only needs to pump the blood into the ventricle.

The pressure in the aorta does go up and down, but you will see that it never goes very low. This is important, because the blood in the aorta has only just left the heart. It needs to have enough pressure to get round the body and back to the heart again.

## Questions

1 Explain why the wall of the left ventricle is thicker than the wall of the right ventricle.
2 Use Figure 5 to estimate the heart rate, in beats per minute, for this person.
3 Figure 5 shows the pressure changes in the left side of the heart. Suggest where the line showing pressure in the right ventricle would be.

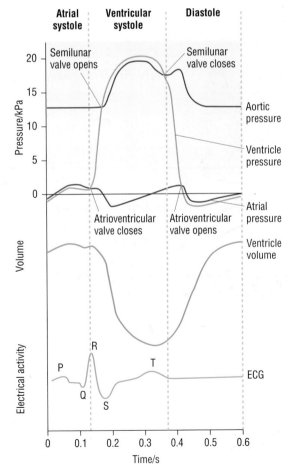

**Figure 5** Pressure and volume changes in the heart during one cardiac cycle

## Cardiac muscle

The wall of the heart is made up of a very special kind of muscle called **cardiac muscle**. It is special because it does not need any stimulation from a nerve to make it contract. We say that it is **myogenic**.

There is a group of specialised cardiac cells in the wall of the right atrium called the **sino-atrial node** (**SAN**). These cells generate electrical impulses that pass rapidly across the walls of the atria from cell to cell. As a result, the atrial walls contract, causing **atrial systole**.

The impulses cannot pass straight on from the atria to the ventricle walls, because there is a ring of fibrous tissue preventing this. The only way that impulses can pass from the atria to the ventricles is by a group of specialised muscle cells called the **atrio-ventricular node** (**AVN**), which acts as a relay point. There is a slight delay here, allowing enough time for the atria to empty completely.

From the AV node, impulses pass very quickly down heart muscle fibres called the **bundle of His** that spread down the septum between the two ventricles. This means that impulses soon reach the *bottom* of the ventricles. After this, the fibres divide into right and left branches at the tip (base) of the ventricles. They then spread throughout the muscular walls in Purkyne tissue. The impulse causes the muscular ventricle walls to contract. This is ventricular systole.

After this, there is a short time when no impulses pass through the heart muscle. This allows the muscle to relax and diastole to occur.

### Recording an electrocardiogram

An **electrocardiogram** (**ECG**) is used to monitor heart function. A cardiology technician will ask the patient to remove his clothes from the waist upwards. Then the technician will place electrodes on the arms, legs and chest. A special ECG cream is used between the electrodes and the skin. The patient is asked to lie down and remain completely relaxed, because any movement will interfere with the recording. The machine records for about 5 minutes. It gives a recording from each electrode.

Look at Figure 3a. This shows a normal ECG for a person with a healthy heart. The line represents the electrical activity in the heart during the **cardiac cycle**.

You will see that the *P wave* occurs shortly before the pressure in the atria increases. This means that the P wave represents the impulses passing from the SAN to the AVN, through the walls of the atria, leading to atrial systole.

The *QRS wave* occurs just before the pressure in the ventricle increases. This shows you that the QRS wave shows the electrical activity in the ventricles that results in ventricular systole. In other words, the QRS wave shows the electrical impulses passing down the bundle of His and along the Purkyne fibres.

The *T wave* is a short phase that occurs as the ventricles recover.

### Abnormal ECGs

Sometimes people have a problem with the electrical activity in the heart. In these cases, the ECG produced will change.

Look at Figure 3. The ECG in 3b shows ventricular **fibrillation**. You will see that this ECG looks very different. There is no P wave and no QRS wave. This is because the muscle in the heart wall is not contracting in a coordinated way. It is likely that a person with an ECG like this has had a **myocardial infarction** (heart attack) and they will almost certainly be unconscious. This person needs urgent medical attention or he will die.

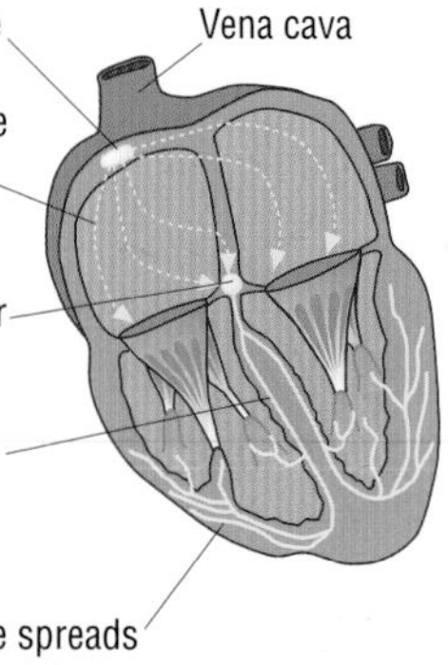

**Figure 1** Electrical activity in the heart

Sinoatrial node / Vena cava / Excitation wave spreads over atria / Atrioventricular node / Purkyne tissue carries wave down septum / Excitation wave spreads up walls of ventricles

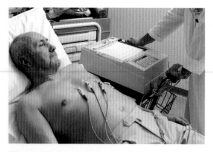

**Figure 2** Recording an electrocardiogram

The ECG in 3c shows atrial fibrillation. There is a small and unclear P wave. The deep S wave in 3d indicates ventricular **hypertrophy**, which is an increase in muscle thickness.

A heightened P wave can indicate an enlarged atrium. A raised S-T segment can indicate a myocardial infarction (see spread 2.4.1.1).

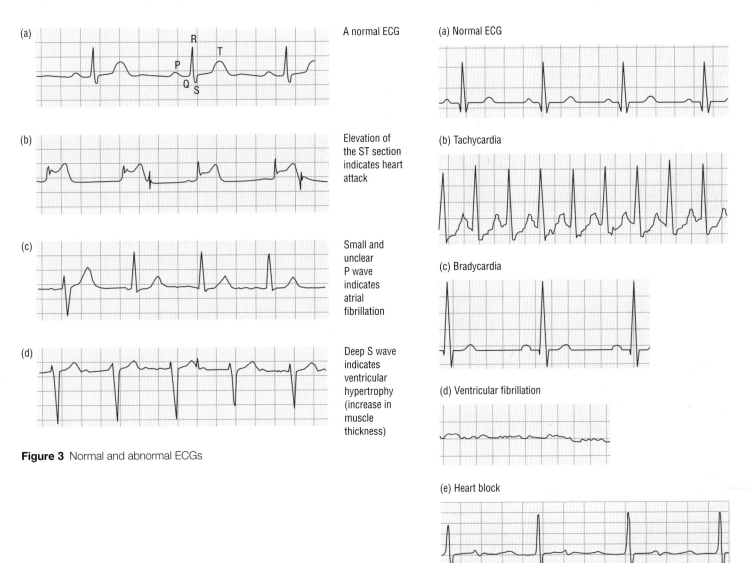

(a)                                                    A normal ECG

(b)                                                    Elevation of the ST section indicates heart attack

(c)                                                    Small and unclear P wave indicates atrial fibrillation

(d)                                                    Deep S wave indicates ventricular hypertrophy (increase in muscle thickness)

**Figure 3**  Normal and abnormal ECGs

(a) Normal ECG

(b) Tachycardia

(c) Bradycardia

(d) Ventricular fibrillation

(e) Heart block

0.2s    Timebase for all ECGs

**Figure 4**  ECGs

# Questions

1  Explain the advantage of impulses passing very quickly down the bundle of His and Purkyne tissue.
2  Explain why there is a short interval between the P wave and QRS wave in a normal person.
3  Use the scale on Figure 4 to calculate the heart rate in beats per minute of the ECGs in (a), (b) and (c).

## Changes during exercise

The **stroke volume** is the volume of blood pumped out of the left ventricle in one cardiac cycle. This is normally 60–80 cm$^3$.

The **cardiac output** is the volume of blood pumped out of the left ventricle in one minute. This figure is normally 4–8 dm$^3$ min$^{-1}$.

Therefore, cardiac output = stroke volume × heart rate.

Another effect of strenuous exercise is that body muscles contract more strongly. The muscles compress the veins and this increases the rate at which deoxygenated blood returns to the heart. The vena cava contains more blood than before, so the heart rate increases. The increased volume of blood in the heart also stretches the heart muscle. This makes the heart muscle contract more strongly. The result of this is that the stroke volume increases. If an athlete undergoes training, her stroke volume will be permanently higher than that of a non-athlete.

Look at Figure 1. You can see that a trained athlete has a greater stroke volume than a non-athlete when their heart rates are the same.

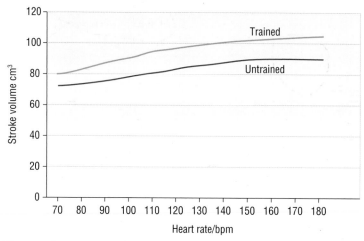

**Figure 1** Graph showing the effect of training on stroke volume

### Measuring pulse rate

The heart rate can be measured by counting how many 'pulses' are felt in an artery per minute. This is because every time the left ventricle contracts, a 'pulse' of blood pushed out into the aorta.

Your pulse can be measured anywhere where you can press an artery against a bone. What you can feel is the expansion of the artery wall as the pulse of blood (the pressure wave) passes through. Pulse rate is usually measured using the 'radial' pulse in the wrist.

If you are measuring the pulse rate of another person, wash your hands carefully before you start.
- Find the position of the radial artery.
- Press firmly against the radial artery with the second and third fingers (because your thumb and first finger have a pulse).
- Count the number of pulses in 1 minute. Alternatively, count the number of pulses in 30 seconds, then multiply by 2 to give a pulse rate in beats per minute.

There are other places where you can measure a pulse. You can see some of these in Figure 2.

There are also meters that give you a reading of your pulse rate. These can be useful to measure your pulse rate before, during and after exercise.

Table 1 shows the normal resting pulse rate for people of different ages.

| Person | Resting heart rate/beats per minute |
| --- | --- |
| Babies 0–12 months | 100–160 |
| Children aged 1–10 | 60–140 |
| Children aged 10+ and adults | 60–100 |
| Highly trained athletes | 40–60 |

**Table 1**  Normal resting pulse

## The effect of exercise on heart rate

The effects of exercise on the heart rate of two boys were investigated. Both boys started to ride on an exercise bike at 2 minutes and stopped exercising at 7 minutes. They wore heart rate monitors. The graph in Figure 3 shows how their heart rate changed before, during and after exercise.

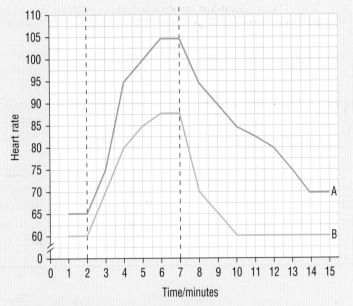

**Figure 3**  Effects of exercise on the heart rate of two boys

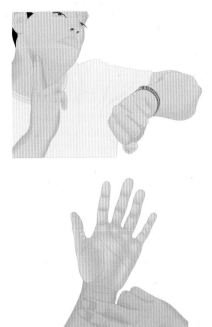

**Figure 2**  Places where you can measure a pulse

- Name the units that should be present on the *y*-axis.
- One of the boys is a highly trained athlete. Which boy is it? Explain your answer.
- How could you improve the reliability of this investigation?

## Questions

1 A person has a stroke volume of 74 cm³ and a cardiac output of 5700 cm³ when he is at rest. Calculate his heart rate in beats per minute.
2 Following vigorous exercise, the same person has a heart rate of 195 beats per minute and a cardiac output of 18 900 cm³. Calculate his stroke volume.
3 Suggest why a trained athlete has a greater stroke volume than a non-athlete when their heart rates are the same.
4 Explain why it is an advantage to an athlete to have a greater stroke volume than a non-athlete when their heart rates are the same.

## Types of blood vessel

**Arteries** carry blood away from the heart. Away from the heart, they divide into smaller vessels called **arterioles**. In turn, arterioles lead into tiny blood vessels called **capillaries**, which exchange materials between the blood and the tissues. Blood leaving the capillaries drains into vessels called **venules**. Venules join up to form **veins**, which carry blood back to the heart.

### Arteries

Look at Figure 1a. This shows the structure of an artery. You can see that it has a thick wall with smooth muscle and elastic tissue in it. This is because the arteries carry blood away from the heart. The blood is under high pressure. Every time the heart beats, a surge of blood passes through the artery. This causes the artery wall to bulge a little. As this happens, the thick elastic layer allows the artery wall to stretch and spring back. This process is called **elastic recoil**. This helps to keep blood pressure high and helps to smooth out blood flow. When you measure your pulse, you can feel these surges in pressure.

There is also a tough fibrous layer around the outside of the artery. This protects the artery from damage as we move around and our skeletal muscles contract and relax. The **lumen** is the space through which the blood flows. You can see that it is relatively small, keeping the blood under pressure. The lining of the artery, the **endothelium**, is a thin, smooth layer. This helps the blood to flow along with as little friction as possible.

### Veins

Figure 1b shows a vein. You can see that it has a much thinner layer of muscle and elastic tissue than the artery. This is because the blood in the vein is under much lower pressure. The lumen of the vein is larger than the lumen of the artery, so the blood is under less pressure and it flows more slowly. The vein also has a thick fibrous layer to protect it from damage. It has a smooth endothelium to reduce friction as blood flows along it.

### Key definition

Humans and many other animals have a closed blood system in which the blood is carried only inside blood vessels. Blood never leaves this system of blood vessels.

### Examiner tip

Pay attention to the *tissues* and the key words that go with them. *Smooth muscle* can contract and relax to regulate the pressure by changing the lumen diameter. *Elastic tissue* stretches to accommodate the pulse of blood from ventricular systole and recoils to maintain the blood pressure. *Get the right words with the right tissue!* A common error is for candidates to suggest that smooth muscle 'smoothes' the bloodflow by preventing friction. This is the function of the *endothelium*.

Terminology is key to good marks. For example, a common mistake is to state that when smooth muscle in an artery wall contracts, the artery gets smaller. It doesn't, the lumen gets smaller.

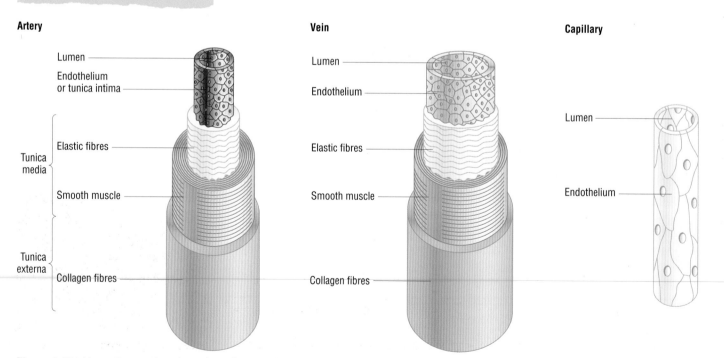

**Figure 1** Structure of **a** an artery, **b** a vein and **c** a capillary. The diagram of the capillary is drawn to a much larger scale

A special feature of veins is that they contain **valves**. You can see how these work in Figure 2. Valves help the blood to keep flowing in one direction only, back to the heart. This happens even in veins in the legs and arms, where blood flows against the force of gravity. As our skeletal muscles contract, they squeeze the veins. This raises the pressure in the veins, which shuts the valves behind and opens the valves ahead, so making sure that the blood keeps flowing back towards the heart.

### Capillaries

Look at Figure 1c. This shows a capillary. It is not drawn to the same scale as the artery and the vein. A capillary is a tiny blood vessel, only 7–10 μm wide. This is actually about the same diameter as a **red blood cell**. Red blood cells can pass through capillaries, but they have to be squeezed through. This means they pass through one at a time, allowing more efficient exchange between the blood and the tissues.

The capillary wall is made of a single layer of thin, flattened endothelium cells. In fact, the capillary wall is like the layer of cells that lines the artery and the vein. Because the capillary wall is very thin, it allows exchange between the blood and the tissues. There are also tiny gaps between the endothelium cells. These also allow substances to be exchanged between the capillaries and the tissues, and allow phagocytic white cells (**phagocytes**) to migrate into tissues.

### Arterioles and venules

Look at Figure 3. You can see that the arterioles have a thin wall, mainly of muscle fij46
bres, but with some elastic fibres. When this muscle contracts, it makes the lumen of the arteriole narrower. When it relaxes, the lumen becomes wider. This means that arterioles can increase or decrease the flow of blood to particular tissues. This is also one means of regulating blood pressure.

Venules have a very thin wall of muscle and elastic tissue. They are like small veins, and carry blood from the capillaries back to the veins.

(a) Pressure falls behind the valve. Pockets fill and close the valve

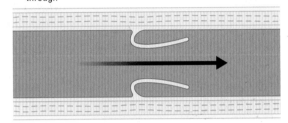

(b) Pressure builds from blood being pushed by the heart. Higher pressure behind the valve pushes it open and blood flows through

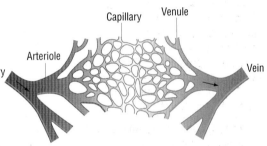

**Figure 2** The action of valves

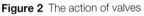

**Figure 3** Relationship between different kinds of blood vessel

| Artery | Capillary | Vein |
|---|---|---|
| Transports blood away from the heart | Links arteries to veins. Allows exchange of materials between the blood and tissues | Transports blood towards the heart |
| Thick wall with muscle and a great deal of elastic tissue | No elastic or muscle fibres | Relatively thin wall with only a small amount of muscle and elastic fibres |
| Bloodflow rapid | Bloodflow slowing | Bloodflow slow |
| High blood pressure – blood flows in pulses | Pressure of blood falling – not in pulses | Low blood pressure – not in pulses |

**Table 1** The main differences between arteries, veins and capillaries

# Questions

1  The lumen of an artery is narrower than the lumen of a vein, although arteries carry the same volume of blood as veins do. Explain why their lumens are different sizes.
2  Look at Figure 4.
   **(a)** Explain how it is possible to identify which vessel is the artery and which vessel is the vein.
   **(b)** A vein is really round in cross-section. Suggest why the section through the vein in this figure is not round.
   **(c)** Use the magnification shown to calculate the thickness of the wall, in μm, in the artery and the vein.

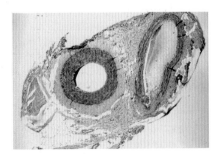

**Figure 4** Photo of artery and vein as seen under microscope (×30)

## The human circulatory system

The human circulatory system transports materials, such as oxygen, round the body by **mass transport**. Mass transport is when everything is moving in a stream in one direction. The cells, **plasma** and dissolved substances are all moving together in the blood.

Very small organisms, with very few cells, do not need a circulatory system as each cell in the organism is very close to the medium in which they live so oxygen and nutrients can be absorbed over their whole body surface by **diffusion**. Larger organisms like humans cannot do this because the diffusion path would be so long that substances wouldn't move fast enough. We need a blood system to carry substances such as oxygen and glucose to respiring cells.

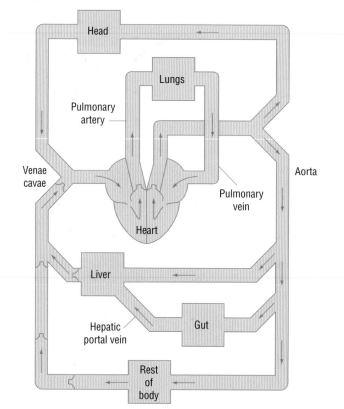

**Figure 1** The human circulatory system

Look at Figure 1. This shows the human circulatory system. You will see that it is a **closed system**. This means that blood stays in the blood vessels all the time. Blood does not leave the blood vessels at any time, except when the body is injured.

The human circulatory system is also a **double circulation**. This means that there are two 'circuits'. The **pulmonary circulation** goes from the heart to the lungs and back to the heart. The **systemic circulation** goes from the heart to the body organs and then back again. When blood passes round the body once, it goes through the heart twice.

Deoxygenated blood that has come from the tissues returns to the right **atrium** of the heart via the vena cava. It passes into the right **ventricle**, and gets pumped to the lungs through the pulmonary artery. Oxygenated blood from the lungs returns to the left atrium of the heart. It passes into the left ventricle. From here, it is pumped to the body in the aorta.

Measurement of pressure

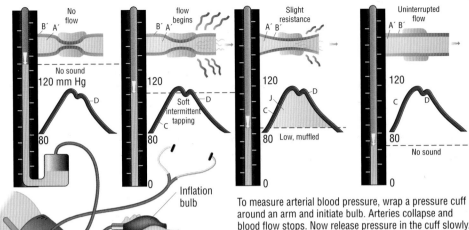

To measure arterial blood pressure, wrap a pressure cuff around an arm and initiate bulb. Arteries collapse and blood flow stops. Now release pressure in the cuff slowly. The pressure where sounds first occur corresponds to the pressure where the artery is just barely able to open for a moment. It is a systolic pressure. Continue releasing pressure until the sounds muffle; the pressure is a diastolic pressure. The sounds arise from the turbulent blood flow through the narrowed (partially collapsed) artery under the cuff.

**Figure 2** Korokov sounds heard in a stethoscope when measuring blood pressure

## How to measure blood pressure

- Make the person sit down for at least five minutes beforehand. Ensure he is relaxed and not moving.
- Ensure that the person is not wearing any tight clothing on the arm.
- Place the cuff placed around the upper arm (*not* over clothing). Then place a stethoscope over the brachial artery, on the inside of the elbow, so that you can hear the brachial pulse.

**Figure 3** A medical professional measuring blood pressure

- Pump air slowly into the cuff so that it inflates. This tightens around the upper arm and restricts the flow of blood to the brachial artery.
- When you cannot hear the pulse any more, very gradually release the air in the cuff.
- As soon as you can hear the pulse again in the stethoscope, record the reading in the cuff. This means that when ventricular systole is occurring, the pressure is just enough to squeeze past the cuff. So the pressure in the artery at systole and the pressure in the cuff are about equal at this point.
- Allow the cuff to deflate slowly until there is no sound in the stethoscope. Again record the pressure in the cuff. This is the diastolic pressure. This means that even when diastole is occurring, there is enough pressure for blood to get past the cuff.
- You can also measure blood pressure using a digital sphygmomanometer.
- Make the person sit down for at least five minutes. She should be relaxed and not moving.
- Ensure that the person is not wearing any tight clothing on the arm.
- Place the cuff around the upper arm (*not* over clothing). The indicator mark should be over the brachial artery.
- The cuff will automatically inflate and deflate again. The blood pressure reading will be shown on a digital display.
- Repeat twice more.

## Questions

1 A person's blood pressure should be measured when they are resting and relaxed. Explain why.
2 People with high blood pressure sometimes have swelling in their legs and feet. This is because tissue fluid has accumulated there. Explain how this happens.
3 Somebody with very low blood pressure will feel faint and may become unconscious. Explain why.
4 Nicotine in cigarette smoke causes the muscle fibres in arteriole walls to constrict. Explain how this causes blood pressure to increase.

## The structure of the lungs

Look at Figure 1. This shows the structure of the lungs and breathing system. When we breathe in, air passes down the **trachea** into the two **bronchi**. The bronchi branch into smaller **bronchioles**, which end in clusters of air sacs called **alveoli**.

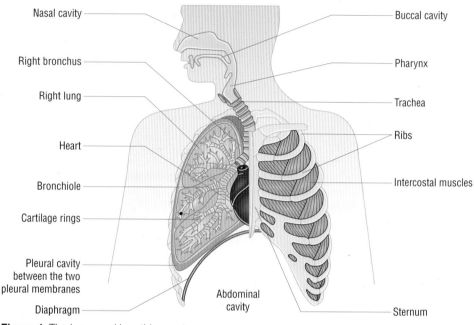

**Figure 1** The lungs and breathing system

### The alveoli

The alveoli are the actual site of gas exchange. They are tiny hollow sacs made of thin, flattened cells called **squamous epithelium**. An 'epithelium' is a lining tissue.

Figure 2 shows squamous epithelium. You can see that the cells are thin and flat. This means that there is only a short distance between the air in the alveoli and the blood in the capillary, so gas exchange is as efficient as possible.

Squamous epithelium is an example of a **tissue**.

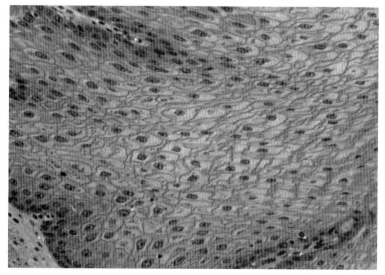

**Figure 2** Squamous epithelium

> **Key definition**
>
> A **tissue** is a group of similar cells specialised to carry out the same function.

> **Examiner tip**
>
> Cells need not always be the same type in a tissue. For example, **ciliated epithelium** tissue contains ciliated cells and mucus-secreting cells. Epithelial tissues are one of four tissue types. There is also muscle tissue and **connective tissue** (e.g. cartilage). The fourth type of tissue is nervous tissue, which you will meet in A2. In each case you should be able to point out how the properties of the tissue and its component cells adapt it to its function.

## The trachea

The trachea has rings of cartilage around it to make sure that it stays open when you breathe in and out. It is lined with a layer of **ciliated epithelium** cells and **goblet cells**. You can see this in Figure 3.

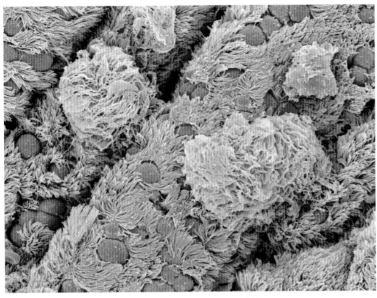

**Figure 3**  Ciliated epithelium cells and goblet cells from the lining of the trachea

The goblet cells are so-called because they are shaped like a goblet. These cells produce large amounts of mucus, which is a **glycoprotein**. The ciliated epithelium cells have many tiny hairs called **cilia**. These beat together in a rhythm and move the mucus back up the trachea to the throat. Dirt and bacteria in the air that is breathed in gets trapped in the mucus. When the mucus reaches the throat, it is swallowed. This means that the dirt and bacteria are destroyed by the acid and the enzymes in the stomach.

## The lungs

Look at Figure 4. This shows a photo of lung tissue. You can see that it consists of many **alveoli**, which are lined with squamous epithelium cells. It also contains many blood capillaries. The lung is an example of an **organ**. An organ is a structure made up of different kinds of tissue. The lung is an organ because it contains different kinds of tissue, such as squamous epithelium and *elastic tissue* (a kind of connective tissue).

# Questions

1  Compare the role of the squamous epithelium cells in the capillaries and in the alveoli of the lungs.
2  Goblet cells make large amounts of mucus, which is a glycoprotein. Name the organelles that would be involved in producing this mucus.
3  Cigarette smoke damages the cilia on the ciliated epithelium cells lining the trachea. Explain why people who smoke develop a persistent chesty cough and are prone to lung infections.

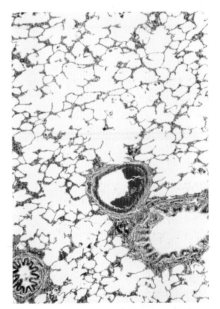

**Figure 4**  Photomicrograph of lung tissue showing squamous epithelium and capillaries

A single-celled organism can take in molecules, such as oxygen, all over its surface. But the cells in a multicellular organism cannot take in oxygen through their surface unless they are on the outside. Look at Figure 1. You can see that the cells in the middle of an organism are too far from the outside. They cannot obtain the oxygen they need from the environment.

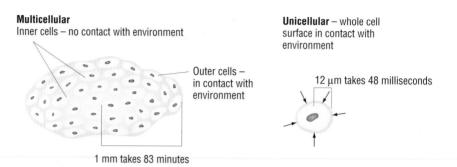

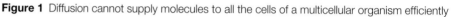

**Figure 1** Diffusion cannot supply molecules to all the cells of a multicellular organism efficiently

## Surface area and volumes

The amount of diffusion between a multicellular organism and its environment depends on:

- its *surface area* – in other words, the number of cells in contact with the environment
- its *volume* – in other words, the space occupied by all the cells that need to be supplied with molecules.

As the number of cells increases, the volume increases. The surface area also increases, but not as much. Look at Table 1. To make it easier, we can assume that an organism is shaped like a cube. Notice how the surface area:volume ratio goes down sharply as the size of the organism increases.

| Length of side/cm | Surface area/cm² | Volume/cm³ | Surface area: volume ratio |
|---|---|---|---|
| 1 | 6 | 1 | 6 |
| 2 | 24 | 8 | 3 |
| 3 | 54 | 27 | 2 |
| 4 | 96 | 64 | 1.5 |
| 5 | 150 | 125 | 1.2 |
| 6 | 216 | 216 | 1 |
| 7 | 294 | 343 | 0.86 |

**Table 1** Surface area:volume ratios

### Gas exchange in the alveoli

Look at Figure 2. This shows the structure of an **alveolus** in the lungs.

The alveolus wall contains some elastic fibres. These allow the alveoli to expand when breathing in, and **recoil** easily when breathing out. The alveoli are also lined with a watery liquid. This contains a detergent-like substance called **surfactant**. It lowers the surface tension of the alveoli, and therefore reduces the amount of effort needed to breathe in and inflate the lungs. It also has an antibacterial effect, helping to remove any harmful bacteria that reach the alveoli.

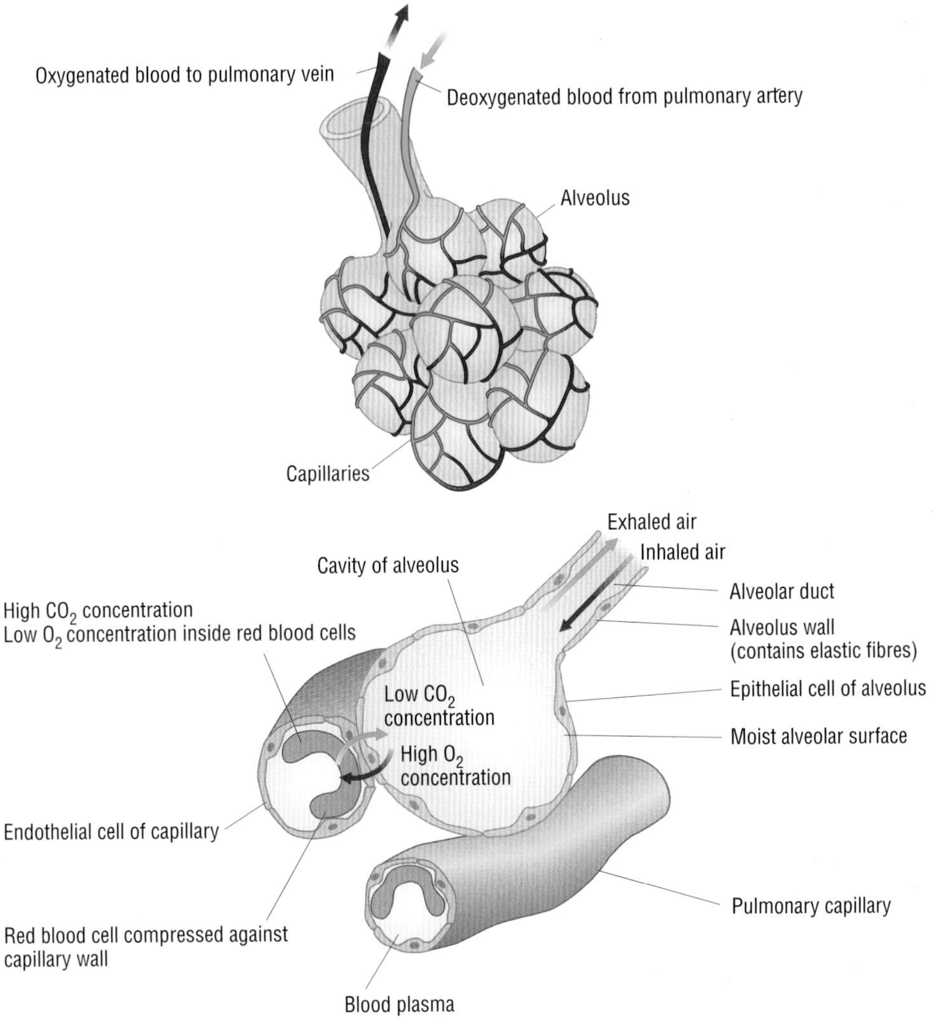


**Figure 2** The structure of an alveolus (not drawn to scale)

A good exchange surface has a large surface area, a thin surface and a steep diffusion gradient.

The lungs are adapted for efficient gas exchange because of a number of factors.

*A large surface area*

- The bronchioles are highly branched, giving a large number of pathways for air to enter and leave the lungs.
- There are millions of alveoli in each lung.
- The alveoli are highly folded, giving an even greater surface area.

*A thin surface*

- The **squamous epithelium** cells in the alveoli are only 0.1–0.5 μm thick, which allows rapid diffusion across them.
- The capillary wall is also made of a single layer of thin, flattened cells.

*A steep diffusion gradient*

- The blood circulation carries oxygenated blood away from the alveoli, and brings deoxygenated blood to the alveoli.
- Ventilation brings air rich in oxygen into the alveoli, and air with increased carbon dioxide is removed from the alveoli.
- The capillaries surrounding the alveoli are narrow, slowing down the bloodflow and allowing plenty of time for efficient gas exchange.

### Examiner tip

When answering a question on how the lungs are adapted for efficient gas exchange, read the question carefully. Does it ask how the alveoli are adapted? Then there is more you could say. The dense capillary network is in close contact with the alveoli. The movement of blood through the capillaries maintains the steepness of the gradient. The narrow width of the capillaries means that erythrocytes are pressed close to the capillary wall and so close to the alveolar wall. This reduces the distance for gas exchange and so speeds it up. And don't forget, if the question is about the lungs, then features such as the cartilage in the trachea and bronchi keeping the airways open are important.

## Questions

1. A premature baby may suffer from respiratory distress syndrome. This happens because the baby's lungs do not contain enough surfactant. Suggest the symptoms of respiratory distress syndrome.
2. Describe in detail the path taken by a molecule of oxygen as it passes from the air in the alveoli to the haemoglobin in a red blood cell.

## Lung capacity

The volume of air breathed in and out of the lungs depends on a number of factors. These include our level of activity, the size of our lungs and how healthy we are.

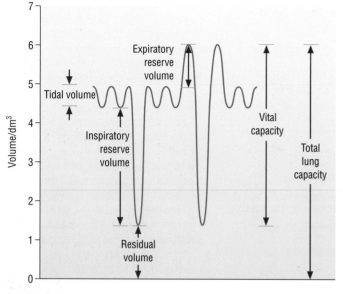

**Figure 1** Lung capacities

Look at Figure 1. This shows some of the volume changes that take place in the lungs during breathing. The actual volumes vary from one person to another.

- The **tidal volume** is the volume of air breathed in and out with a normal breath. This is usually about $0.5\,dm^3$.
- If you breathe out as much air as possible, and then breathe in as much air as possible, you will breathe in about $3.5\,dm^3$. This volume is called the **vital capacity**.
- Even when you have breathed out as much air as possible, there is still about $1.5\,dm^3$ of air left in the lungs. This is called the **residual volume**. It is important that there is some air left in the lungs when you breathe out, or the walls of the alveoli would stick together and it would be difficult to re-inflate the lungs.

### Using a spirometer

Lung volumes can be measured using a piece of apparatus called a **spirometer**. You can see a spirometer in Figure 2.

### Flow rates

Medical practitioners may also wish to measure the rate at which air can be expelled from the lungs when a person forcibly breathes out. This can help them to diagnose and monitor conditions such as asthma.

- The **forced expiratory volume per second (FEV$_1$)** is the volume of air that can be breathed out in the first second of forced breathing out.
- The **peak expiratory flow rate (PEFR)** is the maximum rate at which air can be forcibly breathed out through the mouth.

These volumes can be measured using a **peak flow meter**.

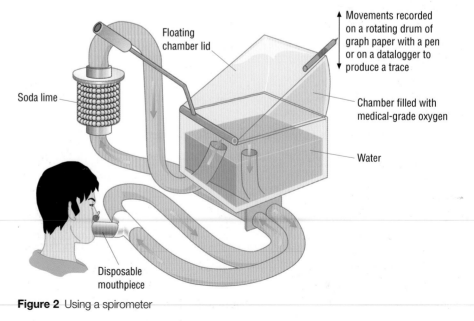

**Figure 2** Using a spirometer

## Using a spirometer

A person breathes in air through a tube connected to a container of oxygen that floats in a tank of water. You can see this in Figure 2.

The floating container rises and falls as the person breathes in and out. The rise and fall is related to the volume of air the person is breathing in and out.

The container has an arm attached to it, with a pen on the end. The pen draws a trace on some graph paper on a rotating drum. You can see a trace in Figure 3.

The air breathed out passes through a chamber containing soda lime, which absorbs the carbon dioxide in the air breathed out before it returns to the oxygen chamber. This stops the person re-breathing carbon dioxide, which would increase their breathing rate.

As the oxygen in the chamber is gradually used up, the volume of oxygen in the chamber reduces gradually. You can see this happening as the trace gradually goes down in Figure 3.

Note that you can also measure lung volumes using a digital spirometer.

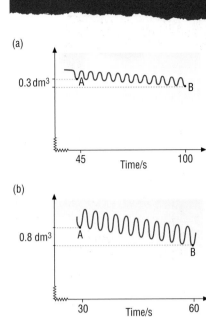

(a)

(b)

**Figure 3** Spirometer traces of a subject at **a** rest and **b** during exercise

### Using a peak flow meter

- Stand up straight and make sure that the indicator is at the bottom of the meter.
- Take a deep breath and fill your lungs completely with air.
- Place the mouthpiece in your mouth and close your lips firmly around it.
- Blow air out of your mouth into the meter as hard as you can in one blow.
- Record the reading on the meter.
- Re-set the meter and take two more readings. Record the highest reading.

Some peak flow meters and digital spirometers will also measure $FEV_1$.

### Normal values for PEFR

Figure 5 shows the normal range of values for PEFR for males and females of different ages.

**Figure 4** Using a peak flow meter

## Questions

**1** Look at Figure 5.
  **(a)** Estimate the person's tidal volume at rest using Figure 3.
  **(b)** Estimate this person's normal breathing rate in breaths per second.

**2** Asthma is a condition in which the bronchi and bronchioles constrict. Some people with asthma find that their condition is made worse when they are exposed to house dust mites. Figure 6 shows a graph of a person's PEFR over several days.
  **(a)** Explain the evidence that this person's asthma is caused by exposure to house dust mites.
  **(b)** Explain why PEFR is lower when a person is having an asthma attack.

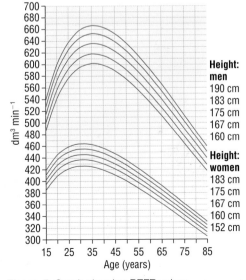

**Figure 5** Graph showing PEFR values

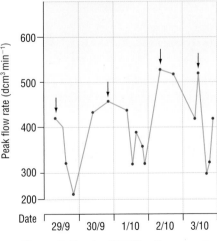

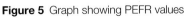

**Figure 6** Graph of PEFR, with exposure to house dust mites indicated by arrows

**Examiner tip**

Whenever breathing has stopped or the pulse is weak, **cyanosis** occurs. The word comes from the Greek word for dark blue – cyan. Cyanosis describes the bluish appearance of the skin especially around the lips. This is due to the build up of deoxygenated haemoglobin.

## Causes of respiratory arrest

Respiratory arrest is when a person stops breathing. It is possible for a person to stop breathing even though her heart is still beating. Many things can cause respiratory arrest, including:

- a respiratory disorder, e.g. severe asthma or pneumonia
- an obstruction in the trachea or bronchi, e.g. caused by choking on food or a child putting a small object into their mouth
- overdosing on drugs that suppress the respiratory system, e.g. heroin, barbiturates.

Respiratory arrest may also occur following cardiac arrest (when the heart stops beating).

### Respired air resuscitation

This is a first aid procedure that you should carry out on a person who is not breathing but who still has a pulse. It is sometimes called 'rescue breathing'.

First of all, dial 999 for an ambulance. If there is another person with you, send them to get help while you start rescue breathing. If you can, wear latex gloves and use a breathing mask.

- Roll the person on to his back, being very careful not to twist the person's neck, head or spine. Then pull his head back and lift the chin as in Figure 1. This will open the person's airway.
- Ensure nothing is blocking the person's airway.
- Gently pinch the person's nose shut, using the thumb and index finger. Then place your mouth over the person's mouth, making a seal.
- Breathe slowly into the person's mouth, and watch their chest to see if it rises. Pause between each breath to let the air flow out.
- If the person's chest does not rise, tilt the head back again and try again.
- After giving two breaths, check for a pulse. If the person has a pulse, continue rescue breathing. You should give one breath every five seconds.
- If the person's pulse stops, you should perform CPR (cardiopulmonary resuscitation, also called Basic Life Support). You can read about this in spread 2.4.1.1.

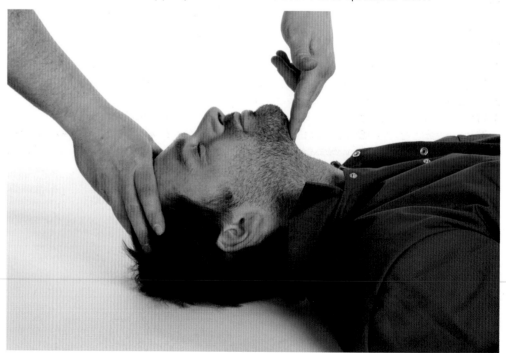

**Figure 1** Opening the airway

### Rescue breathing in children

This is very similar to the procedure for adults, but there are a few differences.

- To open the airway of an infant or child, you do not need to tilt the head so far back.
- Children or infants need one slow breath every three seconds.
- On a baby, you should use your mouth to make a seal over the baby's mouth and nose at the same time.
- Check for a pulse after a minute of rescue breathing (about 20 breaths).

**Figure 2** Rescue breathing

## Questions

1 Suggest why you should use latex gloves and a breathing mask if these are available.
2 Suggest why you should avoid twisting the person's head, neck or spine.
3 Suggest why it is important to check regularly for a pulse.
4 When you do rescue breathing, you are blowing exhaled air into the person's lungs. This contains about 16% oxygen, although fresh air has 21% oxygen. Also, expired air has 4% carbon dioxide, while fresh air has 0.04% carbon dioxide. Explain why it is still helpful to blow expired air into the lungs of a person who is not breathing.

## Self-check questions

**Fill the blanks**

1 The human heart has four chambers. .................... blood entering the heart from the vena cava enters the right .................... . From here, the blood passes through .................... valves into the right .................... . The right ventricle contracts to pump the blood to the .................... via the pulmonary .................... . .................... blood from the lungs enters the left .................... through the .................... veins. This blood passes through the .................... valves into the left .................... . When the left ventricle contracts, blood passes out of the left ventricle along the .................... .

2 Atrial .................... is when the atria ...................., forcing blood from the atria into the .................... . During ventricular systole, the .................... contract. The pressure in the ventricles during systole causes the atrio-ventricular valves to .................... and the semi-lunar valves to .................... . This forces blood out of the heart along the pulmonary .................... and the .................... . During .................... the atria and the ventricles relax.

3 Heart muscle is .................... which means that it contracts without being stimulated by a nerve. The .................... node in the wall of the .................... atrium initiates the heartbeat. From here, impulses pass across the walls of the .................... to the .................... node between the atria and the ventricles. These cause the atria to contract. From the atrio-ventricular node, impulses pass along the .................... tissue to the ventricles. This causes the ventricles to contract.

4 .................... volume is the volume of blood pumped out by the heart during one cardiac cycle. Cardiac output = .................... × .................... rate. During exercise, both stroke volume and heart rate .................... .

5 .................... have a thick muscular wall, containing a great deal of .................... tissue. The diameter of the .................... is comparatively small. They transport blood at .................... pressure away from the .................... have thinner walls, containing some muscle and a little .................... tissue. The diameter of the .................... is comparatively large. They carry blood at .................... pressure back to the heart. They contain .................... to prevent backflow. .................... have no .................... or elastic tissue in their walls. They have permeable walls and supply materials from the blood to cells.

6 Squamous .................... cells in the .................... of the lung are an example of a ...................., because they consist of a group of .................... cells carrying out the same function. A lung is an example of an .................... because it consists of several different kinds of tissue, all working together to carry out the same function. The respiratory tract is lined with .................... epithelium cells. These move mucus that has been secreted by .................... cells back towards the throat. In this way, dirt and .................... are kept out of the lungs.

7 The alveoli are efficient at gas exchange because they have a .................... surface area. The squamous epithelium cells are very .................... and flat, giving a short .................... pathway. The alveoli are well supplied with blood .................... and .................... keeps bringing fresh air into the lungs. This maintains a steep .................... gradient.

# Summary questions

1  The diagram shows some cells from the lining of the respiratory tract.

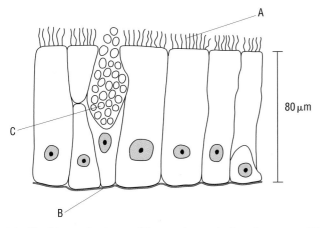

(a) (i)  Name the type of tissue shown in the diagram.  [1]

   (ii)  Names structures **A**, **B** and **C**.  [3]

(b)  Calculate the magnification of the drawing. Show your working.  [2]

(c)  Explain how the tissue shown in the diagram may help to protect against lung infections.  [3]

2  The drawing shows some structures involved in gas exchange in a human.

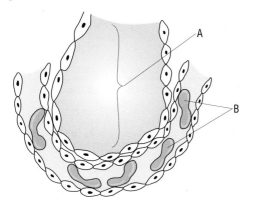

(a)  Name structures **A** and **B**.  [2]

(b)  Surfactant is found lining the walls of structure **A**. Explain the role of surfactant.  [2]

(c)  Describe how a spirometer may be used to measure tidal volume.  [4]

3

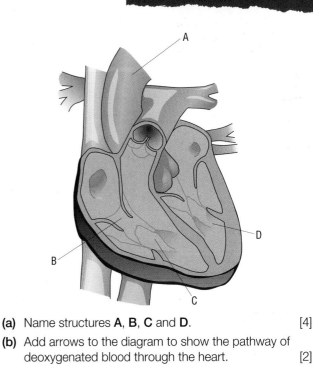

(a)  Name structures **A**, **B**, **C** and **D**.  [4]

(b)  Add arrows to the diagram to show the pathway of deoxygenated blood through the heart.  [2]

(c)  Explain why the wall of structure **D** is thicker than the wall of structure **C**.  [2]

(d)  Some children are born with a 'hole in the heart'. This is a hole in the septum between the left and the right atria. Suggest the effects that this would have on a person with a 'hole in the heart'.  [3]

## Question 1

Lymphocytes can be identified in a blood smear by using a differential stain.

**(a)** Describe the structure of a lymphocyte as seen in a blood smear examined using a light microscope.     [3]

Lymphocytes can also be examined using an electron microscope. A diagram of mature B lymphocyte is shown below.

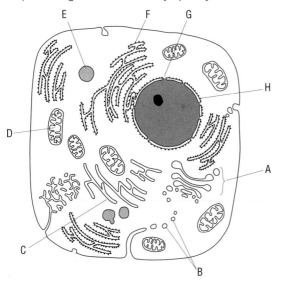

**Figure 1.1**

**(b)** Identify the structures by completing the table using EITHER the appropriate letter OR the name of the organelle.     [6]

| Letter | Name |
| --- | --- |
| A | |
| | smooth endoplasmic reticulum |
| D | |
| E | |
| F | |
| | nuclear envelope |

The function of mature B lymphocytes is to produce antibodies. Antibodies are glycoproteins which are secreted into the blood plasma.

**(c) (i)** What features of the cell in Figure 1.1 suggests that its main function is antibody production?     [3]

     **(ii)** Which structure in the cell will convert the antibody proteins into glycoproteins?     [1]

**(d)** Describe how antibody molecules will be secreted from the mature lymphocyte.     [4]

## Question 2

Fill in the missing words in the following passage about erythrocytes and haemoglobin.

**(a)** Erythrocytes contain a solution of the ...................... protein haemoglobin. Haemoglobin consists of

............................ polypeptide chains. Each chain is

attached to a ............................ group – a

non-protein group which in this case is ...................... .

The chains consist of a sequence of amino acids linked by

.......................... bonds which form their primary

structure. The sequence of amino ...............................

is determined by DNA in the nuclei of stem cells in the

............................................. . The nucleus is lost as the

cells mature and are released into the plasma.     [6]

**(b)** In the formation of a blood clot, erythrocytes become trapped in a network of insoluble protein fibres.

Outline how soluble plasma proteins are converted into insoluble protein fibres when blood clotting occurs.     [4]

**(c)** Antithrombin is a small protein molecule that inactivates enzymes in the clotting cascade. Antithrombin and enzyme molecules bind and the active site of the enzyme is blocked. The substrate cannot enter the active site and a product cannot be formed.

Using examples from the clotting cascade, identify an enzyme, its substrate and its normal product.     [3]

# Question 3

(a)

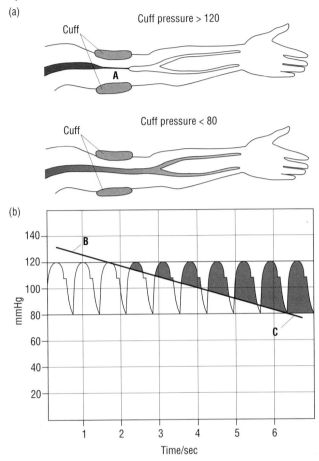

**Figure 3.1**

(a) Figure 3.1 (a) represents what happens in the arm when a blood pressure measurement is taken.

(i) Name the instrument which is being used to measure the blood pressure. [1]

(ii) Name the TYPE of blood vessel which is labelled **A**. [1]

(iii) Indicate the direction of blood flow in this vessel by placing an arrow on the diagram. [1]

(b) Line **B** to **C** on the graph indicates the pressure exerted by the cuff around the arm. This is inflated and then deflated slowly. A stethoscope is placed over the blood vessel. The noises which are detected are called the Korotkov sounds.

(i) Explain why no sounds are heard at one second. [2]

(ii) Between two and four seconds, a regular tapping noise is heard. Using the information on the graph, describe the blood flow responsible for this sound. [2]

(iii) Between four and six seconds, the sounds become softer. Using the information on the graph, explain why. [2]

(iii) Using the information on the graph – give the blood pressure reading that would have been recorded. [2]

# Question 4

A spirometer was used to measure lung volumes. The measurements were carried out at two different breathing rates. The data collected was used to calculate the pulmonary ventilation rate.

(a) State two precautions that should be taken when carrying out investigations with a spirometer. [2]

(b) Pulmonary ventilation is the volume of air breathed per minute. If the pulmonary ventilation rate was 6000 cm³ min⁻¹ and the tidal volume was 200 cm³, what was the breathing rate during the experiment? [2]

(c) The volume of air arriving at the alveolar surface was only 1500 cm³. With reference to the structure of the gas exchange system, comment on the implications of so little of the air inhaled reaching the alveoli. You will gain credit for the use of figures in your answer.

Credit will also be given for good organisation and the use of technical terms. [5]

## DNA structure

DNA stands for deoxyribonucleic acid. DNA is the molecule that stores genetic information. It is present in the chromosomes, inside the nuclei of cells.

DNA is made up of many repeated units, called **nucleotides**. A DNA nucleotide contains
- a five-carbon sugar, deoxyribose
- a phosphate group
- an organic nitrogenous base, which may be adenine, thymine, cytosine or guanine.

Adenine and guanine are both **purine bases**, while thymine and cytosine are **pyrimidine bases**. You can see the structure of a nucleotide in Figure 1.

The DNA nucleotides join together by a series of **condensation reactions** to form a **polynucleotide** chain.

A *molecule* of DNA is made of two polynucleotide chains, with one chain upside-down compared to the other. The two chains or *strands* link together by hydrogen bonds between the organic bases. However, the bonding between the organic bases is highly specific. A purine will pair only with a pyrimidine. However, it is even more specific than this. Adenine will pair only with thymine, forming two hydrogen bonds, while cytosine will pair only with guanine, forming three hydrogen bonds. This is called **complementary base-pairing**. You can see this in Figure 2a.

The DNA molecule twists up to form a shape that is known as a **double helix**. It looks like a twisted rope-ladder. The sugar-phosphate 'backbone' forms the uprights of the 'ladder', and the base-pairs form the 'rungs'. You can see this in Figure 2b.

### Examiner tip

When answering questions on this topic, be sure to make it clear whether you are talking about the whole molecule (two chains) or one strand of it (one chain).

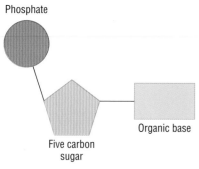

**Figure 1** The structure of a DNA nucleotide

### Examiner tip

Note that it is new *nucleotides* that pair up with the exposed bases, not new bases. Note that a purine (adenine and guanine) always pairs with a pyrimidine (cytosine or thymine). Adenine always pairs with thymine because two hydrogen bonds are formed between them. Cytosine always pairs with guanine, forming three hydrogen bonds.

This means that when DNA replicates during S phase of the cell cycle (see page 64), each strand can be copied accurately. This is the reason DNA is double stranded.

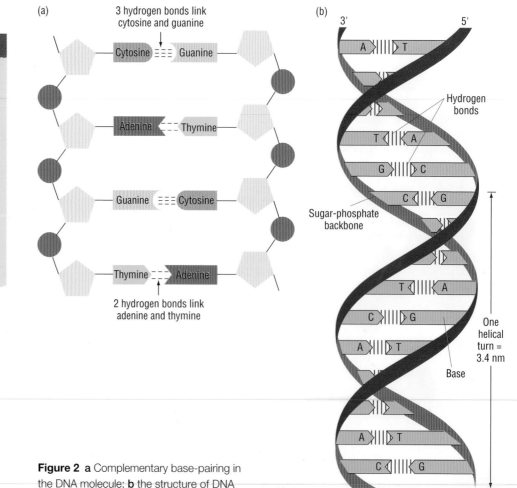

**Figure 2 a** Complementary base-pairing in the DNA molecule; **b** the structure of DNA

## DNA replication

As you know, DNA carries the genetic information that your cells need to function properly. Every cell in your body has the same DNA. It is very important that every time a cell divides all the DNA copies itself accurately. This makes sure that every new cell has exactly the same information.

The way that DNA replicates is called **semi-conservative replication**. This happens during the S-phase of the cell cycle. You will learn more about this on spread 2.1.3.3. You can see how this happens in Figure 3.

- The DNA molecule unwinds and 'unzips'. The enzyme DNA helicase causes the hydrogen bonds between the base-pairs to break.
- New DNA nucleotides pair up with the exposed bases, according to the complementary base-pairing rules. Adenine pairs with thymine, and cytosine with guanine. Note that a purine base always pairs with a pyrimidine base.
- The new nucleotides are joined together by the enzyme DNA polymerase. A phosphodiester bond forms between the sugar of one nucleotide and the phosphate of the next.

Notice how the two new molecules have exactly the same sequence of bases as each other. This is very important. The order of bases in the DNA molecule carries important information. If there was a mistake in this base sequence, it could cause serious problems.

There are enzymes that 'proof-read' the DNA. These check that the DNA has been copied accurately. If there is a mistake, the enzyme corrects it.

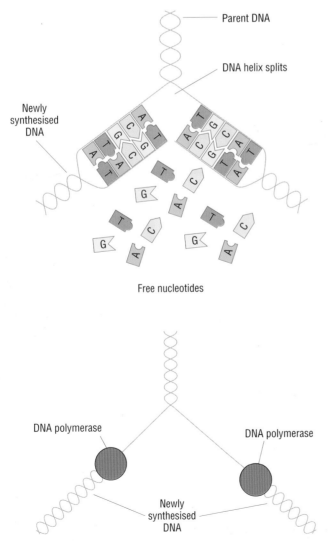

**Figure 3** Semi-conservative replication of DNA

## Questions

1 The bases on one strand of DNA are ATTCGA. Write down the sequence of bases on the opposite strand of DNA.

2 A sample of DNA contains 20% guanine nucleotides. What percentage of nucleotides in this DNA contain adenine?

3 DNA nucleotides join together by a condensation reaction to form a **phosphodiester bond**. What does this mean?

## Cell division

**Mitosis** is a kind of cell division that is used for growth. It produces new cells to increase the size of an organism during growth, or to produce an adult multicellular individual from a single-celled **zygote**. New cells are also needed to replace damaged cells and repair tissues.

When cells divide, it is very important that the new cells are exactly the same as the previous cell. This means that the genetic information in the daughter cells must be an exact copy of the genetic information in the mother cell.

As you know, **DNA** is present in **chromosomes**. These are inside the nucleus of **eukaryotic cells** such as human cells. Each chromosome consists of a long double helix of DNA wrapped around special proteins called histone proteins, together with some other proteins. The single DNA double helix in an average human chromosome contains about 150 million nucleotide pairs. The nuclear envelope isolates the chromosomes from the rest of the cell, but it disintegrates in the early stages of nuclear division (mitosis and meiosis). If the DNA were not part of chromosomes, its long threads would get tangled up and damaged as the replicated DNA separates into new daughter nuclei during nuclear division.

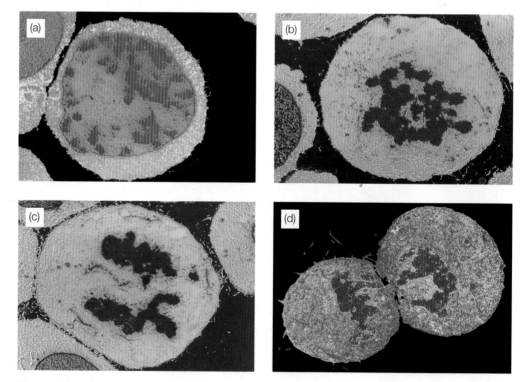

**Figure 1** Stages of mitosis: **a** prophase; **b** metaphase; **c** anaphase; **d** telophase

### Interphase

Interphase is the time when the cell is not dividing or undergoing nuclear division. Cells spend most of the time (usually about 90%) in65
interphase, because this is when the cell is carrying out its functions, growing, and copying its chromosomes in preparation for cell division. On the next spread you will learn more about the different stages of a cell's life – the **cell cycle**.

When you look at a normal cell under the microscope, you cannot see chromosomes. Chromosomes become visible only when the cell is about to divide. During interphase, they are relatively uncoiled and spread out as long, very thin threads. This gives the nucleus a speckled appearance.

## Prophase

Prophase is the first stage of mitosis. By the end of interphase the DNA that was already in the cell has replicated. Each chromosome now consists of two **chromatids** held together by a **centromere**. Each chromatid is one of the two new DNA *molecules* that have been synthesised. The chromosomes are coiling up or condensing, and are now becoming visible as darkly-staining structures.

Human cells contain 23 pairs of chromosomes. This is because we receive one set of chromosomes from our mother and another set from our father. Cells containing pairs of chromosomes are called **diploid**. In humans, the diploid number of chromosomes is 46. The **nuclear envelope** is present for most of prophase. At the end of prophase, the nuclear envelope breaks down, leaving the chromosomes free inside the cytoplasm.

To make it easier, the cell shows only four chromosomes.

## Metaphase

There are two tiny structures inside the cell, but just outside the nucleus, called **centrioles**. During prophase these move to opposite poles of the cell. During metaphase they send out tiny protein threads called **microtubules**. These microtubules form a structure called the **spindle**. The centromeres of the chromosomes attach to the spindle fibres and are pulled to the middle or *equator* of the cell. The position of the equator – and hence the orientation of the two daughter cells – depends on the original orientation of the centrioles. This can be important for cells that will be part of body tissues.

## Anaphase

The centromeres divide and the two 'sister' chromatids of each chromosome separate, pulled apart by the spindle fibres. Remember that each chromatid is an exact copy of the original DNA molecule, so now we can call the chromatids 'daughter' chromosomes. The daughter chromosomes are pulled to the opposite poles of the cell by the spindle fibres.

## Telophase

The spindle starts to break down, and a new nuclear envelope starts to form around each group of chromosomes. You can see this in Figure 2. The DNA in the chromosomes starts to unwind again, so the chromosomes become less distinct.

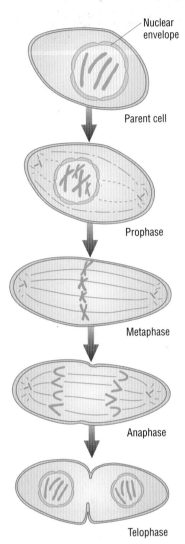

**Figure 2** The stages of mitosis: prophase; metaphase; anaphase; telophase

## Cytokinesis

Following telophase, the cytoplasm divides and two new cells are formed. Each daughter cell receives approximately half of the cell organelles of the original cell.

The time spent in each phase of the cell cycle and mitosis/cytokinesis can be calculated by counting the number of cells in each stage and then calculating this as a percentage of the total cell number. The cells then are assumed to spend this percentage of their time in that stage of the cell cycle. So if 80% of the cells are in interphase and the cells divide every 48 hours, then they spend approximately 38 hours in interphase.

### Examiner tip

A common mistake is to mix up the words 'strand' and 'molecule' and this is never more obvious than when students describe what a 'chromatid' is. Each chromatid is a DNA molecule that has been replicated. Each chromatid will have one old and one new STRAND of DNA in that molecule as a result of semi-conservative replication.

## Questions

1 What is the diploid number of the cell in Figure 2?
2 Distinguish between
   (a) a chromosome and a chromatid.
   (b) a centromere and a centriole.
3 In Figure 1 the spindle fibres cannot be seen. Suggest a reason for this.

# The cell cycle and apoptosis

## The cell cycle

**Mitosis** is just a small part of the **cell cycle**. As you learned on spread 2.1.1.2, cells spend most of the time in interphase.

**Interphase** is sometimes called a resting phase, but it is far from that. Look at Figure 1. You can see that interphase is divided into three parts:

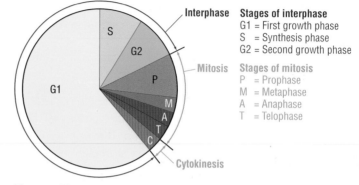

**Interphase**

**Stages of interphase**
G1 = First growth phase
S  = Synthesis phase
G2 = Second growth phase

**Stages of mitosis**
P  = Prophase
M  = Metaphase
A  = Anaphase
T  = Telophase

Mitosis

Cytokinesis

**Figure 1**  The cell cycle

- **G₁**, or the first growth phase. During this stage proteins are made. These proteins are used to build up cell **organelles** and more cytoplasm is made.
- **S**, or synthesis stage. This is the stage when DNA is replicated. You learned about this in spread 2.1.1.1.
- **G₂**, or the second growth phase. During this stage, cell organelles grow and divide. The cell also builds up its energy stores.
- **M**, or mitosis phase. This is when the nucleus undergoes **mitosis**, and then the cell divides (**cytokinesis**).

### Control of the cell cycle

The timing and duration of the cell cycle are controlled by a series of protein complexes made up of activating proteins called cyclins and enzymes called cyclin-dependent kinases (CDKs). The cyclins activate the CDKs. These affect the cycle at specific checkpoints. The three most important checkpoints are:

- the G₁ checkpoint, which late in G₁ triggers the initiation of S phase and the replication of organelles
- the G₂ checkpoint, which late in G₂ triggers processes that lead to mitosis
- the M (metaphase) checkpoint, which controls the entry into anaphase.

This control system is rather like the control on a washing machine, in which electrical signals trigger the machine to switch from washing to rinsing or spinning, or to stop altogether. Instead of electrical signals, internal and external signals cue the assembly or dissassembly of the cyclin-CDK complexes.

Cyclins, and so CDKs, interact with signals from both inside and outside the cell. Outside factors include hormones, drugs and other chemicals that bind to protein receptors on the outer surface of the plasma membrane.

An example of this is if the DNA is damaged by radiation during G₁, a protein p21 binds to two G₁ CDKs. This stops them being activated, so stopping the cell cycle. This allows the DNA to be repaired. It is these CDKs that are disrupted in cancerous cells. In some forms of breast cancer there is over-stimulation of one of the CDKs (Ck4), which then overstimulates cell division. Another major protein that inhibits cell division is p53, which leads to the synthesis of p21 and so blocks cell division. More than half of all human cancers contain a defective p53 gene.

## Syndactyly

*Syndactyly* is when two or more digits on a hand or foot are 'webbed' or fused together. You can see this in Figure 3. In a fetus, the hands and feet start out as paddle-shaped. Normally, apoptosis occurs so that the cells separating the toes and fingers die. However, in some case this does not happen properly.

The condition can be corrected by surgery. Surgeons usually make zig-zag cuts between the fingers, and maybe add new skin from another part of the body.

## Apoptosis

Cells die because they are killed by toxic chemicals or injury. However, they may die because they are triggered to 'commit suicide'. This is called **apoptosis**. Like cell division, this process can be controlled by signals from either inside or outside the cell.

You can see apoptosis in Figure 2. Apoptosis is a very orderly process, quite different from when a cell is killed. In apoptosis

- the cell shrinks
- the DNA and protein in the nucleus breaks down
- the mitochondria in the cell break down
- the cell breaks up into small fragments, wrapped in membrane
- a particular **phospholipid**, phosphatidylserine, which is normally present on the inside of the cell membrane, is placed on the surface
- phagocytic cells such as **macrophages** have protein **receptors** in their plasma membranes that recognise phosphatidylserine. When phosphatidylserine binds to these receptors, it causes the phagocytic cells to engulf these membrane-bound fragments.

### Why do cells commit apoptosis?

There are two main reasons why cells commit apoptosis.  One reason is to destroy cells that are a risk to the organism. For example, apoptosis occurs in

- cells that are infected with viruses
- cells with DNA damage, as these could cause cancer. In a fetus, cells with DNA damage could cause a birth defect. Cells that have damaged DNA increase the production of a protein called p53. This protein induces apoptosis. You will learn more about p53 on spread 2.1.1.4.

Apoptosis is also important in normal development. For example:

- in a fetus, the fingers and toes are formed by removal of the tissue between them. This happens by apoptosis.
- shedding the lining of the uterus (the **endometrium**) during menstruation involves apoptosis
- forming synapses, or connections, between neurones in the brain involves apoptosis of the surplus cells around them.

(a)
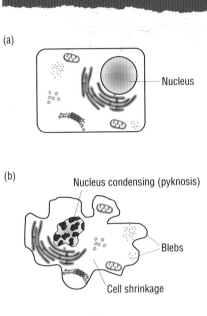
Nucleus

(b)
Nucleus condensing (pyknosis)
Blebs
Cell shrinkage

(c)
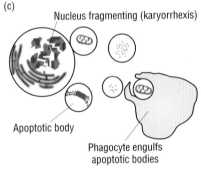
Nucleus fragmenting (karyorrhexis)
Apoptotic body
Phagocyte engulfs apoptotic bodies

**Figure 2** Apoptosis

## Questions

1  A student examined a sample of cells that were dividing by mitosis under a microscope. She counted the number of cells in each stage. The results are shown in the table.

| Stage | Number of cells |
| --- | --- |
| Interphase | 810 |
| Prophase | 105 |
| Metaphase | 15 |
| Anaphase | 6 |
| Telophase | 24 |

(a) Calculate the percentage of the time that these cells spend in interphase.
(b) Which stage of mitosis lasts the longest? Explain your answer.

2  There are 12 units of DNA in a cell during prophase. How many units of DNA will there be:
(a) In the cell during $G_1$?
(b) In the cell during $G_2$?
(c) In one of the nuclei formed at the end of telophase?

3  Explain why phagocytic cells such as macrophages do not engulf healthy cells.

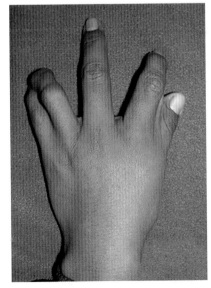

**Figure 3** Syndactyly

## What is cancer?

In a healthy person, mitosis only happens when new cells are needed for growth or repair. This means that the cell cycle is highly controlled by factors both within the cell and by growth factors outside the cell.

Cancer may develop when the cell cycle is not controlled. A tumour is a mass of abnormal cells that has grown when there is no need for growth or repair. Some tumours are **benign**. This means that they tend to grow in one place and do not spread to other parts of the body. If a benign tumour is removed by surgery, it is unlikely to grow back. Examples of benign tumours are moles and uterine fibroids. This does not mean that benign tumours are harmless, however. A benign tumour may press on important nerves or blood vessels and cause harm. For example, a benign tumour in the brain could be very harmful if not removed.

Cancer is when malignant tumours form. These tumours tend to spread into neighbouring tissues. Cells from the tumour may break off and spread to other parts of the body via the blood or lymph systems. This is called **metastasis**. Here, they can start secondary tumours, or **metastases**.

You can see how cancer develops in Figure 1.

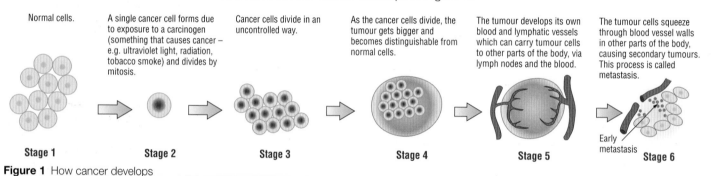

| Normal cells. | A single cancer cell forms due to exposure to a carcinogen (something that causes cancer – e.g. ultraviolet light, radiation, tobacco smoke) and divides by mitosis. | Cancer cells divide in an uncontrolled way. | As the cancer cells divide, the tumour gets bigger and becomes distinguishable from normal cells. | The tumour develops its own blood and lymphatic vessels which can carry tumour cells to other parts of the body, via lymph nodes and the blood. | The tumour cells squeeze through blood vessel walls in other parts of the body, causing secondary tumours. This process is called metastasis. |

Stage 1     Stage 2     Stage 3     Stage 4     Stage 5     Early metastasis  Stage 6

**Figure 1** How cancer develops

### How is normal cell division controlled?

Genes called **proto-oncogenes** stop a cell from dividing too often. They work in two different ways.

- They may carry the genetic code for a receptor protein in the cell membrane. This receptor protein is the right shape for a specific growth factor to bind to it. When the growth factor binds to the receptor protein, the genes needed for DNA replication are 'switched on'.
- They may carry the genetic code for a growth factor, the cyclins and CDKs mentioned in the previous spread.

However, changes may occur in these proto-oncogenes. When a change occurs in DNA, we call it a **mutation**. A mutated oncogene can lead to uncontrolled cell division.

- An oncogene may produce a receptor protein that triggers DNA replication even when an extracellular growth factor is not present. An example of this is the Ras oncogene. In normal cells, it only triggers cell division when it is stimulated by, for example, a hormone. When it mutates, it can permanently trigger cell division even when the growth factor is not present.
- An oncogene may cause excessive amounts of growth factor to be produced.

Cells also contain genes called **tumour suppressor genes**. These code for proteins that stop the cell cycle. They cause cells with mutated DNA to undergo apoptosis. One tumour suppressor gene is the p53 gene. If the p53 gene mutates, cells with damaged DNA may not be detected. This means that cells with damaged DNA continue to divide, passing on their mutations to their daughter cells.

## What causes mutations?

As you learned on spread 2.1.1.1, DNA copies itself very accurately. There are also enzymes that 'proof-read' the DNA and correct any mistakes that are made. However, DNA copies itself so many times that very occasionally a mistake is made that is not corrected by these enzymes. This leads to a mutation.

However, there are factors that increase the rate of mutation. These include ultraviolet radiation and some chemicals in cigarette smoke. Factors that cause damage to DNA are called **carcinogens**.

Ultraviolet light is high-energy radiation. It cannot penetrate very far into the body, but it can penetrate skin cells. Here, it can damage bonds in the DNA. One of the genes that are very likely to be damaged by ultraviolet light is the p53 gene. If this mutates, it can lead to cancer. This is why exposure to ultraviolet radiation can lead to skin cancer.

Cigarette smoke contains several carcinogens. One of these is benzopyrene. This is absorbed by the cells lining the respiratory tract. Here, it is converted to another chemical, BPDE. BPDE binds to the p53 gene and mutates it, so that it no longer works. With the p53 gene not working, cells can divide in an uncontrolled way, leading to the development of a tumour.

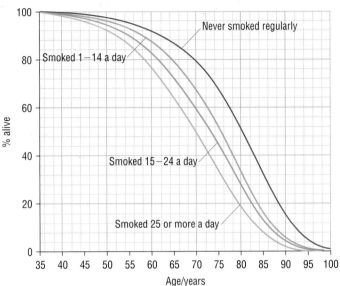

**Figure 2** Graph showing life expectancy of people related to the number of cigarettes they smoke in a day

## The cell cycle and cancer

If we make an analogy with driving a car, oncogenes can be seen as the accelerator, and tumour suppressor genes as the brake.

Retinoblastoma protein (RB) inhibits the cell cycle. In other words, the gene coding for RB is a tumour suppressor gene. RB acts as a 'switch' in the cell cycle. RB is inactivated if it is phosphorylated (has a phosphate group attached). This phosphorylation is carried out by CDKs 2 and 4 during G1. The CDKs act by binding to a second type of protein called cyclins.

So, for the cell to pass from the G1 phase to the S phase:

- cyclins d and e need to be synthesised
- cyclin d activates CDK4 and cyclin e activates CDK2
- the CDKs phosphorylate RB and inactivate it
- the cell progresses into the S phase and is committed to DNA replication and then cell division (G2 and mitosis).

For the cell to pass from S to G2 depends on cyclins in a similar manner.

p53 is a **transcription factor**. It codes for a protein that blocks the interaction of CDK and cyclin, so the RB normally stays active and blocks progression on to cell division. p53 protein binds to DNA and causes another protein, p21, to be made.  If p21 is made it binds to CDK2 and stops it binding to cyclin d.  So the gene for p21 is another tumour suppressor, as is the gene for p53.

If the p53 gene mutates, this means that p21 protein is not made. In turn, this means that cyclin and CDK can bind and inactivate RB. As a result, the cell will divide.

The Ras proto-oncogene has a different mode of action. Ras codes for a receptor on the cell surface. Growth factors, such as hormones, can bind to it. As a result, Ras is switched on and cell cycle inhibition stops. As a result, the cell starts to divide (moves from G1 to S).  If the Ras proto-oncogene mutates, it becomes an oncogene. This means that it is permanently switched on even if growth factors are not present.

## Smoking and lung cancer

Figure 2 shows the life expectancy of people according to the number of cigarettes they smoke per day.

- Give two conclusions that may be drawn from these data.
- Do these data prove that smoking causes lung cancer? Explain your answer.

## Questions

1 A benign tumour in the brain could be particularly dangerous. Explain why.
2 Explain why people have a better chance of being cured of cancer if it is detected early.
3 Most cancer occurs in older people. Use your knowledge of DNA and cancer to suggest a reason for this.

## Stem cells

You will remember that every cell in the body has exactly the same DNA. Yet a liver cell is clearly very different from a nerve cell or a red blood cell. The reason that these cell types look different is because some of the genes they contain are 'switched off' as the cell matures. This process is called **differentiation**. Once a cell has differentiated into a particular kind of cell, it loses the ability to divide and produce more cells of a different kind.

But some cells remain undifferentiated. This means that they can divide and form different kinds of cell. Cells that can do this are called **stem cells**. Scientists believe that stem cells will be very useful in the future to help cure many medical conditions. For example, a person who has broken their spinal cord is paralysed because the nerve cells cannot repair the damage. In the future it may be possible to use stem cells to repair a broken spinal cord.

### Obtaining stem cells

Scientists believe that the most useful kind of stem cells come from embryos that are just a few days old. At this stage, an embryo is a tiny ball of undifferentiated cells. A newly fertilised egg is said to be **totipotent** because it can divide to form a whole new organism. Cells from an embryo, once it has reached about 50–100 cells, are described as **pluripotent** because they have the ability to become any cell type, but not a whole new organism. This is because some cell differentiation has already started.

Adults also have stem cells. These have the ability to become one of several cell types, although they are not pluripotent. Stem cells are found in human bone marrow, inside certain bones. These stem cells are described as **multipotent** because they can form several, but not all, different cell types. The stem cells in the bone marrow can produce all the many different kinds of blood cell, but they cannot become any other kind of cell.

### Potential uses of stem cells

Scientists believe that in the future there could be many uses for stem cells.
- They could be used to treat medical conditions such as Parkinson's disease, Alzheimer's disease, heart disease, stroke, arthritis, diabetes, burns and spinal cord damage.
- They could be used to test the effects of experimental drugs.
- They can be studied to improve our knowledge about how diseases develop.

However, people have concerns about the use of stem cells in medicine.
- Although embryonic stem cells are taken from embryos that are only four or five days old, many people think these embryos represent a potential human life. They believe that it is wrong to use embryos in this way.
- Some scientists think it is possible that stem cell therapy might lead to a virus being passed on to someone, quite unintentionally.
- The media used to grow stem cells use nutrients from animals. There is a small risk that diseases could be passed from animals to humans in this way.
- Some people are worried that stem cells might become cancerous.

The Human Fertilisation and Embryology Authority is the UK's independent regulator overseeing safe and appropriate practice in fertility treatment and embryo research.

It licenses and monitors centres carrying out IVF, donor insemination and human embryo research. It also provides a range of detailed information for patients, professionals and the Government. You can find out more about its role at http://www.hfea.gov.uk.

Most scientists would prefer to use stem cells from an adult rather than from an embryo. This would also mean that the stem cells would come from a person's own body, so they would not be rejected by the person's immune system. But very few stem cells are found in an adult, and at the moment it is difficult to culture enough cells. Also, adult stem cells do not grow into any kind of cell. Their main role is to repair the tissue in which they are found. So they differentiate only into cells found in that tissue.

### Stem cells in bone marrow

Stem cells in the bone marrow produce the many different kinds of blood cell. These stem cells are called haemocytoblasts. You can see how they produce blood cells in Figure 1.

Red blood cells are produced from haemocytoblasts. Each stem cell divides by **mitosis**, giving 8 or 16 proerythroblasts. These stem cells are now committed to differentiate. As the red blood cell develops

- haemoglobin gradually builds up
- the cell nucleus breaks down and is extruded from the cell
- cell organelles are lost, such as mitochondria
- the cell gradually gets smaller.

An immature red blood cell is called a 'reticulocyte' because of the large amount of endoplasmic reticulum that it contains.

Stem cells in the bone marrow give rise to committed cells called monoblasts. These then develop into **monocytes** in the bone marrow. Monocytes spend a few hours in the blood before returning to the tissues. Here they mature into **macrophages**.

Stem cells also give rise to committed cells called lymphoblasts. Some of these cells remain in the bone marrow, and others migrate to the **thymus gland** and other lymphoid tissue. Here they divide and mature into lymphocytes.

Stem cells also give rise to committed cells called myeloblasts, which mature into other kinds of **leucocytes** (white blood cells) such as neutrophils.

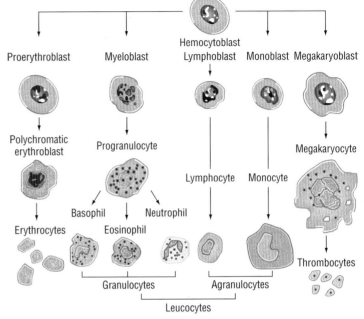

**Figure 1** Blood cell production in the bone marrow

### Examiner tip

You need to be able to describe the key differences in appearance between stem cells and the mature blood cells described earlier – and this is a good opportunity to revise the functions of those organelles!

### Case study: CellStore

*Umbilical cord stem cell storage is where cord blood (which is rich in stem cells) from your baby's umbilical cord is safely collected at birth, using the collection kit supplied by CellStore plc. The midwife who delivers your baby will do this. The cord blood is frozen and stored safely.*

*Stem cells allow the body to regenerate damaged cells and repair organs. This means that stem cells can be used to treat leukaemia, anaemia and heart problems. CellStore gives you the chance to protect your child's future health. In the future, other health conditions may be treatable using stem cells.*

*The cost of the kit and storage of the cord blood for 20 years is only £1,600. We believe this is a small price to pay for your child's future wellbeing.*

Do you think it would be a good idea to pay a company such as CellStore to store umbilical cord blood from a newborn baby? Set out your answer as 'pros' and 'cons'.

## Questions

1 Suggest why cells taken from an embryo after 4–5 days are not useful as stem cells.

2 Explain why reticulocytes contain large amounts of endoplasmic reticulum.

3 Red blood cells only last about four months. Use your knowledge of the structure of a red blood cell to suggest why it has a short lifespan.

You learned on spread 2.1.1.4. that some factors increase the chances of developing cancer. We are going to look at some of these in more detail.

## Types of radiation

Some kinds of radiation can increase the chances of developing cancer. Types of radiation that are high in energy can damage the bonds in DNA. This leads to **mutations**. You will remember from spread 2.1.1.4 that ultraviolet radiation can cause skin cancer. Other kinds of radiation that can cause cancer include X-rays and gamma rays.

### Chemical carcinogens

You will remember from spread 2.1.1.4 that some chemicals cause changes in DNA. These chemicals are called **carcinogens**. Because DNA can be repaired by cells to some extent, it often takes a number of attacks by carcinogens before cancer is triggered. So a carcinogen can be regarded as increasing the likelihood of cancer occurring. Several different chemicals in cigarette smoke cause cancer. Some other examples of chemical carcinogens and the types of cancer with which they are associated are shown in the table.

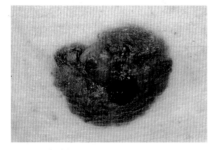

**Figure 1** Melanoma, a form of skin cancer

| Chemical carcinogen | Type of cancer |
| --- | --- |
| Benzene | Leukaemia |
| Hair dyes | Bladder cancer |
| Arsenic | Lung cancer, skin cancer |
| Soot and mineral oil | Skin cancer |
| Asbestos | Lung cancer, mesothelioma |
| Alcohol | Oesophageal cancer, oropharyngeal cancer |

**Table 1** Carcinogens and associated cancers

### Ageing

A few kinds of cancer, such as leukaemia, are found mainly in younger people. But most types of cancer, such as colon cancer, become more common as people age. This happens because older people have had more years of contact with factors that cause cancer, such as ultraviolet radiation. They have had more time to accumulate damage to their DNA. Scientists believe that several different mutations have to occur to DNA for cancer to develop. It is likely to take some years for several different mutations to occur.

### Heredity

Some people inherit **genes** that make them more likely to develop cancer. Some forms of breast cancer seem to be linked to particular gene mutations. Two genes, BRCA1 and BRCA2 are examples of genes that can easily mutate to cause breast cancer. Women who have one of these mutated genes have about a 60% lifetime risk of developing breast cancer. They also have a significantly higher risk of developing ovarian cancer. Some kinds of colon cancer can also be caused by inheriting a particular mutated gene.

### Viruses

Some cancers have been shown to be caused by viruses. Viruses infect cells by inserting new genetic material into cells, so they cause changes in a cell's DNA. One example of a cancer caused by a virus is cervical cancer. Most cases of cervical cancer are caused by the Human Papilloma Virus.

## Case study: processed meat and bowel cancer

The European Prospective Investigation into Cancer and Nutrition (EPIC) looked at the dietary habits of over 500 000 people across Europe over 10 years. They found that bowel cancer risk was a third higher for those who regularly ate more than two 80 g portions of red or processed meat a day, compared to less than one a week. The researchers defined red meat as beef, lamb, pork and veal. Processed meat was mostly pork and beef that were preserved by methods other than freezing. They include ham, bacon, sausages, liver pate, salami, tinned meat, luncheon meat and corned beef.

Since the study began, 1330 people have developed bowel cancer. The study also found a low-fibre diet increased the risk of bowel cancer.

Eating poultry had no effect on the risk of bowel cancer. However, the risk for people who ate one portion or more of fish every other day was nearly a third lower than those who ate fish less than once a week.

There are several theories about why red meat should increase the risk of bowel cancer. One explanation is that haemoglobin and a similar protein, myoglobin, which are found in red meat, trigger a process called nitrosation in the gut, which leads to the formation of carcinogens. Alternatively, the problem might be caused by compounds called heterocyclic amines. These are carcinogens created in the cooking process. But these compounds are also found in poultry, which has not been linked to an increased cancer risk.

**Figure 2** The risk of bowel cancer is higher for people who regularly eat processed meat

### Examiner tip

Correlations do not 'prove' a cause but they do establish probable links between two factors, in this case smoking and the risk of cancer. With two sets of data, such as number of cigarettes smoked and incidence of lung cancer, a 'correlation coefficient' can be calculated. This can be compared to statistical tables to see if the figure obtained is 'significant'.

Significance is an important concept in evaluating the strength of the evidence. It takes into account a set of results arising by 'chance' rather than due to a link. Normally results are said to be significant if they could occur by chance less than 5% of the time.

## Evaluating epidemiological evidence

Studies such as EPIC collect vast amounts of data. The data collected is then assessed using statistical techniques which look at the strength of the correlation between a factor and the increased risk of a cancer. If the evidence is sufficiently strong, the results will be published and used, for example, to inform government policy. Look at Table 2 below:

| | | |
|---|---|---|
| Cigarette smoking | Sufficient: | Lung, oesophagus, larynx, pharynx, oral cavity, pancreas, bladder, nasal cavity and sinuses, stomach, liver, kidney, cervix and myeloid leukaemia |
| Passive smoking | Sufficient: | Lung |
| Pipe and cigar smoking | Sufficient: | Lung, oesophagus, larynx, pharynx, oral cavity, liver, bladder, bowel, stomach and pancreas |
| Chewing tobacco | Sufficient: | Oral cavity |

**Table 2** Strength of evidence for an increased risk of cancer due to tobacco consumption

The evidence that links tobacco smoke, even passive smoking, to lung cancer was sufficiently strong to persuade governments across Europe to introduce a ban on smoking in public places.

## Questions

1  How would you carry out a study like EPIC?
2  What are the problems in carrying out a study like this?
3  What are carcinogens? What problems are there in trying to find out whether a chemical is a carcinogen?

Several different techniques are used to detect cancer.

## X-rays

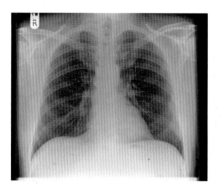

**Figure 1** Chest X-ray

X-rays are a kind of high-energy radiation. X-rays give off photons. X-ray photons pass through the softer parts of the body quite easily, but denser parts of the body, such as bones, tend to absorb the photons. In an X-ray machine, X-rays are produced by an electrode on one side of the patient. A camera on the other side of the patient contains photographic film which records the pattern of X-rays passing through the body. Dense parts of the body, like bones or tumours, appear whiter on the film, and softer parts of the body appear darker. This is because they allow more X-rays through. You can see this in Figure 1.

### Mammography

Mammography is the use of X-rays to detect small tumours in the breast. Low-dose X-rays are passed through the breast. If any unusual lump shows on the X-ray, further tests will be done. Mammography is offered on the NHS to all women over the age of 50. You can see how this is done in Figure 2.

**Figure 2** Mammography

### CT scans

CT scans are also known as CAT scans, standing for computerised (axial) tomography. The patient lies on a platform which slowly moves through a hole in the CT machine. The machine is ring-shaped and it turns around the patient, taking X-ray pictures from all different angles. A computer uses these pictures to build up a three-dimensional image of the body. This makes it much easier to see where a tumour is. Figure 3 shows a CT scan.

## Thermography

Thermography uses cameras that are sensitive to infrared radiation. Warmer parts of the body give off more infrared radiation. This means that the resulting image shows which parts of the body are warmest. You will remember that cancer cells divide more quickly than other cells. This means that they respire more and release more heat. So tumours show up as warmer areas. You can see this in Figure 4.

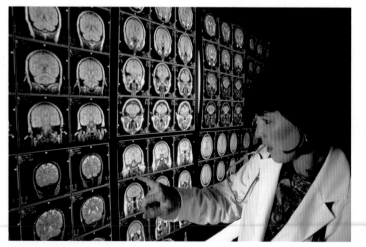

**Figure 3** CT scan results

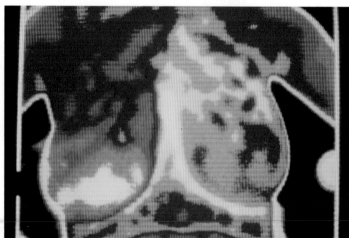

**Figure 4** Thermography

# Ultrasound

Ultrasound uses sound waves to build up an image of parts of the body that may contain tumours. It is particularly useful to detect tumours in soft parts of the body such as the liver. It is much cheaper and more portable than most other methods of detection. Figure 5 shows an ultrasound scan showing a testicular tumour.

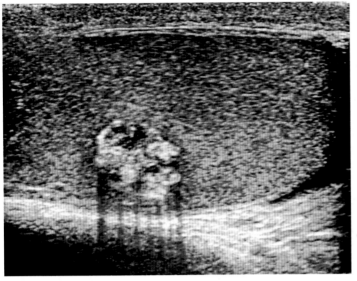

**Figure 5** Ultrasound

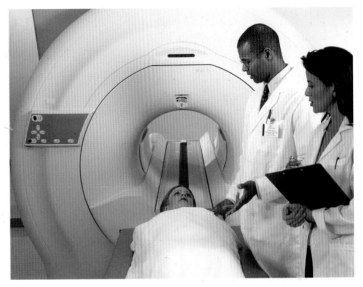

**Figure 6** MRI scanner

# MRI scans

This stands for magnetic resonance imaging. The patient lies inside a very large magnet. The machine measures the magnetic field in different parts of the body. A computer puts these measurements together to build up a three-dimensional image. The image is more detailed than a CT scan, but the MRI machine is very expensive. Although the procedure is not painful it can be uncomfortable, as people have to lie very still for a long time in the small space inside the magnet, while the machine makes quite a lot of noise. You can see an MRI scanner in Figure 6.

# PET scans

PET stands for positron emission tomography. The patient is injected with a radioactive substance that breaks down and releases gamma rays. These gamma rays are detected, and the pattern of gamma rays emitted is used to build up an image of the body. The most metabolically active parts of the body give off more gamma rays. PET scanners are very expensive and there are only a few available in the country.

# Questions

1 Radiographers who carry out X-ray investigations on patients stand behind a screen when the X-ray is taken. This prevents the radiographer from absorbing X-rays. Explain why this is important.
2 Explain why tumours emit more gamma rays than healthy tissue when a PET scan is carried out.
3 CT scans are normally carried out only when a person is suspected of having cancer. One reason for this is that it is expensive. Suggest another reason for this.

Epidemiologists are scientists who study the distribution of diseases. This can help us to learn about the factors that cause disease. Epidemiologists study the **prevalence** of disease. They also study the incidence of a disease. Incidence is the number of new cases of a disease occurring in a population at a given time.

### Key definitions

**Prevalence** is the number of existing cases of a disease in a given population at a given time. For example, the prevalence of COPD in the UK population could be as high as 4000 cases per 100 000.

**Incidence** is the number of new cases of a disease in a given population in a year.

## The incidence and prevalence of breast cancer

Look at Figure 1. This shows the incidence of breast cancer in the UK in 2000.

Table 1 shows the risk of a woman developing breast cancer at different ages.

You will see from the graph in Figure 1 that most cases of breast cancer occur in women who have reached the menopause. This is the stage in life when a woman's menstrual cycle stops. The woman is no longer fertile and her periods stop. Most women reach the menopause at about the age of 50.

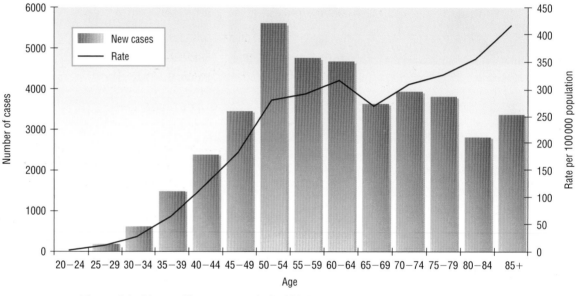

**Figure 1** Incidence of breast cancer in the UK, 2000

While a woman is menstruating, her body produces high levels of oestrogen. After the menopause, oestrogen levels fall. We know that oestrogen protects against the development of breast cancer. However, men can also develop breast cancer, though it is much rarer than in women.

Apart from being a woman over 50, there are several other risk factors for breast cancer:
- starting puberty early
- having a late menopause
- having a family history of breast cancer
- drinking a large amount of alcohol
- being obese.

### Treating breast cancer

Often **surgery** is carried out. This means that the tumour is cut out. Often, a **lumpectomy** is carried out. This means that the surgeon removes the tumour and a border of breast tissue around it. Sometimes the surgeon decides that **mastectomy** is a better option. This involves removal of the whole breast. Following this, the woman will

| Age | Risk |
|-----|------|
| Up to age 25 | 1 in 15 000 |
| Up to age 30 | 1 in 1900 |
| Up to age 40 | 1 in 200 |
| Up to age 50 | 1 in 50 |
| Up to age 60 | 1 in 23 |
| Up to age 70 | 1 in 15 |
| Up to age 80 | 1 in 11 |
| Up to age 85 | 1 in 10 |
| Lifetime risk (all ages) | 1 in 9 |

**Table 1** Breast cancer risk and age

usually be offered reconstructive surgery. Sometimes lymph nodes are removed as well. This is done when it is likely that cells from the breast tumour have spread into the lymph nodes.

**Chemotherapy** is also used to treat breast cancer. This means using drugs to treat breast cancer. These drugs are usually chemicals that are toxic to dividing cells. You will remember that cancer cells are dividing very rapidly. This means that chemotherapy should kill cancer cells while having a much smaller effect on normal cells. Some cells in the body divide particularly quickly, for example, cells in hair follicles and the cells in the bone marrow that produce blood cells. This is why cancer drugs can sometimes have side effects such as reducing the number of white blood cells or hair loss. Chemotherapy may be used:

- before surgery, to reduce the size of the tumour
- after surgery, to make sure that all the cancer cells have been removed
- to treat breast cancer that has spread, or come back.

**Radiotherapy** is another breast cancer treatment. This uses ionising radiation to destroy cancer cells. Ionising radiation destroys actively growing cells more than other cells. The radiation can be targeted very accurately at the tumour, to minimise the damage to healthy cells. Like chemotherapy, this may be used before surgery to reduce the size of a tumour, or after surgery to destroy any remaining cells.

**Tamoxifen** is a hormone treatment for breast cancer. Scientists have found that the hormone oestrogen encourages breast cancer cells to grow and divide. Tamoxifen works by preventing oestrogen getting into breast cancer cells. It does this by blocking the oestrogen receptors on the surface of the breast cancer cells. Tamoxifen is not useful for all kinds of breast cancer, but if a doctor prescribes it for a patient, they usually take tamoxifen for five years.

**Immunotherapy** is another kind of breast cancer treatment. An example of this is the drug **herceptin**. This is useful for 15–20% of breast cancer patients. You will remember from spread 2.1.1.4 that there are receptors for growth factors on the cell surface. Some breast cancer patients have a particular receptor on their cancer cells that cause the cells to divide rapidly. Herceptin is an antibody that is the right shape to bind specifically to these protein receptors and block them so that they cannot cause rapid cell division.

You will remember from spread 2.1.2.1 that some viruses can cause cancer. An example is Human Papilloma Virus (HPV), which can cause cervical cancer. A vaccine has been developed that protects women against HPV. Other kinds of immunotherapy are being developed. Scientists hope that, in the near future, it may be possible to produce vaccines that help the body's immune system to destroy cancer cells.

**Complementary therapies** can also be used as part of the treatment for cancer. Complementary therapists usually work with the person as a whole, not just the part of the body with the cancer. This is called a holistic approach. Complementary therapies can make a person feel better while they are having cancer treatments, and reduce the side effects such as tiredness, anxiety, sleeping problems, constipation, diarrhoea and sickness. Different kinds of complementary therapy are available, including relaxation therapies, hypnotherapy, art therapy, therapeutic touch, reiki, meditation and aromatherapy.

> **Examiner tip**
>
> One role of our immune system is to destroy cancerous cells. You will study the immune system later in this course. You will need to be aware of the links between immunity and cancer.

# Questions

1 Suggest why ionising radiation destroys rapidly dividing cells more than normal cells.
2 Explain why herceptin will not work for all kinds of breast cancer.
3 A vaccine against HPV will be most effective if it is given to girls aged 13 or 14. However, there are ethical issues in giving this vaccine to girls of this age. Suggest what these ethical issues may be.

## The need for testing

It is very important that new drugs are tested before they are used. It takes several years to develop a new drug before any kind of testing starts. A new cancer drug may need to be tested for about six years in the laboratory before it reaches the stage of **clinical trials**. During this time, drugs may be tested on animals and on cell cultures. Very few drugs even get to the stage of clinical trials. It is estimated that of every 1000 new drugs that are created, only one will ever reach the stage of clinical trials.

Clinical trials are needed to check that the drug is actually effective. It is also necessary to find out if there are any side effects that may be harmful, and to find out what dose is effective. Doses for different drugs may also vary between adults, children and elderly people.

### Clinical trials

- Phase 1 trials are the earliest trials in the life of a new drug or treatment. They involve small numbers of people (usually fewer than 30) and are carried out to find the safe dose range, any side effects, how the body copes with the drug, and whether the drug is effective. These trials take a long time. The first few patients are given a very low dose of the drug. If this goes well, the next few people will be given a slightly higher dose. This continues until the scientists find the right dose to give. The people are studied in detail. People in phase 1 cancer trials often have advanced cancer and have had all the available treatments. This is because some people will benefit from the new drug being tested, but most people will not.

- Phase 2 trials are used to find out if the treatment works well enough to be tested in bigger phase 3 trials. Phase 2 trials of cancer drugs are used to find out what kinds of cancer the drug is effective against, and to find out more information about the best dose and any possible side effects. These trials usually use more people than phase 1 trials. About 50 people may be involved. If these trials show that the drug is as effective as existing treatments, or better, then it moves into a phase 3 trial.

- Phase 3 trials compare the new treatment with the best currently available treatment. These trials usually involve much larger numbers of people. This is because a new treatment might be only slightly better than the existing treatment, so you need to use large numbers of people to make this small difference noticeable. These trials may involve thousands of patients in many different hospitals in different countries.

- Phase 4 trials are carried out after the drug has been given a licence for doctors to prescribe it. These trials are done to find out more about possible side effects, and what the long-term benefits and problems are. It also gives more information about how the drug works, as it is being used on larger numbers of people.

Phase 3 trials are often **randomised**. This means that people are chosen at random to go into a group. One way you might pick a name at random is to pull names out of a hat. One group will be given the new treatment, and the other group will be given the existing treatment. By choosing people at random, this ensures that, on average, the two groups will be as similar as possible to each other.

In many studies some people are given a **placebo**. This means that a person is given an injection or a tablet that looks exactly like the real drug being tested, but it does not contain the drug. In other words, it is a 'dummy' treatment. This is done to make sure that it is really the drug that is making the person improve. Some people are biased, and will say they feel better even if they have a tablet made of something like sugar.

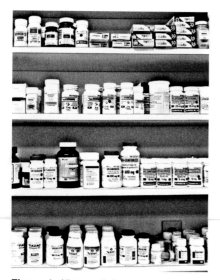

**Figure 1** All prescription drugs are thoroughly tested and trialled

However, the trial may not always use a placebo. For example, if a phase 3 trial is being carried out on people with cancer, it would not be ethical to give some patients a cancer treatment and other patients a dummy drug. In this case, the new drug is used with one group, and the standard treatment is used with the other group. In this way, everybody is receiving a treatment for their cancer.

Many studies are **blind** trials. This means that the person taking part does not know which treatment they are getting. They could be receiving the new drug, the standard treatment or the placebo. Studies may also be **double-blind**. This means that neither the scientists carrying out the study nor the patients know which treatment they are getting. This is done by giving every patient a code number. A computer randomly allocates the code numbers to treatment groups. The list of patients and their code numbers is kept secret until the end of the study.  This means that the scientists are not biased and also prevents any psychological effects on the patients of knowing they have or have not been given the drug – such effects might affect the results.

### The role of NICE

The National Institute for Health and Clinical Excellence (NICE) evaluates the effectiveness of new drugs, and provides guidelines for the NHS. In the UK, drugs can only be prescribed on the NHS using NICE guidelines. This is done so that public money is not wasted on prescribing ineffective drugs.

**Figure 2** Would you know which of these drugs was a placebo?

New drugs that are being developed may be effective, but they often cost thousands of pounds every month for just one patient. The NHS is paid for out of taxation and people do not want to pay more tax than is absolutely necessary. It is becoming increasingly difficult to fund the full cost of healthcare for everyone. There are many different opinions about this. Some people's ideas are listed below.

- The NHS should not have to pay for treating illnesses caused by a person's lifestyle – for example, treating lung cancer in a person who smokes.
- We should all be prepared to pay more tax if this provides full healthcare for everyone.
- We should all have to pay the cost of our own healthcare by buying medical insurance. This means that people who are more likely to need treatment will have to pay higher premiums – but that is fairer than the current system.
- At the moment, patients with some conditions like breast cancer can get expensive treatment on the NHS because people are sympathetic to their cause. But it means that other patients, like elderly people with Alzheimer's disease, are not having enough money spent on them. There are other factors to consider as well, such as the difficulty in evaluating the benefits of extending a particular person's life by a few years when a complete cure is not possible.

## Questions

1 Explain why drugs are tested on cell cultures and animal models before being used on humans.
2 Suggest why clinical trials need to be carried out on humans, rather than animals.
3 Suggest why only one out of every 1000 new drugs created ever gets to clinical trials.
4 What are the advantages and disadvantages of carrying out phase 1 trials of a new cancer drug on terminally ill cancer patients?
5 Suggest the advantages of carrying out a blind or double-blind study.

# Self-check questions

## Fill the blanks

1 The cell cycle is made up of interphase, ..................... and cytokinesis. ..................... is the longest phase of the cell cycle. It is subdivided into G1, S and G2. G1 and G2 are growth phases, and S is when ..................... takes place. Mitosis consists of four phases. ..................... is when the chromosomes appear. Each chromosome consists of two ..................... joined together by a ..................... In ..................... the nuclear ..................... has broken down, and a system of microtubules called the ..................... has been set up by the ..................... . The chromosomes line up along the ..................... of the spindle, attached to the fibres by their ..................... In ....................., the centromeres divide and the chromatids, which can now be called daughter ....................., move to opposite ..................... of the cell, pulled by the spindle fibres. In ....................., the daughter chromosomes reach the poles of the cell. They de-condense, the spindle fibres ..................... and a new nuclear ..................... re-forms. ..................... is when the cell divides into two. Each daughter cell is ..................... identical to the parent cell.

2 DNA stands for ..................... acid. It is a polymer made up of many repeated units called ..................... A nucleotide is made up of ..................... sugar, a ..................... group and an ..................... base. There are four organic bases in DNA: adenine and ..................... which are purines, and cytosine and ..................... which are pyrimidines. The sugar of one nucleotide joins to the phosphate group of the next nucleotide by a ..................... reaction. The two polynucleotide strands join together by ..................... base-pairing between the bases. Adenine always pairs with ..................... and cytosine pairs with .....................

3 DNA replicates by..................... replication. The DNA double helix 'unzips' because the ..................... bonds between the organic bases break. New nucleotides line up alongside the exposed bases, according to the complementary ..................... -pairing rules. An enzyme called ..................... joins the nucleotides together to form two new strands. Each new molecule of DNA has one 'old' strand and one 'new' .....................

4 ..................... is a very orderly process in which cells 'commit suicide'. It happens to destroy cells that are a risk to the organism, This includes cells that are infected with ....................., or cells with ..................... damage, as these can lead to cancer. Cells with damaged DNA produce large amounts of a protein called ..................... This protein induces ..................... Cell deletion by ..................... and cell addition by ..................... are essential for normal growth and repair.

5 ..................... cells are undifferentiated cells that have the potential to develop into many different kinds of cell. Stem cells in the bone ..................... may ..................... to form erythrocytes and many different kinds of .....................

6 There are many different factors that can increase the chances of developing cancer, including some kinds of radiation, ..................... such as chemicals in tobacco smoke, ageing, viruses and ..................... ..................... is the study of the distribution and causes of disease. There is epidemiological evidence to link ..................... with lung cancer, ..................... with bowel cancer, and mutations in the BRCA1 gene with ..................... cancer.

## Independent assortment of chromosomes and genetic variation

With *two* pairs of chromosomes, the number of possible combinations is 4. You can see this in the diagram below:

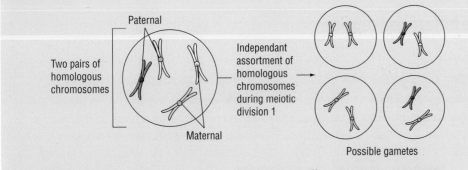

There are $2^2$ combinations. So wth 23 pairs in humans, there are $2^{23}$ combinations = 8388608!!

# Questions

1  Look at Figure 2. Describe what is happening in the cell.

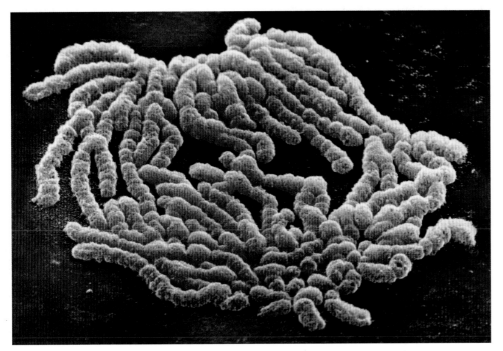

**Figure 2** Meiosis

2  The same parents can produce many children that are all different (unless identical twins are born). Explain why.

3  The bar chart in Figure 3 shows the DNA content in a cell at different stages of meiosis. Describe what is happening between:

   **(a)** Stage A and stage B.

   **(b)** Stage B and stage C.

   **(c)** Stage C and stage D.

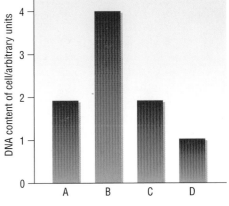

**Figure 3** Bar chart showing DNA content of a cell at different stages of meiosis

## Care before conception

Preconceptual care is the care that a mother should take of herself before she even becomes pregnant. It is very important that the mother is healthy before she becomes pregnant, because the fetus will be growing inside her uterus for nine months.

If the mother smokes, it is very important that she gives up smoking even before she becomes pregnant. This is because smoking can cause serious harm to the baby's development. The baby's father should also stop smoking, as passive smoking can harm both the mother and the baby. There is also evidence that men who do not smoke produce healthier sperm than men who smoke.

Women who are trying for a baby, as well as women who are already pregnant, should avoid drinking alcohol. Too much alcohol during pregnancy can cause problems in the developing baby. Women who drink more than two units of alcohol a day (two small glasses of wine or a pint of low-strength beer) are more likely to have babies with language, attention and hyperactivity problems. It is best not to drink alcohol at all, but if a pregnant woman does drink alcohol, she should consume no more than 1–2 units of alcohol once or twice a week.

Before becoming pregnant, it is important for women to check whether they are immune to rubella (German measles). If the mother catches the rubella virus during pregnancy, it can cause serious problems in the unborn baby, such as deafness, brain damage and eyesight problems. Women can check whether they are immune to rubella by having a simple blood test at their GP surgery. If a woman is not immune to rubella, she can have the vaccination. However, a woman should avoid becoming pregnant within three months of the vaccination, because the vaccine contains a live form of the virus.

Toxoplasmosis is a parasitic infection found in cats. It can spread to humans via cat faeces. If a pregnant woman catches toxoplasmosis, there is a high risk that she will pass the parasite on to her unborn baby. The parasite does not cause serious problems in adults. But in an unborn baby it can cause brain damage, hearing and eyesight problems, or epilepsy. In extreme cases it can cause miscarriage or stillbirth. To avoid getting this parasite, pregnant women should avoid contact with cat litter trays and take care when gardening that they do not come into contact with cat faeces on the ground or buried in the soil.

The food poisoning bacteria *Listeria* and *Salmonella* can harm the unborn baby. *Listeria* bacteria may be present in soft, unpasteurised cheeses and ready meals. *Salmonella* may be present in raw or undercooked meat and raw eggs. Pregnant women can safely eat pasteurised dairy produce, hard-boiled eggs and thoroughly cooked meat.

Some fish, such as shark and swordfish, should not be eaten by pregnant women as they may contain high levels of mercury. Tuna may be eaten, but only in small amounts.

## Care after conception

Postconceptual care is the care that a woman receives once she has become pregnant. This is often called **antenatal care**.

The first appointment happens early in the pregnancy, at around 11 or 12 weeks. The pregnant woman will be asked questions about her general health, and whether she has had children before. This is so that the medical staff can decide if any special care is needed during pregnancy. The pregnant woman will be given advice about pregnancy services, advice on diet and exercise, information about maternity benefits and the routine screening tests that will be done. A sample of urine will be tested to see whether protein or glucose is present. Protein in the urine may be a sign of an infection or kidney disease or of high blood pressure. Glucose in the urine may be a sign that gestational diabetes is developing, although this does not necessarily lead to diabetes after the baby is born. The woman's

**Figure 1** Drinking alcohol and smoking can cause serious harm to the development of the fetus

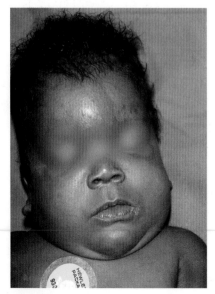

**Figure 2** Photo of a child with fetal alcohol syndrome

weight and height will be noted, and her blood pressure will be taken. This is to ensure that the mother is gaining the right amount of weight, and to check that she is not developing **pre-eclampsia**. Pre-eclampsia is a very dangerous condition characterised by high blood pressure. It is the cause of maternal and infant deaths.

### Later appointments
Pregnant women will have appointments every four weeks after week 12 of the pregnancy, every two weeks from week 32, and every week during the last three or four weeks. These may be at the hospital, at the local clinic, or in the woman's own home. At each appointment, the woman's blood pressure and weight will be measured. The doctor or midwife will examine the woman's abdomen to check that the baby is growing at the right rate. They will also listen to the baby's heartbeat. They will ask the mother how she is feeling, and about the baby's movements. A sample of the mother's urine will be tested. The doctor or midwife will also check that the mother does not have any swelling in the legs, arm or face, as this could be a sign of pre-eclampsia.

## Diet during pregnancy
A woman should expect to gain between 10 kg to 15 kg during pregnancy. She should not try to lose weight or avoid weight gain at this time, as she must make sure she is eating enough nutrients for the developing baby. But she does not need to eat very much more than usual. However, it is important that the woman should eat a healthy balanced diet. A pregnant woman is advised that she should consume, every day:

- four to six portions of carbohydrates (rice, pasta, cereals, bread, potatoes)
- at least five portions of fruit and vegetables
- three portions of protein (meat, poultry, fish, pulses, eggs)
- at least one portion of dairy products (milk, cheese, yoghurt)
- at least two litres of fluids such as water and fruit juices. This helps the body to get rid of toxins and waste products, and can also help to prevent constipation and nausea.

Pregnant women and women who intend to become pregnant are advised to take folic acid supplements. Folic acid reduces the risk of problems with the baby's spine and brain, such as spina bifida. Pregnant women should also eat plenty of green leafy vegetables, bread and breakfast cereals, because these contain folic acid.

Sometimes blood tests in pregnancy show that the pregnant woman is anaemic. This means that she is short of iron. The doctor may prescribe iron tablets. The woman should also eat plenty of green leafy vegetables and wholemeal bread, as these contain iron.

Pregnant women should also take regular gentle exercise, such as swimming, to ensure they are in good health and that they are prepared for labour. They should also relax frequently and have plenty of rest.

## Tests during pregnancy
There are several routine tests that are offered to all pregnant women, as well as some tests that are only offered to women with specific risk factors. A blood sample may be tested in order to
- find out the mother's ABO blood group, in case she needs a blood transfusion when the baby is born
- find out the mother's Rhesus blood group. This is because Rhesus negative mothers will need an injection after the baby's birth to protect their next baby from a serious kind of anaemia if the baby is rhesus positive
- check for the virus that causes hepatitis
- check whether the mother has HIV, the virus that causes AIDS. Pregnant women can pass this virus to their unborn baby, although medical treatment can reduce this risk
- check whether the mother has syphilis
- check whether the mother is immune to rubella.

Ultrasound scans can be carried out, as well as other tests for genetic defects. You will learn more about these on spread 2.2.1.5.

**Examiner tip**

You need to revise the immune system to really understand this section. The blood test for rubella is looking for **antibodies** to the virus and the vaccine will have the same **antigens** as the virus. The structure of viruses and bacteria are also relevant here!

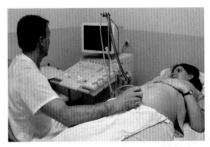

**Figure 3** Pregnant woman undergoing ultrasound

## Questions
1 Explain the difference between preconceptual and postconceptual care.
2 Explain why it is important that pregnant women do not become anaemic.

**Case study**

Kayleigh, who lives with her partner Chris, was delighted to hear she was pregnant after two miscarriages. Once the pregnancy got past 12 weeks, Kayleigh thought that nothing could go wrong. She had what the doctors called a 'low risk' pregnancy, and her scans were normal. But her son Jack was stillborn.

Doctors had failed to spot signs of fetal growth restriction, one of the main factors in full-term stillbirth. Almost 4000 babies were stillborn in Britain in 2006 according to the latest government figures. Despite falling since the 1960s, the rate of stillbirths has remained stubbornly unchanged for more than a decade and the rate (per 1000 births) actually went up from 5.3 in 2001 to 5.7 in 2003. Experts believe that many of these still births would be avoided if more trained staff and midwives were available.

## Diet in pregnancy

A pregnant woman should be very careful to eat a healthy balanced diet during pregnancy. In the UK, dietary reference values (DRVs) are published for various nutrients. These are guidelines for populations as to the amount of energy and nutrients needed by different groups of healthy people in the UK.

The table gives some DRVs. It also gives the EARs (estimated average requirements) for energy and RNIs (Reference nutrient intake) for protein and other nutrients. RNI levels would be sufficient for 97.5% of the population.

| | Energy/ MJ | Protein/% of energy intake | Protein/g | Calcium/ mg | Iron/ mg | Vitamin A/µg | Vitamin C/µg | Folic acid |
|---|---|---|---|---|---|---|---|---|
| Female 15–18 years | 8.83 | 15 | 45 | 800 | 14.8 | 600 | 40 | 400 |
| Female 19–50 years | 8.10 | 15 | 45 | 700 | 14.8 | 600 | 40 | 400 |
| Pregnancy | +0.8 * | ** | +6 | ** | ** | +10 | +10 | 600 |

\* = in last three months   \*\* = no extra amount needed

**Table 1** Dietary reference values and estimated average requirements

Proteins are needed for the growth of the baby, the placenta and the uterus and synthesis of haemoglobin, antibodies and enzymes. Calcium is needed to strengthen bones and teeth, and to maintain levels in the plasma for muscle contraction and nervous system function. Iron is needed for the formation of haemoglobin. Vitamin A is needed to make the pigment rhodopsin, which is present in the rod cells of the retina in eye. Vitamin C is needed for several functions including collagen formation.

### Measuring fetal growth during pregnancy

An ultrasound scan can be used to measure fetal growth. A machine sends out sound waves into the body. These are reflected back and used to produce an image on a screen. A gel is spread on the mother's abdomen. Then the scanner is passed over the uterus and an image of the fetus and the placenta can be seen.

The scan can show whether the mother is having a single baby or twins. From 18 weeks onwards, most organs can be seen and a scan will show if these are developing properly. Looking at the blood flow in the umbilical cord can show whether the baby is receiving enough oxygen and nutrients. The scan will show the position of the placenta. In some women, the placenta is lying abnormally close to the inside of the cervix. This can cause serious bleeding during labour, and a Casearean delivery may be needed.

The baby will also be measured. This will help the midwife to determine when the baby will be due. It is also used to tell whether the baby is the right size for its age. Two main measurements are carried out

- Crown-rump length. This is the length of the fetus from the top of its head to its bottom (see Figure 1).
- Biparietal diameter is the width of the head at its widest. This is usually measured only after 12 weeks.

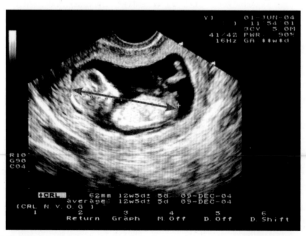

**Figure 1** Measuring the fetus

### The effects of smoking

When a pregnant woman smokes, carbon monoxide in the cigarette smoke combines with haemoglobin in both the mother's and the baby's blood. This stops the haemoglobin from carrying oxygen, reducing the oxygen supply to tissues in mother and baby. Nicotine reduces the diameter of blood vessels, including those in the placenta and in the fetus. This reduces the fetus's supply of oxygen. It also makes the fetus's heart beat faster.

The combined effects of cigarette smoke mean that the baby is likely to be born prematurely. The baby's lungs are likely to be less well developed and the risk of a still birth or death in early infancy is higher.

### Alcohol

Alcohol passes across the placenta into the baby's blood. Mothers who drink more than two units of alcohol a day (equivalent to a pint of ordinary strength beer or lager) are more likely to have babies who have problems in developing speech, language and attention span. Women who drink six units of alcohol a day, or more, are at risk of having babies with fetal alcohol syndrome. These children have mental and physical retardation, behavioural problems, and facial and heart defects (see page 84).

## Questions

1  Give a use of calcium in the body. (Clue: think about blood clotting).
2  Suggest why placenta praevia, *or a low-lying placenta*, may cause problems.
3  The table shows the crown-rump length of an average fetus during the first few weeks of pregnancy.
   (a)  Plot a graph of these data. Remember to include a title, use a sensible scale, and label the axes with the correct units.
   (b)  Use the graph to estimate the age of a fetus that has a crown-rump length of 110 mm.
   (c)  Use your graph to estimate the crown-rump length of a fetus at 15 weeks.
4  The graph shows the biparietal diameter of a fetus at different stages of pregnancy.
   (a)  Suggest reasons why it is important to measure biparietal diameter during pregnancy.
   (b)  A fetus has a biparietal diameter of 50 mm. Use the graph to estimate the age of the fetus.
   (c)  Suggest why normal babies can vary in their biparietal diameter.
5  Suggest why a pregnant woman who eats a healthy diet does not need to increase her intake of iron during pregnancy.

| Time after woman's last period/weeks | Crown-rump length/mm |
|---|---|
| 6 | 45 |
| 6.5 | 65 |
| 7 | 92 |
| 7.5 | 130 |
| 8 | 150 |
| 8.5 | 190 |
| 9 | 220 |
| 9.5 | 270 |
| 10 | 310 |
| 10.5 | 360 |
| 11 | 410 |
| 11.5 | 470 |
| 12 | 520 |
| 12.5 | 600 |
| 13 | 650 |
| 13.5 | 740 |
| 14 | 790 |

**Table 2** Average fetus size during pregnancy

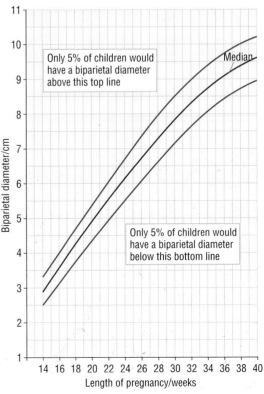

Only 5% of children would have a biparietal diameter above this top line

Median

Only 5% of children would have a biparietal diameter below this bottom line

Biparietal diameter/cm

Length of pregnancy/weeks

**Figure 3** Graph showing biparietal diameter of a fetus at different stages of pregnancy

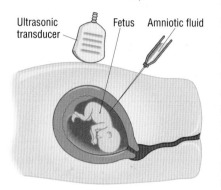

**Figure 1** Amniocentesis

## Problems in meiosis

You will remember from spread 2.2.1.2. that **meiosis** is used to form **gametes** with half the number of chromosomes of the parent cells. Each gamete should contain just one from each pair of **homologous chromosomes**. One pair of chromosomes determines the sex of the offspring in humans. They are called the sex chromosomes. Females have two identical X chromosomes. You can see this in Figure 1. Males have one X chromosome and one Y chromosome. Normally, every female gamete contains one X chromosome. However, half of all sperm contain an X chromosome and the other half contains a Y chromosome. If an egg is fertilised by an X sperm, the baby will be a girl, but if it is fertilised by a Y sperm, the baby will be a boy.

On rare occasions the sex chromosomes do not separate correctly in anaphase 1 or 2 of meiosis. In these cases, there will be one gamete without a sex chromosome at all, and another gamete with two sex chromosomes. When chromosomes do not separate correctly in meiosis, it is called **non-disjunction**.

### Turner's Syndrome

If a gamete with an X chromosome is fertilised by a gamete without a sex chromosome, the baby will have only one sex chromosome. We write this as XO. The child will have Turner's Syndrome. You can see a female with Turner's Syndrome in Figure 2.

### Klinefelter's Syndrome

If a gamete with an X chromosome is fertilised by a gamete with both an X and a Y chromosome, the baby will be XXY. The child will have Klinefelter's Syndrome and you can see this in Figure 3.

### Amniocentesis

**Amniocentesis** is a test that can be carried out to sample the fluid that the **fetus** is growing in. Some cells from the fetus are present in the fluid. A test called karyotyping can be carried out on these cells to see if the baby's chromosomes are normal. You can see the test in Figure 5.

**Ultrasonography** (ultrasound scanning) is used to find out where the placenta and the fetus are. The healthcare practitioner inserts a hypodermic needle into a suitable place

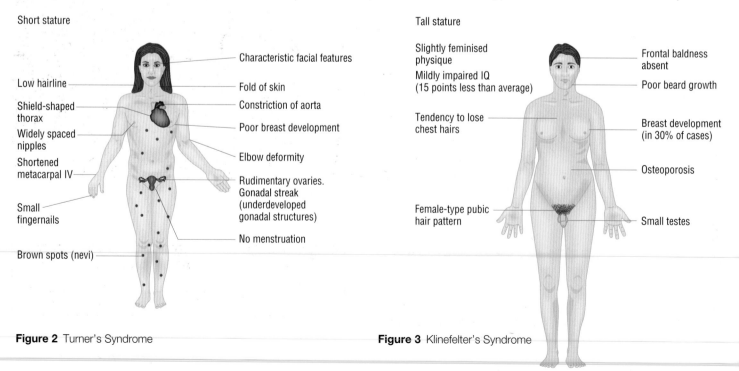

**Figure 2** Turner's Syndrome          **Figure 3** Klinefelter's Syndrome

**Normal**
Sperm with 23 chromosomes
including an X or Y chromosome

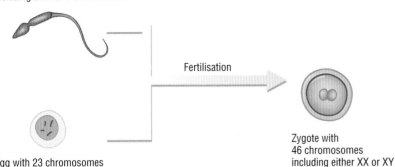

Fertilisation

Egg with 23 chromosomes
including an X chromosome

Zygote with
46 chromosomes
including either XX or XY

**Figure 4** How Turner's Syndrome and Klinefelter's Syndrome occur

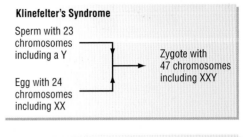
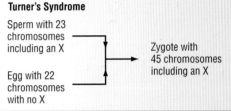

**Klinefelter's Syndrome**
Sperm with 23 chromosomes including a Y

Egg with 24 chromosomes including XX

Zygote with 47 chromosomes including XXY

**Turner's Syndrome**
Sperm with 23 chromosomes including an X

Egg with 22 chromosomes with no X

Zygote with 45 chromosomes including an X

on the mother's abdomen, away from the placenta and fetus. She then withdraws a sample of amniotic fluid using a syringe. This test is carried out only after 15–16 weeks of pregnancy.

## Chorionic villus sampling (CVS)

This test can be carried out after about 10 weeks of pregnancy. Again, ultrasound is used to find the location of the fetus and placenta. A tiny needle is used to remove a sample of the placenta. A disadvantage of both amniocentesis and CVS is that they slightly increase the risk of miscarriage. Chorionic villus sampling also carries a very small risk of deformity in the fetus.

## Karyotyping

A **karyotype** is a picture of all the chromosomes in a cell. The chromosomes are photographed during metaphase of **mitosis** and then sorted into pairs. You can see a karyotype in Figure 5.

Cells from the placenta or from the amniotic fluid are cultured in a special medium. A chemical is added that stimulates them to divide by mitosis. Then a chemical called colchicine is added, which stops the spindle being formed. This means the cells stop dividing at the beginning of metaphase.

The cells are placed in a dilute salt solution so they swell up, and a stain is added to make the chromosomes stand out. The chromosomes are then photographed.

When cells are obtained by CVS, a karyotype can be made within 10–14 days, but when the cells come from amniocentesis it takes a little longer. This is because more fetal cells are sampled using CVS.

**Figure 5** A karyotype

## Questions

1 Copy and complete the table to show the advantages and disadvantages of amniocentesis and CVS to investigate whether a fetus has a chromosome defect.

| | Advantages | Disadvantages |
|---|---|---|
| Amniocentesis | | |
| Chorionic villus sampling | | |

2 The karyotype in Figure 6 is taken from an unborn baby. Its mother wants to know if the child has a genetic defect. What can you tell her?

## Measuring infant growth

It is very important to measure a child's growth in order to ensure that it is growing normally. There are several ways of doing this.

### Monitoring changes in weight

The nurse places the baby on scales that are specially designed for weighing babies. The baby is supported so he cannot fall out. The nurse removes the baby's clothes to make sure the measurement is accurate, and records its weight. The baby is re-positioned and weighed again. In all, three readings are taken. The mean of the two readings that are closest together is calculated. If one of the readings is very different from the others, a further reading is taken as a check.

Sometimes babies are very distressed and will not lie quietly on the scales. If this happens, the mother stands on adult scales. The scales are set to zero, then she holds the baby and stands on the scales again. This is also repeated three times.

### Measuring infant length

This is shown in Figure 1. The baby is laid down on a special piece of measuring equipment. It is based on the same principle as the gauge used to measure your feet in a shoe shop. The head rests against one end of the equipment. The baby's legs are stretched out and a slide is brought up so that it touches the baby's feet.

### Measuring head circumference

This is measured by placing a tape measure around the baby's head and measuring the circumference at the part where the head is widest.

### Infant growth charts

The measurements are compared against a chart showing 'normal' growth. These charts are compiled by taking the measurements from very large numbers of children and plotting them on a graph.

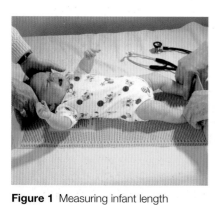

**Figure 1** Measuring infant length

| Nutrient | Needed for |
|---|---|
| Carbohydrates | Release of energy in cellular respiration |
| Lipids | Making cell membranes, to provide energy in cellular respiration, and to insulate nerve cells |
| Essential fatty acids | Two polyunsaturated fatty acids which cannot be made in the body, linoleic acid and alpha-linolenic acid, which must be eaten in the diet |
| Proteins | Forming new cells and tissues. Also, proteins are needed to make haemoglobin, antibodies, enzymes and other proteins in the body |
| Essential amino acids | Amino acids that cannot be made in the body. There are twenty different amino acids needed to make proteins in humans. Many of these can be made from other amino acids, but eight of these cannot be made from other amino acids. This means they must be eaten in the diet |
| Calcium | The development of healthy bones and teeth |
| Iron | Forming haemoglobin |
| Phosphorus | The development of healthy bones and teeth |
| Vitamin A | Making the pigment rhodopsin. This is present in the rod cells of the eye, which allow us to see in dim light |
| Vitamin C | Wound healing and the formation of collagen, a protein important in the formation of healthy skin, tendons, bones, and support tissues |
| Vitamin D | The formation of healthy bones and teeth |

**Table 1** Infants need a healthy balanced diet

### Nutrients needed by infants

Infants need a healthy, balanced diet if they are to grow properly. Table 1 shows some important nutrients that an infant should eat, and what the nutrients are needed for. Remember, the amount required will be given by the DRV values. These vary with age.

## Babies overfed to meet flawed ideal

New research is beginning to confirm what many mothers have long suspected – that the most commonly used growth charts, based on babies fed high-protein formula milk, wrongly classify lean but healthy babies as underweight. What's more, by encouraging mothers to overfeed their babies, the charts may be setting perfectly healthy children on the path to obesity.

The most popular growth chart, produced by the US National Center for Health Statistics (NCHS), has been used for nearly 30 years to provide a reference against which to judge the growth of a new baby. Introduced in 1977, when obesity was not yet common, its main aim was to make sure babies didn't suffer from malnutrition. What is now being increasingly recognised is that these charts were based on babies that were atypically heavy: almost all of them had been bottle-fed and came from white, middle-class families in Ohio.

The charts were revised by the US Centers for Disease Control and Prevention (CDC) in 2000 to include more breast-fed infants, but this has little bearing on the preceding 23 years. Scientists believe that these charts have led to overfeeding of infants for decades.

Last week, evidence was produced showing that babies fed high-protein formula milk put on weight far faster and more extensively than those fed breast or low-protein formula milk. Previous studies reached the same conclusion, but these relied on simply watching how a group of babies turned out, without being able to isolate the effect from other factors, such as wealth or the smoking habits of parents. In the new study, 1000 infants in five European countries randomly received breast milk, low-protein formula milk or high-protein formula milk and were monitored until the age of 2.

At any given age, babies in the high-protein group weighed around twice as much above the norm as the highest-scoring babies in other groups. The scientists are now following the babies to school age to see whether those who were heaviest during the first two years of life are at greater risk of obesity later on, as earlier studies have suggested.

There is growing acceptance that the NCHS charts are out of date. Last month, representatives from 31 European countries met in Italy to discuss whether to adopt new charts. These are based on a study coordinated by the WHO (World Health Organisation) of 8500 children in six diverse countries, all of whom were breast-fed and reared in optimal circumstances, free from poverty, illness and malnutrition.

*Based on an article in* New Scientist, *26th April 2007.*

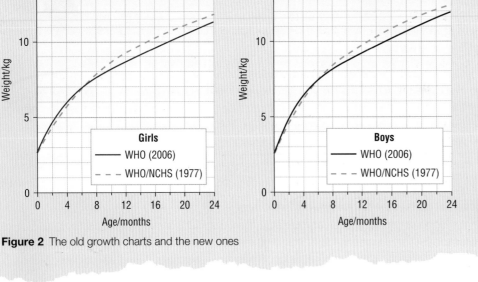

**Figure 2** The old growth charts and the new ones

- How useful are growth charts for infants?
- Give as many reasons as you can to suggest why the old charts are unsuitable.
- The new study shows that babies receiving high-protein milk gain more weight than other children. Explain why it was important that:
  - (a) a large number of children was studied.
  - (b) three different methods of feeding were investigated.
  - (c) the children were randomly assigned to each group.
- Do you think the new charts based on a study coordinated by the WHO are better? Give reasons for your answer.

## Questions

1 Suggest why it is important to weigh a small baby three times.
2 Explain why it is better to use infant scales, than to weigh the mother holding the baby after accounting for the mother's weight.
3 Suggest why infant growth is measured by three different methods, and not just one.
4 Suggest reasons why a healthy baby might be a little smaller than average.

## Growth during the human life cycle

A graph of the mass of a human being plotted against age would show the **absolute growth** of a human. This is useful because it shows the overall growth pattern in humans. But it is also possible to plot a graph of the change in height or mass against time. This produces an **absolute growth rate** curve. You can see this in Figure 1a. This kind of graph shows how the rate of growth changes during the time of the study. It shows when the rate of growth is highest.

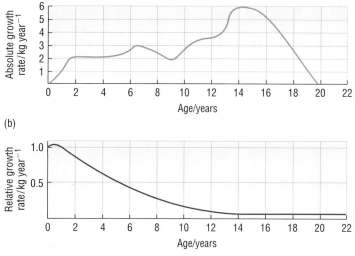

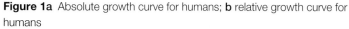

**Figure 1a** Absolute growth curve for humans; **b** relative growth curve for humans

> ### Key definition
>
> The **absolute growth rate** is the change in mass or height divided by the time period. For example, a child who weighs 13.6 kg on one birthday, and 15.4 kg on his next birthday, has an absolute growth rate of 1.8 kg year$^{-1}$.

> ### Key definition
>
> The **relative growth rate** is the change in weight or mass, divided by the weight or mass at the beginning of the time period. For example, a child who weighs 13.6 kg on one birthday, and 15.4 kg on his next birthday, has a relative growth rate of 0.13 kg year$^{-1}$.

Figure 1b shows the **relative growth rate** of a human. This shows the efficiency of growth, because it compares the mass or height with the measurement at the beginning of the time period.

Human growth is divided into four phases:
- *Fetal growth* when the growth rate is very high
- *Infancy and childhood* when the growth rate is very rapid at first and then the rate of growth slows down to a steadier rate
- *Adolescence* when there is a rapid growth spurt due to puberty
- *Adulthood* when the growth rate falls to zero, as the person's body mass or height stops increasing.

### Different parts of the body grow at different rates

In multicellular animals such as humans, the newly fertilised egg divides into a ball of cells (**blastocyst**) (see spread 4.2.1.3) and ultimately an embryo. The newly divided cells are genetically identical, but in time most of them become specialised into different kinds of cells with different functions. These cells in turn are grouped into different **tissues**, which go on to make up different **organs**. The organs themselves are organised into **organ systems** (for example, the circulatory system, which consists of the heart and blood vessels, and the digestive system, which comprises the stomach and intestines).

**Gradients** of chemical concentrations and complex pathways of chemical signalling that interact with cells in different parts of the embryo and with the immediate environment are involved in this process of **differentiation**. Remind yourself about plasma membrane receptors and stem cells on spreads 1.1.1.4 and 2.1.1.5.

Not all the organ systems develop and grow at the same rate. The head of a fetus grows rapidly and reaches adult size early. This is because humans have large brains to learn complex skills such as language and using tools. Large-brained infants are able to learn complex skills from an early age. By contrast, the reproductive organs develop much later. This is so that humans do not reproduce until they have learned the complex skills they need to look after a very dependent baby.

Figure 2 shows the different growth rates of different organ systems in a human. You will see that the nervous system grows very rapidly in young children. This is mainly because of the growth of the brain.

The **lymphatic system** is involved in the immune system. A fetus does not need an immune system because it is protected by the mother's **antibodies**, which cross the placenta. Similarly, a newborn baby receives antibodies from its mother in breast milk. But the lymphatic system develops very rapidly, and in childhood the immune system is very active in producing antibodies to common childhood **pathogens**.

The last organ system to develop is the reproductive system. This is because the reproductive organs are not needed until the child reaches puberty. In girls, this happens between 9 and 15 years, whereas boys enter puberty between 10 and 16 years.

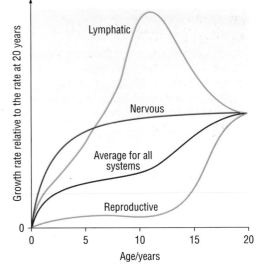

## Questions

1 Suggest an advantage of using relative growth rate rather than absolute growth rate to monitor the rate of growth of a child.

2 (a) Describe the pattern of growth for boys and girls shown in Figure 3.

**Figure 2** Different growth rates of different organ systems in a human

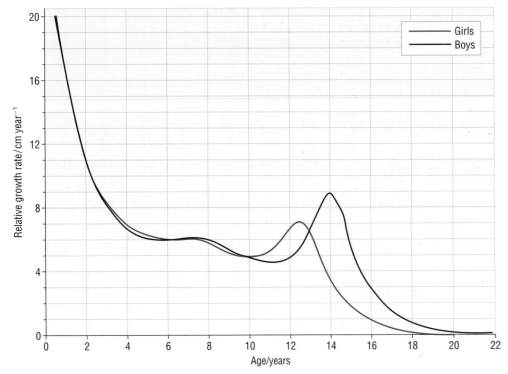

**Figure 3** Relative growth rate of boys and girls

(b) Explain the differences between the pattern of growth shown for boys and girls.

# Self-check questions

## Fill the blanks

1  ................... is a type of cell division that produces ................... gametes from diploid cells. It is a two-stage division resulting in ................... haploid cells at the end of the ................... division. In prophase 1, ................... chromosomes pair together forming ................... . The ................... of homologous chromosomes cross over and exchange genetic information at the ................... . In metaphase 1, the bivalents line up on the ................... of the spindle. The homologous pairs of chromosomes separate in ................... . At the end of ..................., two haploid nuclei are formed. The second division of meiosis is very similar to ................... . It results in ................... haploid cells being formed.

2  Another source of variation is meiosis is independent assortment of ..................., which happens during ..................., and independent assortment of ..................., which happens during ................... . As a result of independent assortment and ..................., all the gametes formed are genetically ................... from each other.

3  During ..................., the mother should eat a healthy ................... . She will need to ensure that she eats enough protein so that the fetus develops properly. ................... is important for the development of bones and ..................., and iron is needed for the formation of ................... in red blood cells. ................... is needed to make the pigment rhodopsin, which is present in the rod cells of the retina in the ................... . Vitamin C is needed to form ..................., a protein that holds cells together. It is important in bones, teeth, gums and blood vessels.

4  During pregnancy, the growth of the fetus can be monitored using ................... scanning. The ................... diameter is measured, which is the diameter of the head at the widest point. ................... length is also measured.

5  During pregnancy, the mother should avoid ................... . This is because carbon ................... in cigarette smoke combines with ................... in the mother's and baby's blood, which stops the haemoglobin from carrying ................... ....... reduces the ................... of blood vessels, including those in the placenta. This reduces the fetus's supply of ................... .

6  Women who drink more than two units of ................... a day (two small glasses of wine or a pint of low-strength beer) are more likely to have babies with language, attention and ................... problems.

7  The ................... system grows very rapidly in young children. This is mainly because of the growth of the ................... . The................... system is involved in the immune system. A fetus does not need an immune system because it is protected by the mother's ................... that cross the placenta. Similarly, a newborn baby receives antibodies from its mother in ................... milk. However, the lymphatic system develops very rapidly, and in childhood the immune system is very active in producing antibodies to common childhood pathogens. The last organ system to develop is the ................... system. This is because the reproductive organs are not needed until the child reaches ................... . In ..................., this happens between 9 and 15 years, whereas ................... enter puberty between 10 and 16 years.

# Summary questions

**1** **(a)** State the importance of iron in the diet of a girl.  [1]

**(b)** A woman who is planning to become pregnant is advised by her doctor to supplement her diet with folic acid.

**(i)** State why the increased intake of folic acid is advised.  [1]

**(ii)** Her doctor will also ask her if she has had german measles or the rubella vaccination.

Explain why this information is important.  [4]

**2** **(a)** State and explain the nutritional requirements of a pregnant woman, from the time of conception to the birth of her baby.  [6]

**(b)** Describe how large amounts of alcohol ingested by the mother may affect the developing fetus.  [5]

**3** One method of measuring fetal growth is to measure, from an ultrasound scan, the length of the back from the crown of the fetus to its rump.

**(a)** Outline how the technique of ultrasound scanning works.  [2]

**(b)** The table shows the changes in the mean fetal crown to rump length during pregnancy.

| Gestational age of fetus/ weeks | Mean crown to rump length/mm |
|---|---|
| 12 | 57 |
| 16 | 112 |
| 20 | 160 |
| 24 | 203 |
| 28 | 242 |
| 32 | 277 |
| 36 | 313 |
| 40 | 350 |

Using the data in the table, calculate the percentage increase in mean crown to rump length from 36 to 40 weeks. Show your working and give your answer to the nearest whole number.  [2]

**4** The graph shows the growth curves of different parts of the human body compared with the overall growth curve.

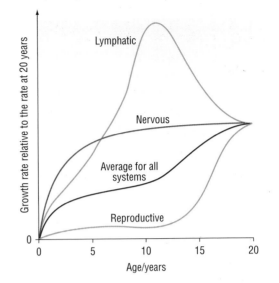

**(a)** Suggest biological advantages for the pattern of growth shown by

**(i)** the brain and skull;

**(ii)** the reproductive organs.  [3]

**(b)** Distinguish between the terms *absolute growth rate* and *relative growth rate*.  [3]

**5** The diagrams below show animal cells at various points in the first division of meiosis. The diagrams are not in the correct sequence.

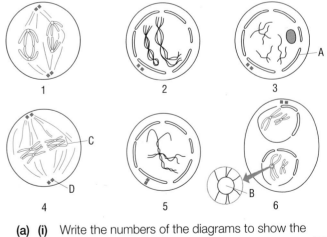

**(a)** **(i)** Write the numbers of the diagrams to show the correct meiotic sequence.  [3]

**(ii)** Identify the structures labelled A to D.  [4]

**(b)** Describe how meiosis results in gametes that are genetically different from each other.  [6]

## Infectious diseases

Infectious diseases are caused by a range of **pathogens**, including bacteria, viruses, fungi and protoctists.

Some infectious diseases are **endemic**. This means that they are always present in the population. For example, chickenpox is endemic in the UK because there is always someone, somewhere in the UK, who has the disease. But sometimes there are **epidemics** of a disease. This is when the disease spreads suddenly so that it affects a large number of people over a widespread area. If chickenpox is affecting a large proportion of people at the same time, we call it an epidemic. Some diseases become **pandemic**. This term is used to describe the spreads of AIDS (acquired immunodeficiency syndrome). This disease affects large numbers of people all over the world.

### Mycobacterium tuberculosis

*Mycobacterium* causes the disease tuberculosis, also known as TB. It is a bacterium. Every year, 1.6 million people die from this curable disease. Bacteria are very small organisms, with cells that are different from the **eukaryotic** cells seen in plants and animals. Their cells are also much smaller than eukaryotic cells. You can see the structure of *Mycobacterium* in Figure 1.

Bacterial cells are called **prokaryotic** cells. You will see that the structure of these cells is very different from the structure of a eukaryotic cell.

### Key definitions

An **infectious disease** can be spread from one organism to another. These diseases are also known as **communicable diseases**. They are caused by **pathogens**. A pathogen is an organism that causes disease. Most pathogens are microorganisms.

**Endemic** – an infectious disease which is always present in the population or 'prevalence pool'.

**Epidemic** – a sudden increase in the incidence in a specified area such as a city or country.

**Pandemic** – a rise in the incidence of a disease on a global scale.

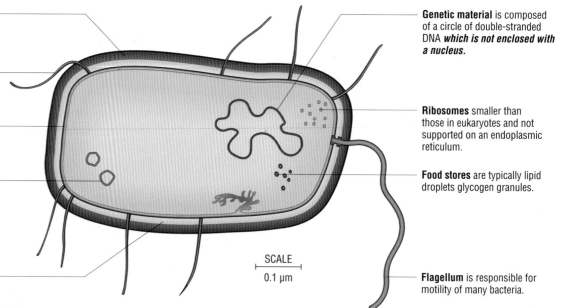

**Capsule** is a layer of mucilage which may unite bacteria into colonies.

**Pili** are protein rods concerned with cell–cell attachment. The *sex pilus* is involved in DNA transfer.

**Plasma membrane** is a typical phospholipid bilayer.

**Plasmids** are small pieces of circular DNA which replicate independently of the main genome.

**Cell wall** has a rigid framework of *murein*, a polysaccharide cross-linked by peptide chains.

**Genetic material** is composed of a circle of double-stranded DNA *which is not enclosed with a nucleus.*

**Ribosomes** smaller than those in eukaryotes and not supported on an endoplasmic reticulum.

**Food stores** are typically lipid droplets glycogen granules.

SCALE
0.1 μm

**Flagellum** is responsible for motility of many bacteria.

**Figure 1** The structure of a prokaryotic cell

### The development of TB

TB is spread when a person who has TB coughs or sneezes droplets of mucus containing the bacterium into the air. These droplets are breathed in by another person, and so *Mycobacterium* infects the lungs. It can also be caught when a person drinks unpasteurised milk from cows that are infected with a related bacterium, *Mycobacterium bovis*. TB can infect almost any organ in the body, but it usually infects the lungs. At first, the person develops a fever and loses weight. He usually has a persistent cough and feels very tired. This is called the **primary infection**, and in people with healthy immune systems the disease may not spread any further. However, if the disease does develop, the person will cough up blood-stained sputum, as their lung tissue is damaged. He will develop chest pain and suffer from night sweats.

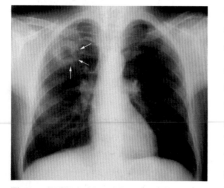

**Figure 2** Chest X-ray showing TB infection indicated by white arrows

**Figure 3** A fund-raising event in aid of the Chelsea Hospital for Consumption (tuberculosis) in 1846. Patients were isolated in large hospitals to prevent the spread of the disease and to allow them to recuperate

TB can be diagnosed by a chest X-ray. Look at Figure 2. You can see that the infected area looks cloudy.

TB can also be diagnosed by taking a sample of sputum from a patient. This can be examined in a laboratory to see if TB bacteria are present.

TB can spread to other organs, such as the heart, bones, pancreas, thyroid and skeletal muscles.

People are more likely to develop TB if they live in damp, overcrowded conditions, if they abuse alcohol and other drugs, or if they are infected with HIV. Also, people who are already suffering from other medical conditions that make the body less able to resist disease are at risk. This includes people with diabetes.

### Treatment of tuberculosis

Like other infections caused by bacteria, tuberculosis can be treated successfully using antibiotics. Isoniazid is often used. But the antibiotics used to treat TB must be taken over six to nine months to be effective. This is a problem, because people are willing to take the drugs at first because they feel ill. The antibiotics destroy the most susceptible strains of *Mycobacterium* at first. The person feels better after this, so they often decide to stop taking the drugs. But the *Mycobacterium* cells that survive in the body are the most resistant and difficult to destroy. If the drug is stopped, these bacteria multiply and spread to others. These bacteria are likely to be resistant to the drugs.

To try to stop resistance developing, people with TB are usually given a 'cocktail' of three or four different drugs. It is likely that at least one of these will be effective. Also, a scheme has been developed called *Direct Observation Therapy* (DOTS). This means that people are watched while they take their medication, to make sure that they complete the course of treatment. DOTS has other advantages too, because it means that the person's health can be monitored.

The best way to prevent TB occurring in a population is **vaccination** using the BCG (Bacille–Calmet–Guerin) vaccine. This is an *attenuated vaccine*. This means that it contains a live but weakened strain of *Mycobacterium bovis*.

## Questions

1  Make a list of the differences between a prokaryotic cell, such as *Mycobacterium*, and a eukaryotic cell.

2  Explain why people who live in overcrowded conditions are more likely to develop TB.

3  The number of cases of TB in the UK is rising at the moment, although the incidence of the disease had dropped to a very low level by 1990. Suggest reasons for this increase.

# The human immunodeficiency virus

Viruses do not have a cell structure and are many times smaller than bacteria. Outside living cells, they show no signs of life. Viruses can replicate only when they are inside a living cell. They can be described as **obligate intracellular parasites**.

The **human immunodeficiency virus**, HIV, causes the disease AIDS – **acquired immunodeficiency syndrome**. It was first recognised as a disease in 1981. At the present time, we have no cure for AIDS. Treatment is available which prolongs life, although the body remains infected by the virus. Without treatment the virus leads inevitably to death from AIDS or from AIDS-related infections.

The structure of the HIV virus is shown in Figure 1.

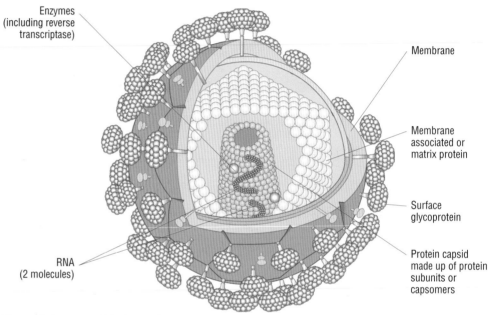

**Figure 1** Structure of the human immunodeficiency virus

### Means of transmission

The HIV virus spreads from one infected person to another person in body fluids. It is usually spread through sexual intercourse if one partner is infected, or by transfer of infected blood from one person to another, for example, when intravenous drug abusers share needles. HIV can also be spread from an infected mother to her baby, either during pregnancy or in breast milk. Blood transfusions are now screened for HIV, but in the past some people became infected by receiving transfusions of infected blood or through being given infected blood products such as **factor VIII**.

Once HIV enters the blood, the virus infects a certain kind of white blood cell, a T helper lymphocyte. The virus usually remains dormant inside these cells for a long time. During this time the person has no symptoms, but their blood contains antibodies against the virus. The infected person is said to be HIV positive.

A person may remain HIV positive for many years, showing no symptoms. However, after a few years HIV starts to replicate. The infected T helper cells release many new viruses and are themselves destroyed. The viruses that are released infect new T helper cells. As T helper cells are destroyed, this makes the person susceptible to **opportunistic infections**, such as TB. This stage of the disease is known as 'full-blown AIDS'. The person usually dies within two years. Apart from TB, other opportunistic infections that can cause death include Kaposi's sarcoma (a form of cancer) and pneumocystis pneumonia. You can see examples of Kaposi's sarcoma in Figure 2.

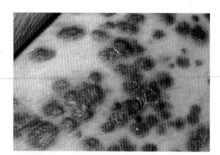

**Figure 2** Kaposi's sarcoma

## Controlling the spread of HIV

Although scientists are searching for a vaccine against HIV, and are trying to develop drugs that will cure AIDS, there is as yet no cure. But there are some drugs that slow down the development of AIDS. The best way to deal with the spread of HIV is to avoid catching it in the first place. There are several ways of doing this:

- screening of blood for transfusion
- using condoms during sexual intercourse
- needle-exchange schemes for injecting drug abusers, so that they have fresh sterile needles available
- educating people about the ways in which HIV is spread
- offering HIV tests to people at risk, e.g. prostitutes and intravenous drug abusers
- encouraging people who are HIV positive to contact people that they might have infected, and advising them to avoid spreading the infection
- encouraging HIV positive mothers not to breast feed
- encourage people to 'take the test' – many people pass on the virus because they are not aware that they are HIV positive (see spread 2.3.2.4 for more information on testing.)

**Examiner tip**

You may be asked to 'describe' the pattern of HIV in a population from tables or a graph. Remember to use figures from the graph or the table. Remember to give units.

Look at Figure 3. This shows the number of cases of HIV and deaths from AIDS in the USA per year.

1 Describe the pattern shown by HIV and by AIDS.
2 Explain why the two lines are different.
3 Suggest reasons why the death rate from AIDS is decreasing.

HIV and AIDS are found in almost every country in the world. It is estimated that about 40 million people are infected worldwide. There were 4.3 million new cases in 2006 alone. One part of the world that is very badly affected is sub-Saharan Africa. Zimbabwe is the worst affected country, where 25% of the adult population are thought to be infected. Life expectancy is only 39 years. This is having a devastating effect on children, many of whom are orphaned as a result of AIDS. Other children are kept off school because their sick parents cannot afford the fees, or need the children at home to look after them.

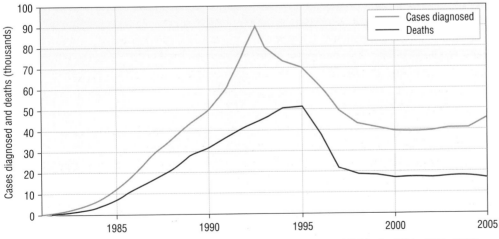

**Figure 3** Graph showing the number of cases of HIV and deaths from AIDS in the USA, 1981–2002

## Questions

1 Some biologists argue that viruses, such as HIV, are not living organisms. Suggest why.

2 In sub-Saharan Africa, many women are HIV positive. Should women in these areas be advised not to breast-feed their babies? Give reasons for your answer.

## Antibiotics

You have already learned that antibiotics can be used to treat diseases caused by bacteria. Figure 1 shows some of the ways in which antibiotics kill bacteria.

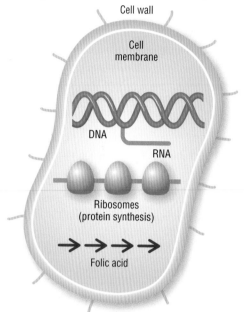

**Inhibition of...**

| Cell wall synthesis | Protein synthesis |
| --- | --- |
| Penicillins | Macrolides |
| Cephalosporins | Chloramphenicol |
| Carbapenems | Tetracycline |
| Daptomycin | Aminoglycosides |
| Glycopeptides | Oxazolidonones |

| DNA synthesis | Folic acid synthesis |
| --- | --- |
| Fluoroquinolones | Sulfonamides |
| | Trimethoprim |

**RNA synthesis**

Rifampin

**Figure 1** Some of the ways in which antibiotics target and kill bacteria cells

## Antibiotic resistance

Antibiotics were first used on a large scale in the Second World War. They seemed to be 'wonder drugs', curing people of infections that had killed people only a few years before. But almost as soon as antibiotics were discovered, some resistant bacteria were found.

Bacterial populations show some variations in their genetic makeup. When antibiotics were used to treat a bacterial infection, most of the bacteria were killed. But by chance some of the bacteria had a gene that gave them some resistance to the antibiotic. For example, some bacteria can make an enzyme called penicillinase, which breaks down penicillin. The bacteria with a resistance gene survived, and multiplied quickly. All their offspring contained the resistance gene.

In some bacteria, mutations occurred. A mutation might cause a bacterium to have resistance to an antibiotic. Once again, these resistant bacteria were able to survive when antibiotics were used. They multiplied and produced a large number of offspring, all resistant to the antibiotic.

Antibiotics have been used so widely that we now have bacteria that are resistant to most antibiotics. The most well-known is MRSA, which stands for methicillin-resistant *Staphylococcus aureus. Staphylococcus aureus* is a bacterium that normally lives harmlessly on the skin or in the nose of many people. However, if it gets inside the body, for instance under the skin or into the lungs, it can cause infections such as boils or pneumonia. People who carry this organism are usually totally healthy and are considered simply to be carriers of the organism. MRSA is a strain of *Staphylococcus aureus* that is resistant to commonly used antibiotics.

MRSA is found in patients in hospitals, but can also be found on healthy people who are not in hospital. Normally, a person who carries MRSA has no ill effects and does not even know she carries MRSA. But if MRSA organisms are passed on to someone who is

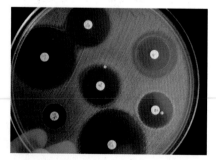

**Figure 2** Agar plate of bacteria with antibiotic discs and inhibition zones

already ill, then a more serious infection may occur in that individual. People who have surgical wounds are especially at risk. When patients with MRSA are discovered in a hospital, the hospital will usually try to prevent it from passing to other patients. This is known as *infection control*.

Some strains of MRSA can be treated using unusual antibiotics that are rarely used, but there are also strains of MRSA that are resistant to all known antibiotics.

We can reduce the chances of antibiotic resistance developing by
- only taking antibiotics when absolutely necessary
- always completing the course of antibiotics.

### Reducing the spread of MRSA

In hospitals, medical staff have to be very careful to avoid spreading bacteria such as MRSA from one patient to another. There are several precautions that they can take:
- apply an alcohol-based hand rub to their hands between every patient contact
- wear disposable gloves when examining a patient or when exposed to body fluids. These should be thrown away after dealing with that patient
- wear disposable plastic aprons before exposure to body fluids, so that bacteria are not transferred to clothing
- visitors should use an alcohol-based hand rub before entering the ward
- equipment should be thoroughly cleaned before a different patient uses it, e.g. mattresses, beds, lockers and curtains
- hospital wards should be cleaned thoroughly to make sure that MRSA is not present on surfaces
- patients with MRSA infections should be kept isolated from other patients. High bed occupancy and lack of spacing between beds can also cause problems.

In September 2007 it was announced that white coats are to be banned in hospitals as part of a series of new measures aimed at reducing hospital infections. Staff will be required to keep their arms bare, and clean, below the elbow. This will mean wristwatches and jewellery are also banned.

Doctors will be discouraged from wearing ties while dealing with patients.

British Medical Association head of ethics Dr Vivienne Nathanson said new dress codes must be practical and sensitive to religious feelings.

She said: 'It is very important to emphasise that clean hands, bare elbows and short sleeves are only one aspect of preventing and controlling infection. A coordinated approach addressing all the relevant factors, for example dress code, bed occupancy, hygiene in hospital and isolation policies, is most likely to be successful.'

1 Make a list of all the ways to control hospital infections that are mentioned in this passage.
2 Explain how each helps to reduce the spread of infection.

## Questions

1 Use Figure 1 to explain why antibiotics cannot be used to treat viral infections.
2 A student was prescribed antibiotics for a throat infection. After 3 days she felt better, so she stopped taking the antibiotic tablets and kept the rest in the bathroom cupboard, saying she might need the tablets another time if she felt ill. Explain why this was an unwise thing to do.

### XDR-TB

XDR-TB is the abbreviation for extensively drug resistance tuberculosis.

TB can usually be treated with a course of four standard or FIRST LINE anti-TB drugs.

The misuse of these drugs – (for example by not completing the course) and the mismanagement of these drugs – (for example by prescribing them at incorrect dosage or by making them available without supervision) has led to the development of MDR-TB – or multidrug resistance TB. This requires a second line of anti-TB drugs to treat it – but these are expensive and can have some side-effects. However, if these are mismanaged or misused, then XDR-TB will develop and the options for treating this will then be very limited. XDR-TB has now been reported in South Africa, Botswana and Mozambique

### DOTS

DOTS is at the heart of the World Health Organisation's strategy to stop the spread of TB – particularly XDR-TB. DOTS stands for DIRECTLY OBSERVED THERAPY SHORT COURSE. The strategy combines accurate DIAGNOSIS of TB, a STANDARDISED multi-drug treatment with HIGH QUALITY anti-TB drugs given as part of a SUPERVISED programme. The hope is that by effective implementation of DOTS, TB can be brought under effective control.

For up to date information on the WHO TB programme visit their website www.who.int/

## The importance of plants

It is estimated that about half the medicines we use in the West contain materials that have come from plants or are synthetic forms of plant products. Many modern medicines have been developed from drugs used in traditional medicine or folk healing. In general, medicines derived from plants are safer than synthetic products, and are often cheaper.

One way to develop medicines from plants is to extract chemicals from plants and test them to see if they have any therapeutic uses. But this is very expensive and time-consuming. A more efficient method is to study plants used in traditional folklore. Many of these plants prove to contain useful medicines.

Curare is a chemical obtained from several species of bitter-tasting vines that grow in the Amazon. It is the source from which the drug d-tubocurarine was developed. This is a muscle relaxant that revolutionised modern surgery. Traditional Amazon tribes used this plant to poison their arrows. It helped them to hunt animals more efficiently. Other tribes use curare to treat certain severe skin infections.

**Figure 1** *Aframomum melegueta*, grains of paradise (Zingiberaceae)

### Antimicrobial compounds

Plants sometimes produce antimicrobial compounds. One example is the Ethiopian pepper, *Xylopia aethiopica.* Local tribespeople use it for many purposes, including as a cough remedy, a tonic given to women after they have given birth, and to treat stomach ache, bronchitis and dysentery. Scientists who have studied this plant have found that its fruit contains an antimicrobial compound that kills a range of bacteria. Grains of Paradise, *Aframomum melegueta*, is a fruit found in many tropical countries. Traditionally, it has been used to treat such conditions as parasitic worms, measles, leprosy, and haemorrhage in women who have given birth. Scientists have found that the fruit contains several essential oils that kill bacteria and fungi as well as the parasitic worm *Schistosoma.*

Table 1 lists some medicines and the plants that they come from.

| Medicine | Use | Plant |
|----------|-----|-------|
| L-dopa | Treatment of Parkinson's Disease | *Mucuna sp.* (velvet bean) |
| Codeine | Painkiller | *Papaver somniferum* (opium poppy) |
| Gossypol | Male contraceptive | *Gossypium spp.* (cotton) |
| Digitalin | Cardiotonic | *Digitalis purpurea* (fox glove) |
| Aesculetin | Anti-dysentery | *Fraxinus rhynchophylla* (Korean ash) |
| Salicin | Painkiller | *Salix alba* (willow) |
| Topotecan | Anti-cancer agent | *Camptotheca acuminata* (Chinese happy tree) |

**Table 1** Medicines and their source plants

### Loss of species

Unfortunately, many plant species in the world are endangered and have even become extinct. Many of the plants that are becoming extinct have not yet been named. It is likely that among the plants being lost are plants containing useful medicines that we do not yet know about. One reason so many plants are becoming extinct is that forests are being cut down to provide timber for buildings, or to clear land to build houses. Forests are also being cleared to create more farmland. Some of the people who are clearing these forests are poor people who are simply trying to make a living for their family.

## Conservation

Kew Gardens in London is involved in a project called *DNA barcoding*. A standard short piece of DNA from a particular **locus** (see spread 5.1.2.2) on a particular chromosome is chosen. Scientists then find the DNA sequence of this piece of DNA for as many plants as possible.

This information can be used:

- to identify different life stages of the same plant, e.g. seeds and seedlings
- to identify fragments of plant material
- in forensic investigations
- in the verifying of herbal medicines/foodstuffs
- in biosecurity and trade in controlled species
- to build up inventories and ecological surveys.

An international breeding initiative has been set up, called the Consortium for the Barcode of Life. Kew is working with 10 other organisations to select a suitable DNA region to be used as a barcode. The project intends eventually to build up barcodes for every land plant.

Kew Gardens has also set up a Millennium Seed Bank (see Figure 2). They are collecting seeds from many plant species all over the world and storing them. Seeds stored in suitable conditions can last for decades and sometimes even hundreds of years. Kew Gardens already have seeds stored from species that are thought to be endangered in the wild.

**Figure 2**  Millennium Seed Bank

### Examiner tip

What do we mean by 'bar codes'?

You will meet this idea again in spread 5.1.2.2 in A2 when you look at DNA profiling.

Quite simply, specific regions of DNA are cut up and separated into bands using a technique called Gel Electrophoresis. This gives a banding pattern. Where a pattern is associated, for example, with production of compounds with medicinal properties, this is a quick way of screening plant species for potential sources of drugs.

### Key definitions

**In situ conservation** – aims to preserve the whole habitat that the plant grows in. We are conserving the plant where it naturally grows.

**Ex situ conservation** – where habitats have been destroyed and the plant is possibly extinct in the wild, specimens may still exist. These may be in botanic gardens or as seeds in seed banks. This is *ex situ* conservation.

## Questions

1. How could you investigate whether a compound from a plant has antimicrobial activity?
2. A new compound from a plant has been shown to have antimicrobial activity. However, pharmaceutical companies have to consider many more factors before they decide whether to develop the compound for sale as a medicine. Suggest some of these factors.
3. Imagine that you are a scientist who has discovered a new medicine from plants. In laboratory tests it appears to be useful in treating cancer. Design an investigation you could carry out to test whether it is effective against cancer in people.
4. When Kew Gardens collects seeds for the seedbank, they try to take samples from different places, rather than collecting all the seeds in one place. Suggest why.
5. Imagine that you work for the Millennium Seed Bank. An article has appeared in a newspaper saying that the seed bank is an expensive waste of money. You have been asked to write to the newspaper justifying its work. What points will you make?

## Non-specific defence mechanisms

The body has a range of defences to protect it against disease-causing organisms, or **pathogens**. Some of these are called the **non-specific responses** because they always work in the same way, regardless of what the pathogen is and whether this is the first or second 'attack'. Many of these defences are simple barriers:

- unbroken skin is a barrier that is difficult for pathogens to penetrate
- if the skin is broken, blood clotting takes place to prevent loss of blood and to seal off the wound
- hydrochloric acid in the stomach and an acid pH in the vagina make it difficult for most pathogens to survive, because their enzymes are denatured in acid conditions
- the epithelium of the respiratory tract is covered with mucus, and there are ciliated cells present. Pathogens stick to the mucus and then the tiny **cilia** beat, moving the mucus back up to the throat. From here, the mucus is swallowed and the pathogens are killed by the acid in the stomach
- the conjunctiva, the membrane covering the front of the eye, is protected by fluid secreted by the tear ducts. This secretion contains an enzyme, lysozyme, which digests bacterial cell walls.

But some pathogens still manage to enter the body, despite these primary defences. The second line of defence involves phagocytic cells (**phagocytes**) and inflammation.

### Phagocytosis and inflammation

Phagocytosis is a process carried out by **neutrophils** and **monocytes**. Monocytes become **macrophages** – the phagocytic cells found in organs such as the spleen and lungs. These cells are known as phagocytes. They are produced in the bone marrow inside long bones, from stem cells. Damaged cells and tissues release chemicals or **cytokines** that attract phagocytes. Phagocytes are able to recognise bacteria, and this process is made easier because the bacteria become coated in **plasma proteins** called **opsonins**. The phagocytes engulf the bacteria, as you can see in Figure 1. The bacterium becomes enclosed in a membrane-coated vesicle called a **phagosome**. **Lysosomes** containing digestive enzymes move towards the phagosome and fuse with it. The digestive enzymes break down the bacterium, releasing soluble products that diffuse into the cytoplasm of the phagocyte.

 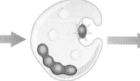 

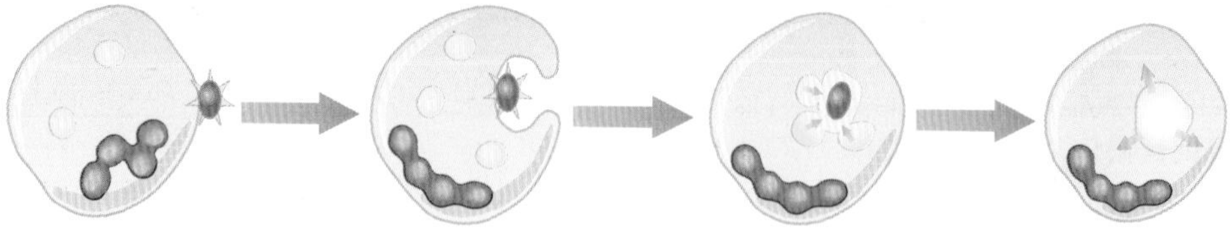

| Pathogen attached to phagocyte by antibody and surface receptors | Pathogen engulfed by infolding of phagocyte membrane | Lysosomes release lysins into phagosome | Harmless end products of digestion are absorbed |

**Figure 1** Phagocytosis

### Inflammation – another non-specific response

You have probably noticed that when an area of the body is wounded, the tissue around the wound becomes swollen, red and painful. This is called *inflammation*. It happens because chemicals, including *histamine*, are released from the damaged tissues. These chemicals cause **vasodilation** in the capillaries. Vasodilation causes more bloodflow in the area, so the temperature of the tissue around the wound rises. This will increase the rate of the chemical reactions involved in blood clotting but higher temperatures will slow down the rate of bacterial growth. The capillaries become more permeable, so more **plasma** escapes from the capillaries into the surrounding tissues. This is what causes the swelling. The plasma contains neutrophils, monocytes and various proteins that help to combat pathogens.

### The immune system

The immune response is called a **specific response**. The cells involved in the immune response attack the pathogen, but cause no harm to other cells and tissues. This is because the cells of the immune system can distinguish the body's own cells (or 'self') from those that are foreign (or 'non-self').

**Antigens** are macromolecules such as **glycoproteins**, on the surface of organisms or parts of organisms such as cancer cells, which the body recognises as non-self, and which therefore trigger an immune response. When a foreign antigen is detected, the immune system produces a specific antibody.

**Antibodies** are proteins synthesised by a particular kind of **lymphocyte**, called **B lymphocytes**. You can see the general structure of an antibody in Figure 2.

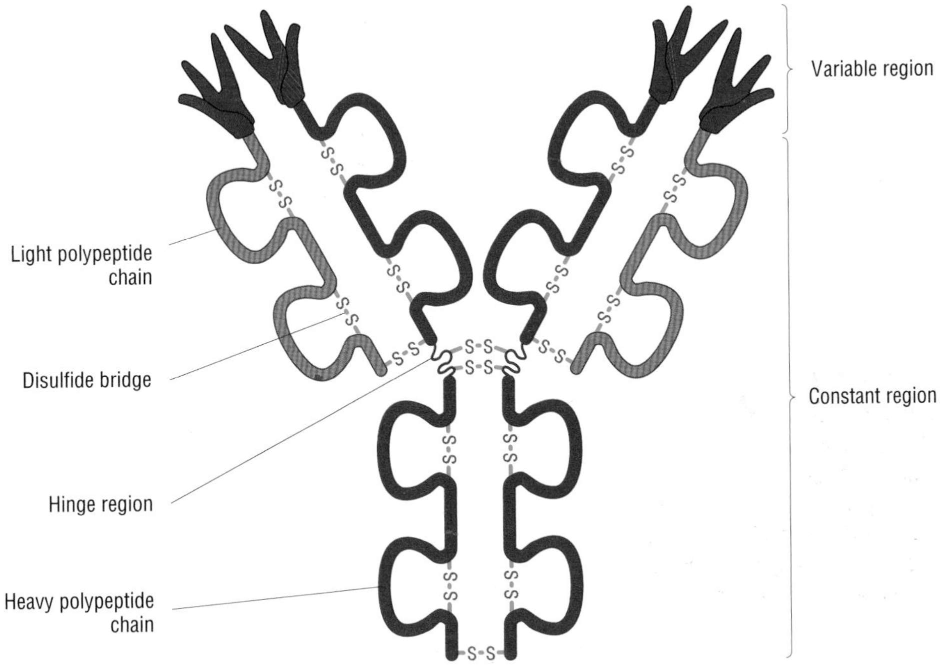

**Figure 2** The structure of an antibody

You will see that each antibody contains four polypeptide chains, two long chains and two short chains. They have two **binding sites** which can fit exactly on to the antigen, rather like a key fitting in a lock – the antigen and the binding site have **complementary** shapes. Each kind of antibody has a different binding site, because it is specific to just one antigen. Therefore the binding sites are called the **variable region**.

Antibodies destroy pathogens in various ways:

- **Agglutination:** antibodies can cause bacterial cells to clump together, which makes it easier for phagocytes to engulf them.
- **Precipitation:** antibodies can cause soluble antigens to precipitate out, so that they can be engulfed by phagocytes.
- **Neutralisation:** antibodies can bind to toxins produced by foreign cells, neutralising them so that they do not cause harm.
- **Lysis:** antibodies can bind to foreign cells. The antibodies then attract enzymes, which attach to the antibodies and digest the foreign cells.
- **Opsonins:** opsonins include antibodies and some other molecules of the immune system. A special attachment site on the antibody's constant region binds to a receptor site on the **plasma membrane** of a phagocytic cell, while its variable region binds to the bacterial antigen. The 'bacterium' is held on the phagocytic cell so it can be engulfed.

**Examiner tip**

Antibodies are another example of a protein with a **quaternary structure.** The shape of the *variable region* changes because the **tertiary structure** of this region changes. The tertiary structure changes because the *sequence of amino acids* (**primary structure**) and hence the *R groups* change in this region.

## Questions

1 People who have serious burns often die of infections. Explain why.
2 Explain why the area around a wound looks red.
3 Use your understanding of antibody structure to explain why antibodies to one strain of a 'flu virus might not recognise a different strain.

## The role of B lymphocytes

**B lymphocytes** are formed from **stem cells** in the bone marrow. They mature in the bone marrow, which is why they are called B lymphocytes.

There are something like 10 million different B lymphocytes, each with a different shape of **receptor** in its **plasma membrane**. Each of these cells is capable of producing a different **antibody**. During fetal development, these B lymphocytes are constantly meeting other cells. The cells they meet will almost certainly be the fetus's own cells, because the fetus is protected against infection by the placenta. Any B lymphocytes with receptors that exactly match the body's own cells are either destroyed or suppressed. This means that, by the time a baby is born, the only B lymphocytes left will be those that are capable of attacking foreign, or *non-self*, antigens.

When a **pathogen** enters the body, one kind of B lymphocyte will have receptors that exactly fit the **antigens** on the pathogen. This B lymphocyte divides by **mitosis** to produce a clone of identical B lymphocytes. You can see this in Figure 1. Some of these cells develop into **plasma cells**. These secrete specific **antibodies** against the pathogen. This is called the **primary immune response**. But some of the cells become **memory B lymphocytes**, also known as memory cells. These last much longer than plasma cells, and can survive for decades. These do not secrete antibodies, but serve as the body's **immunological memory** of the original antigen. If they ever encounter the same antigen again, they divide and rapidly produce plasma cells and more memory cells. This means that the second time the pathogen is encountered, the response is much faster and more intense. This is the **secondary immune response**.

### Examiner tip

B lymphocytes differentiate into plasma cells. This is a good opportunity to review what differentiation is. B lymphocytes contain very little cytoplasm. Differentiation will involve the cell becoming much larger to house the rough endoplasmic reticulum and Golgi apparatus needed to synthesise the antibodies and package them into vesicles for exocytosis. These processes will also require energy so far more mitchondria will be needed. Consequently, plasma cells will look very different to B lymphocytes as they are now adapted to carry out their function – the production of antibodies.

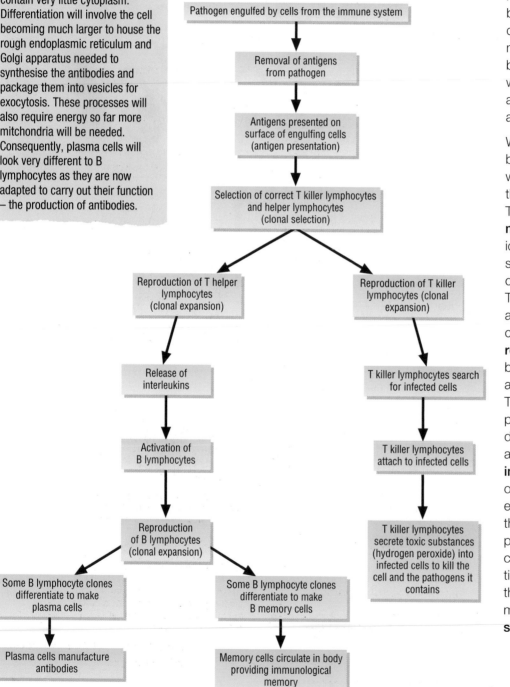

**Figure 1** The role of B lymphocytes and T lymphocytes in the immune response

# The role of T lymphocytes

**T lymphocytes**, like B lymphocytes, are produced from stem cells in the bone marrow. However, they mature in the thymus gland, which is why they are called T lymphocytes.

T lymphocytes, like B lymphocytes, have different shapes of receptors on their cell surface membrane. A **macrophage** that has engulfed a pathogen by **phagocytosis** digests the pathogen and 'presents' the antigen on its cell surface membrane. The T lymphocyte with the right shape of receptor to fit the antigen binds to the macrophage. The T lymphocyte divides to form a clone of T lymphocytes. You can see this in Figure 2.

Some of these T lymphocytes develop into **T killer lymphocytes** that kill any cell carrying the specific antigen. Others develop into **T helper lymphocytes** which secrete chemicals that stimulate phagocytosis by phagocytes and antibody production by B lymphocytes. Both kinds of T lymphocyte produce their own kind of **memory cells**, which can respond to a later infection by the same pathogen. A further kind of T lymphocyte is a **T suppressor lymphocyte**. These 'wind down' the action of the immune system once the pathogen has been destroyed.

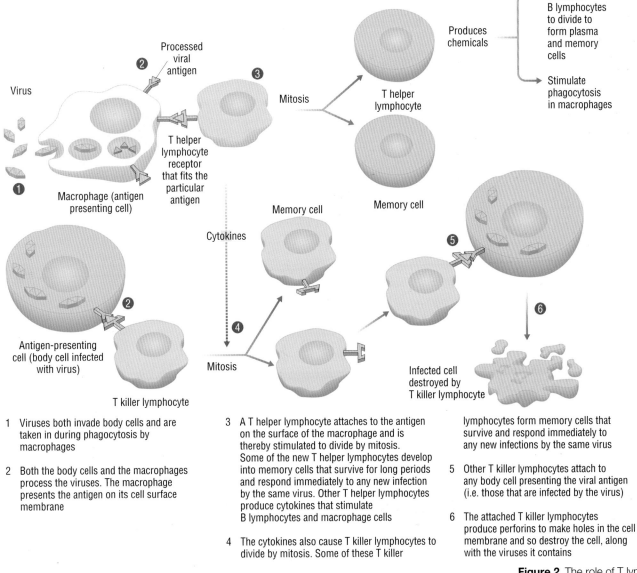

1. Viruses both invade body cells and are taken in during phagocytosis by macrophages

2. Both the body cells and the macrophages process the viruses. The macrophage presents the antigen on its cell surface membrane

3. A T helper lymphocyte attaches to the antigen on the surface of the macrophage and is thereby stimulated to divide by mitosis. Some of the new T helper lymphocytes develop into memory cells that survive for long periods and respond immediately to any new infection by the same virus. Other T helper lymphocytes produce cytokines that stimulate B lymphocytes and macrophage cells

4. The cytokines also cause T killer lymphocytes to divide by mitosis. Some of these T killer lymphocytes form memory cells that survive and respond immediately to any new infections by the same virus

5. Other T killer lymphocytes attach to any body cell presenting the viral antigen (i.e. those that are infected by the virus)

6. The attached T killer lymphocytes produce perforins to make holes in the cell membrane and so destroy the cell, along with the viruses it contains

**Figure 2** The role of T lymphocytes in the immune system

# Questions

1. Describe how an antibody is produced inside the cell of a B lymphocyte.
2. Suggest why it is possible to have a cold many times, even though memory cells remain in the blood after each infection.

## The primary and secondary response to infection

You learned on spread 2.3.2.2 that when the immune system has responded to a **pathogen**, long-lived **lymphocytes** or **memory cells** remain behind. This means that if you are challenged by the same pathogen on a future occasion, your immune system may respond so quickly that you have no symptoms of the disease.

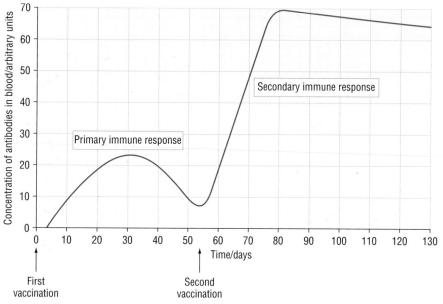

**Figure 1** The primary and secondary response

Look at Figure 1. You will see that when a person first encounters a pathogen, there is a very brief delay before specific antibodies appear in the blood. This is because it takes time for the right **T lymphocyte** to be located and selected and to divide to produce the **T helper lymphocytes**. The specific **B lymphocyte** has also to be located and activated to divide to form a clone of **plasma cells**. However, after this, specific antibodies are produced. This is the **primary response**. The pathogen is destroyed by the immune system and, over time, the number of antibodies in the blood falls.

The second time that a person encounters the pathogen, so long as the antigens are the same, complementary memory B cells are already present. This means they can divide very quickly to produce a clone of plasma cells that produce larger numbers of antibodies. The secondary response is quicker and greater. The pathogen is destroyed, usually before the person has any symptoms of the disease. This is what we mean by the term 'immune'.

### Vaccination

The improved secondary response to an antigen is the basis of vaccination. A vaccine is a preparation that contains antigens. These antigens will be identical to those on a known pathogen and they will stimulate a primary response, so that memory B cells are made. This means that if the person ever encounters the real pathogen, memory B cells will produce a faster secondary response.

There are different forms of vaccine.

- *Live vaccines:* these vaccines consist of the living microorganisms with the same **antigens** as the pathogen. However, these microorganisms have been subcultured many times in a laboratory and are no longer able to cause the disease. This is called an **attenuated strain** of the pathogen. The measles, mumps and rubella vaccines fall in this category.
- *Dead microorganisms* that have been killed by chemicals and heat: although the pathogen is dead, it still carries the antigens that stimulate an immune response. The vaccines for Hepatitis A and cholera are in this category.
- *A fragment of a pathogen* such as a viral coat component: can be used as a vaccine. The HPV vaccine consists of a viral coat protein of the Human Papilloma Virus.

### Vaccinating children

Table 1 shows the recommended schedule of vaccination for children in the UK. You will see that the first vaccinations happen when the child is very young. Some vaccinations are given in more than one dose. This is done to ensure that enough memory B cells are produced to give the child immunity against the disease. These extra vaccinations are called **boosters** because they boost the number of memory B cells.

---

### How many people need to be vaccinated against measles to avoid spreading the disease?

Up to 94% of a population need to be vaccinated to protect the whole population from measles. So, in a population of 13 450 how many people should be vaccinated? To find this out, you need to find 94% of 13 450.

$$\frac{94}{100} \times 13\,450 = 12\,643.$$

| When to immunise | Diseases protected against | Vaccine given |
|---|---|---|
| Two months old | Diptheria, tetanus, pertussis (whooping cough), polio and *Haemophilus influenzae* type b (Hib)<br>Pneumococcal infection | DTaP/IPV/Hib<br>+ Pneumococcal conjugate vaccine (PCV) |
| Three months old | Diptheria, tetanus, pertussis, polio and *Haemophilus influenzae* type b (Hib)<br>Meningitis C | DTaP/IPV/Hib<br>+ MenC |
| Four months old | Diptheria, tetanus, pertussis, polio and *Haemophilus influenzae* type b (Hib)<br>Meningitis C<br>Pneumococcal infection | DTaP/IPV/Hib<br>+ MenC<br>+ PCV |
| Around 12 months | *Haemophilus influenzae* type b (Hib)<br>Meningitis C | Hib/MenC |
| Around 13 months | Measles, mumps and rubella<br>Pneumococcal infection | MMR<br>+ PCV |
| Three years four months to five years old | Diptheria, tetanus, pertussis and polio<br>Measles, mumps and rubella | DTaP/IPV or dTaP/IPV<br>+ MMR |
| 13 to 18 years old | Tetanus, diphtheria and polio | Td/IPV |

**Table 1** Routine childhood immunisation programme in the UK

Some parents worry about having their child vaccinated. Like anything else in life, vaccinations carry some risk. But the risks of vaccination are very much smaller than most people think. For example, with the MMR vaccination there is a one in 1000 chance that the child will suffer from febrile convulsions (fits). But one in 200 children who catch measles will suffer from febrile convulsions and long-lasting damage or even death. So the risk of convulsions is much greater in unvaccinated children. Measles is very common, so unvaccinated children are very likely to catch the disease, especially because the child is infectious for two or three days before symptoms appear. These risks need to be compared with other activities. By comparison, about 5000 children are killed or seriously injured on the roads in the UK every year.

### Herd immunity

If enough people in a population are vaccinated against a disease, it is not possible for the disease to spread in the population, and everybody is protected. This is called **herd immunity**. In 1977 the World Health Authority announced that smallpox had been eradicated. This happened through a careful programme of vaccination around the world. It was estimated that 83–85% of a population had to be vaccinated against smallpox to prevent the disease being spread. The figure is different for different diseases. To prevent diphtheria spreading, 85% need to be vaccinated; for pertussis (whooping cough), 92–94% need to be vaccinated; while for measles, the figure is 83–94%.

If a person in a population that is mainly vaccinated has a disease, s/he may pass the pathogen on to one or two others, but most people will be immune. The idea behind herd immunity is that once a certain proportion of the population is vaccinated, the chances of an infected person meeting a susceptible person and passing on the pathogen are much reduced. If the person does not pass on the pathogen, it cannot multiply in its human host and the pathogen population becomes much smaller.

## Questions

1 The routine childhood immunisation programme in the UK starts when children are only two months old. At this age, the child's immune system is not yet fully developed. Suggest why it is recommended that children are vaccinated when they are only two months old, rather than waiting until they are older.

2 The percentage of people in a population who need to be vaccinated for herd immunity to occur varies according to the disease. Suggest why this percentage varies.

### Examiner tip

Table 1 shows the current guidelines for childhood vaccination in the UK. However, these guidelines can change from time to time, so be careful!

### Measles deaths tumble by 60% worldwide

An international vaccination programme to combat measles has exceeded targets, reducing child deaths from the disease by 60% between 1999 and 2005. The International Measles Initiative launched in 2000 by the World Health Organization and UNICEF had aimed for a 50% reduction in 45 target countries.

WHO announced that deaths have fallen from 873 000 during 1999 to 345 000 by the end of 2005. In Africa, deaths from measles have fallen by 75%. Measles deaths in children under five fell from 791 000 to 311 000 over the same period, globally.

The new figures estimate that, altogether, measles vaccinations have prevented 7.5 million deaths between 1999 and 2005, and 2.3 million of these were attributable to the intensified programme.

Ironically, measles deaths are climbing again in some rich countries, such as the UK, because of lingering but unfounded fears that a combined triple vaccine, which protects against measles, mumps and rubella (MMR), causes childhood autism.

Smallpox has already been eradicated, and polio is now present only in isolated pockets of the world. However, scientists believe it is still too early to talk about eradication of measles.

# Using antibodies and making vaccines

## Blood groups

Sometimes people need a blood transfusion, for example, after a major accident or during a surgical operation. It is essential that they are given blood of the correct blood group.

The ABO blood groups are the best known. There are four blood groups, A, B, AB and O. The blood groups are named according to which **antigens** are present on the red blood cell surface membrane. These antigens may be proteins, carbohydrates, glycoproteins or glycolipids. You can see this in Figure 1. However, there are also corresponding **antibodies** present in the **plasma**. If a person is given a blood transfusion, the antigens on the donor red blood cells should *not* correspond and bind to the antibodies in the recipient's plasma. Usually, health professionals try to match the blood group exactly. But in an emergency, blood of group O can be given to anybody. This is because there are none of the ABO antigens on the red blood cells of a person with group O. People with blood group O are sometimes called 'universal donors'.

If a person is given a donation of blood from the wrong group, the red blood cells from the donor's blood will be clumped or agglutinated by the antibodies in the recipient's plasma. The agglutinated red cells can block blood vessels and stop the circulation of the blood to various parts of the body – particularly in the capillaries in the kidney. The agglutinated red blood cells also break open and their contents leak out. The red blood cells contain **haemoglobin**, which becomes toxic when outside the cell. This can kill the patient.

There are also **Rhesus blood groups**. Many people also have an antigen called a *Rhesus factor* or *antigen D* on the red blood cell's surface. People who have it are called Rh+. Those who do not have the antigen are called Rh–. A person with Rh– blood does not have Rh antibodies naturally in the blood plasma (as one can have A or B antibodies, for instance). However, a person with Rh– blood can develop Rh antibodies in the blood plasma if he or she receives red blood cells from a person with Rh+ blood. The Rh antigens in the donated blood can trigger the production of Rh antibodies. A person with Rh+ blood can receive blood from a person with Rh– blood without any problems.

Rhesus blood groups do not usually cause any problems. However, because a person's blood group is genetically determined, it is possible for a Rhesus negative mother to carry a Rhesus positive baby if her partner is Rhesus positive. Normally, during pregnancy the mother's blood and the baby's blood do not mix. But towards the end of pregnancy or during the birth the placenta can start to break down and a few red blood cells from the baby can pass into the mother's blood. These are treated as 'foreign cells' and stimulate the mother's B lymphocytes to make antibodies against the antigens on the baby's red blood cells. This should not harm the baby, as this happens at the very end of pregnancy.

But if the same mother later has another Rhesus positive baby, there are more serious problems. Antibodies from the mother can pass across the placenta into the baby's circulation. Here, they destroy fetal red blood cells. If this is not treated, the baby will be born with *haemolytic disease* of the newborn, or may even be stillborn. Haemolytic disease of the newborn used to be treated by giving the baby a total blood transfusion at birth, but it can now be prevented. Pregnant women who are Rhesus negative are given an injection of anti-Rhesus globulin early in pregnancy. This stops them making anti-Rhesus antibodies by binding to any Rhesus positive cells and preventing the mother from synthesising her own antibodies. In a sense, the mother is made passively immune to Rh+ red blood cells.

| Blood group | Antigens on red blood cells | Antibodies in plasma |
|---|---|---|
| A | A | anti-B |
| A | B | anti-A |
| AB | A and B | neither |
| O | neither | anti-A and anti-B |

**Figure 1** Antigens on red blood cells and antibodies in plasma of different blood groups

### Examiner tip

One very common mistake is to confuse **agglutination** of red blood cells with *clotting*. Remember, clotting results in the formation of **fibrin** from **fibrinogen**. In agglutination, the antibody has several **variable regions** that bind to several red blood cells simultaneously, so clumping them together.

### Haemolytic disease in literature?

In Arthur Miller's play 'The Crucible', a key character is Mrs Putnam. She has lost seven of her eight children either stillborn or shortly after birth. Her description mentions 'yellowing' and she suspects witchcraft. But Rhesus incompatibility could be the reason. If she is Rhesus negative and her husband is Rhesus positive, haemolytic disease could occur time after time.

### Developing a vaccine against HIV

You will remember that a vaccine contains antigens from a **pathogen**. This causes the body's immune system to produce specific antibodies and memory cells against that pathogen.

It has not yet been possible to produce a vaccine against HIV. This is because the virus keeps mutating. As a result of the **mutations**, antigens with a variety of different shapes are produced. This means that antibodies against one strain of the virus will not bind to antigens of a slightly different shape. Furthermore, the antigens of HIV can change even in one person. Many people think that it is unlikely that we will ever have a vaccine that is effective against HIV.

### Using antibodies to test for HIV and TB infection

There are two main types of HIV test.

One test uses **PCR (Polymerase Chain Reaction)** (see spread 5.1.2). This uses the leucocytes from a blood sample, and makes copies of the DNA present. HIV inserts a DNA copy of its genetic material into T helper lymphocytes. PCR 'amplifies' sections of this 'proviral' DNA. As a 'positive control', a section of the beta haemoglobin gene is normally amplified at the same time to check that the PCR system is working correctly. If PCR finds both viral DNA and beta haemoglobin DNA, the test is positive. These tests are not often used in adults because they are expensive and more complex to carry out. But they are useful to test newborn babies whose mothers have HIV.

Another type of test is the *HIV antibody test*. People who are infected with HIV will have antibodies against the virus, but healthy people will not. You can see how the test works in Figure 2.

You can also test for the presence of *HIV antigens*. These tests use antibodies and enzymes to detect the antibodies or antigens. You can see how the test works in Figure 2.

The *Mantoux test* is used to test people who have been exposed to TB. A small amount of **serum** containing TB antigens is injected under the skin. If the person has been exposed to TB, there will be an **immune response**. The greater the immune response, the more inflammation will be seen on the site of the injection 48–72 hours later. If a person shows a positive Mantoux test, they may be investigated further to see whether they have TB.

### Human Papilloma Virus

Many cases of cervical cancer are caused by strains of Human Papilloma Virus, HPV. This virus is sexually transmitted. Some other strains of HPV cause genital warts. Recently, a vaccine has been produced that protects against most strains of HPV. Doctors hope that this vaccine could greatly reduce the incidence of cervical cancer. But if the vaccine is to be effective, girls should be vaccinated before they become sexually active. There are plans to vaccinate girls aged 11–12 in the UK in the near future.

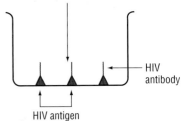

Add blood sample

HIV antibody

HIV antigen

1 Test dish has antigens from HIV attached. Blood sample is added. If HIV antibodies are present they attach to the antigens

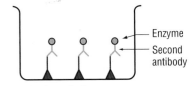

Enzyme
Second antibody

2 The dish is washed to remove any substances that have not bound to the antigens. A second antibody is added. This will bind to the HIV antibody if it is present. It has an enzyme attached to it

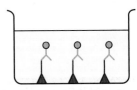

3 The dish is washed again and a colourless dye is added. The enzyme, if it is present, causes the colourless dye to change colour

**Figure 2** Testing for HIV

YORK
COLLEGE
Sim Balk Lane
York
YO23 2BB

## Questions

1 Blood group AB is sometimes called the 'universal recipient'. Explain why.
2 Explain why antibodies against one strain of HIV will not be effective against a different strain of HIV.
3 Explain why a person who has been exposed to TB will have a stronger reaction to the Mantoux test than a person who has not been exposed to TB.
4 Design an antibody and enzyme test, like the one in Figure 2, to test for
  (a) TB antigens
  (b) HIV antibodies

### Key definition

**Mortality** – refers to deaths. The data might be qualified (e.g. infant mortality) and is usually expressed as a rate (e.g. number of deaths per 10 000)

**Morbidity** – refers to people who are affected. Incidence and prevalence data is morbidity data.

## What is epidemiology?

Epidemiology is the study of the occurrence, distribution and control of diseases in populations. When studying disease, epidemiologists use a number of terms. These are listed below.

- **Incidence** – the number of new cases of a specific illness diagnosed or reported during a stated period of time, usually one year.
- **Incidence rate** – the number of new cases of a specific illness diagnosed or reported during a stated period of time, divided by the number of people at risk for the disease. For example, if five children develop measles in a community of 5000 people during one year, the incidence rate of measles in that community is 5/5000 or one per 1000 people per year. Another term for incidence rate is the **morbidity rate**.
- **Prevalence** – the number of current cases of a condition or illness at one time, no matter when it started. The term is usually used to describe conditions that last a long time, or that are chronic.
- **Prevalence rate** – the number of current cases of a condition or illness at one time, divided by the total number of people who may be at risk for the illness or condition. For example, if eight people in a community of 4000 have pancreatic cancer, but only one was diagnosed this year, the prevalence rate of pancreatic cancer in that community is 8/4000 or two per 1000 people.
- **Mortality rate** – the number of deaths from a specific cause per 1000 people in a population, per year. For example, if 850 people in a population of 100 000 died of TB in one year, the mortality rate would be 8.5 per 1000 people.

### Notifiable diseases

Doctors in England and Wales have a statutory duty to notify the Local Authority of suspected cases of certain infectious diseases. The Local Authority is required to inform the HPA Centre for Infections (CfI) every week of details for each case of each disease that has been notified. The Information Management and Technology Department within the CfI has responsibility for collating these weekly returns and publishing analyses of local and national trends.

The infectious diseases that need to be notified to the Local Authority include poliomyelitis, anthrax, cholera, food poisoning, malaria, measles, meningitis, mumps, rabies, rubella, tetanus, tuberculosis, viral hepatitis and whooping cough.

### The global impact of TB and HIV infection

Look at Figure 1. This shows the incidence of TB and HIV worldwide. You will notice that countries with a high incidence of HIV usually have a high incidence of TB as well.

### Past epidemics

Influenza is a disease that causes regular epidemics. The graphs in Figure 2 show the effects of three different influenza pandemics in Liverpool, UK. Graph A shows the death rate from respiratory conditions (pneumonia, influenza and bronchitis) while graph B shows the death rate from all causes.

### Future pandemics

Avian influenza is an infection caused by avian (bird) influenza (flu) viruses. These influenza viruses occur naturally among birds. Wild birds worldwide carry the viruses in their intestines, but usually do not get sick from them. But avian influenza can spread quickly among birds and can kill domesticated birds, including chickens, ducks and turkeys. Most kinds of avian influenza virus do not infect humans, although sometimes they may spread to a few people who have close contact with infected birds.

Humans can be infected by many different kinds of influenza viruses. These different kinds have slightly different antigens on their surface. The antigens of influenza viruses are constantly changing. A bird may also become infected with human influenza viruses at the same time as being infected with an avian influenza virus. If this happens, a new strain of virus may result.

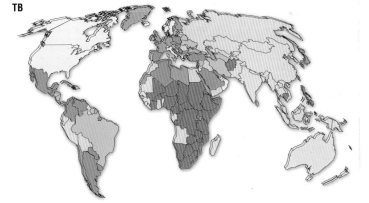

**HIV**

**TB**

**Adult prevalence %**

| | | |
|---|---|---|
| ▰ 15.0 – 34.0 | ▰ 1.0 – < 5.0 | ▰ 0.1 – < 0.5 |
| ▰ 5.0 – < 15.0 | ▰ 0.5 – < 1.0 | ▰ <0.1 |

**TB incidence rates per 100 000 population**

| | | | |
|---|---|---|---|
| Less than 10 | 25 to 49 | 100 to 299 | No estimate |
| 10 to 24 | 50 to 99 | 300 or more | |

**Figure 1** Incidence of HIV and TB worldwide

A type of influenza virus, a strain called H5N1 virus, is a kind of influenza virus that occurs mainly in birds. It is highly contagious among birds and can kill them. H5N1 virus does not usually infect people, but infections with these viruses have occurred in humans. Most of these cases have resulted from people having direct or close contact with H5N1-infected poultry or H5N1-contaminated surfaces. There is concern that if the virus changes slightly, it could become much more infective to humans. There is concern that it could then spread very quickly around the world and maybe cause large numbers of deaths among humans.

Use your knowledge of antigens and the immune system to explain why people can be infected by influenza on many occasions.

Suggest why the graphs in Figure 2 include causes of death other than influenza.

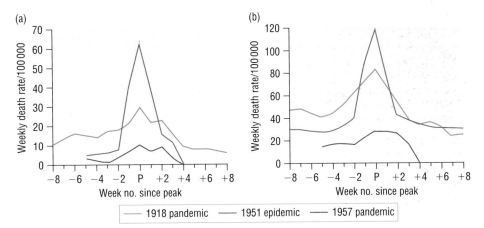

**Figure 2** Effects of three different influenza pandemics in Liverpool

## Questions

1  If the prevalence of a disease is high, but its incidence rate is low, what does it tell you about the disease?

2  Suggest why **(a)** mumps **(b)** food poisoning **(c)** tuberculosis are notifiable diseases.

3  Suggest reasons why the official figures for incidence of some diseases may not be the same as the true incidence of these diseases.

### Controlling and preventing TB and HIV/AIDS

1  Use your knowledge of TB and HIV/AIDS to suggest ways of controlling and preventing **(a)** TB **(b)** HIV/AIDS.

2  Explain why the incidence of these diseases is much higher in some countries than in others.

### Examiner tip

LDCs are the worlds least developed countries. Thirty-three countries in Africa are classed as LDCs. You should be prepared to explain the link between distribution of HIV and TB in terms of ECONOMIC FACTORS (lack of government money to invest in health care and low incomes), SOCIAL FACTORS (overcrowding, migration to cities for work, lack of education) and ETHICAL FACTORS (stigmatisation of HIV victims, low status of women). These are also factors in the difficulties met when planning control measures. In addition, the BIOLOGICAL problems with HIV (mutation rates, the latent period, the lack of early symptoms) will also affect control programmes.

# Self-check questions

**Fill the blanks**

1  An infectious disease can be spread from one organism to another. They are also known as ..................... diseases. They are caused by ..................... A pathogen is an organism that causes disease.

2  ..................... causes the disease tuberculosis, also known as ..................... . It is a ..................... . Bacterial cells are ..................... . This means they do not have a ..................... with a nuclear membrane and they do not have ..................... organelles such as mitochondria. They have circular ..................... and smaller ..................... than eukaryotic cells. Tuberculosis, like other infections caused by ....................., can be treated successfully using .....................

3  The human immunodeficiency virus, ....................., causes the disease ..................... – acquired immunodeficiency syndrome. Viruses do not have a ..................... structure and are many times smaller than bacteria. Viruses can only ..................... when they are inside a living cell. The HIV virus is spread from one infected person to another person in body ..................... . HIV enters the blood and infects ..................... T cells. The virus usually remains ..................... inside these cells for a long time. During this time, the person has no symptoms, but their blood contains ..................... against the virus. The infected person is said to be HIV ..................... . After a few years, HIV starts to replicate. The infected helper T cells release many new ..................... and are themselves destroyed. As helper T cells are destroyed, the person becomes susceptible to ..................... infections, such as TB. This stage of the disease is known as '..................... AIDS'. A person can be tested for HIV infection by examining the blood for ..................... against HIV.

4  Many medicines that we use in the West come from ..................... . Unfortunately, many plant species in the world are endangered and becoming ..................... even before they have been named. It is likely that plants containing useful ..................... that we do not yet know about are among these plants that are being lost. Kew Gardens in London has set up a ..................... Seed Bank. They are collecting ..................... from many plant species all over the world and storing them.

5  When the body becomes infected with a pathogen, the body's ..................... system recognises the pathogen as foreign. This is because the pathogen has foreign proteins on its surface, called ..................... . T-lymphocytes produce T ..................... lymphocytes that destroy pathogens and infected cells, as well as T ..................... and T suppressor lymphocytes. Specific B lymphocytes form clones of ..................... cells, that produce specific ..................... that destroy the specific antigen. They also produce ..................... cells that carry an immunological memory of the antigen.

6  Vaccines contain ..................... . These cause the immune system to produce ..................... cells against the specific antigen, so that if the person later becomes infected with the real pathogen, they will undergo a rapid ..................... response. They will produce large numbers of ..................... so that the pathogen is destroyed before the person becomes ill. In the UK, children are ..................... against many different diseases. If enough children are vaccinated, it is difficult for an infectious disease to spread in a population. This is called ..................... immunity.

# Summary questions

**1** In 1928, Sir Alexander Fleming was the first scientist to observe the effect of an antibiotic on the growth of a bacterium. The bacterium was called *Staphylococcus aureus*. Antibiotics are now widely used in the treatment of infectious diseases.

(a) Suggest why the discovery of antibiotics was so important. [2]

(b) Excessive use of antibiotics has led to the development of resistant strains of bacteria, such as methicillin-resistant *Staphylococcus aureus* (MRSA). Explain how the excessive use of antibiotics has led to the development of resistant strains of bacteria. [5]

(c) When patients are prescribed a course of antibiotics to treat a bacterial infection, they are told that it is important to complete the course, even if the symptoms of the infection have gone. Explain how failing to follow this advice may lead to the development of resistant strains of bacteria such as MRSA. [4]

(d) Outline three precautions which should be taken to reduce the spread of antibiotic resistant bacteria in hospitals. [3]

(e) Why is MRSA infection of particular concern in wards for elderly patients? [2]

**2** Tuberculosis (TB) is a disease caused by the bacterium *Mycobacterium tuberculosis*. Historically, TB has been one of the world's worst fatal diseases and it still kills over 1.5 million people every year.

(a) Describe the role of the Mantoux test in a vaccination programme for TB. [3]

**3** The introduction of a foreign antigen into the blood triggers an immune response.

(a) Explain the meaning of the term *immune response*. [2]

(b) Figure 1 shows the change in antibody concentration in human blood, in response to the introduction of antigen X.

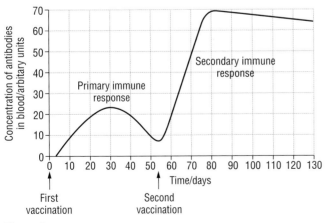

**Figure 1**

Using the information in the figure,

(i) describe **two** differences in the response to the first and second exposure to antigen X; [2]

(ii) explain how a vaccine works. [4]

**4** HIV/AIDS is arguably the most serious disease to spread throughout the world.

(a) Outline the relationship between HIV and AIDS. [3]

(b) Describe three ways in which HIV can be spread by an infected human. [3]

(c) Antibiotics have been used successfully to treat many infectious diseases. Why are antibiotics not used to treat HIV infection? [1]

(d) It is difficult to calculate accurately the number of people in the United Kingdom who are HIV positive. It has been suggested that compulsory HIV screening should be introduced.

(i) State two reasons, other than lack of compulsory screening, why it is difficult to calculate accurately the number of people who are HIV positive. [2]

(ii) Outline **two** ethical problems that could arise from compulsory screening. [2]

**5** (a) Figure 2 shows the structure of the bacterium *Mycobacterium tuberculosis*.

Cell wall made of peptidoglycan

Ribosomes for protein synthesis – these are smaller than eukaryotic ribosomes

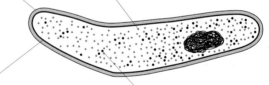

Cell surface membrane – controls entry and exit of substances into and out of the cell

Circular DNA – this carries genes coding for the proteins inside the cell. The DNA is not Complexed with protein

**Figure 2**

(i) Using Figure 2, describe two characteristic features of a prokaryotic cell. [2]

(ii) One-third of the world's population is infected with tuberculosis bacteria.

Describe how the bacteria may be spread. [2]

(b) Explain how the increase in HIV infection is accelerating the spread of tuberculosis (TB). [2]

(c) Outline the actions of B and T lymphocytes in fighting infection. [7]

## Types of disease

You have already learned about **infectious diseases**. These are diseases that are caused by **pathogens**, such as bacteria or viruses. These diseases are spread from one infected person to another, for example, TB or HIV/AIDS. **Non-infectious** diseases are not caused by pathogens, and consequently they do not spread between people. You have already learned about cancer. Most types of cancer would be classified as non-infectious diseases. But there is some evidence that cervical cancer is caused by HPV (Human Papilloma Virus), so 'cancer' should not be treated as one type of disease. **Coronary heart disease** (CHD) is a good example of a non-infectious disease.

### Atherosclerosis

The lining of a healthy artery is smooth so that blood can flow easily through it. But if this smooth **endothelium** becomes damaged, for example by high blood pressure, a fatty plaque or **atheroma** can build up in the wall. You can see this in Figure 1.

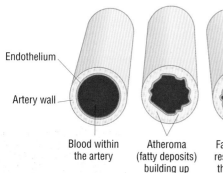

Endothelium

Artery wall

Blood within the artery

Atheroma (fatty deposits) building up

Fat deposits develop restricting blood flow through the arteries

**Figure 1** The development of atheroma

**Phagocytic cells** migrate to the damaged area and start to accumulate lipids, becoming **foam cells**. Fatty deposits build up underneath the damaged endothelium. Smooth muscle cells in the artery lining divide and form fibrous **connective tissue** to repair the damage. Tissue and fatty deposits build up in the area very slowly over many years, and the growing **plaque** narrows the **lumen** of the artery. Eventually, the plaque breaks through the endothelium, which becomes roughened. A blood clot, or **thrombus**, starts to form. This makes the lumen of the artery even narrower.

Eventually the blood clot may grow so large that it blocks the artery. Alternatively, a piece of the clot may break off – an **embolus**. The embolus travels in the blood until it reaches a place where the artery is too narrow for it to pass. This is likely to be somewhere where the artery is already narrowed by atheroma. If the **coronary** artery becomes blocked in one of these ways, a heart attack or **myocardial infarction (MI)** may occur. The coronary arteries supply the heart muscle with glucose and oxygen. If the coronary artery is blocked, some of the heart muscle cells may die and this is what is meant by an MI.

### Angina

When the coronary arteries become narrowed by atheroma, less blood can pass through these arteries to supply the heart muscle. Problems usually occur if the person starts to exercise, for example, climbing stairs. The heart muscle needs more glucose and oxygen for faster respiration, but the coronary artery is too narrow to supply these. This can result in a pain called **angina pectoris**. You can see how this may develop in Figure 2.

People with angina experience pain in the centre of the chest. They often say it feels like something is crushing the chest. The pain starts in the front of the chest but can spread down the left arm, towards the jaw and down towards the abdomen.

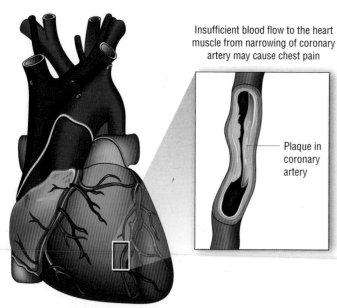

Insufficient blood flow to the heart muscle from narrowing of coronary artery may cause chest pain

Plaque in coronary artery

**Figure 2** Development of angina

### Recognising a heart attack

The symptoms of a heart attack include:

- severe chest pain that does not ease with rest
- breathlessness and nausea
- feeling faint
- severe sweating
- sense of 'doom'
- irregular pulse
- ashen, cold skin and blue lips.

The first thing you should do is make the casualty comfortable. Sit them in a supported position with their knees bent. Then dial 999 and ask for an ambulance, saying that you suspect a heart attack. If the person has any medication for a heart condition, s/he should take it.

Monitor the person's condition, for example, measuring his/her heart and breathing rate regularly. Also note his/her level of consciousness, and reassure him/her. Should the casualty lose consciousness, be prepared to give him/her CPR (see below).

### Cardiac arrest

Cardiac arrest occurs when the heart no longer beats efficiently enough to pump blood round the body. Myocardial infarction is one cause of cardiac arrest. When a cardiac arrest has occurred, no pulse can be felt and the person loses consciousness. The skin goes white or grey and the lips become blue. The first aid treatment for a patient who has suffered cardiac arrest is **cardio-pulmonary resuscitation (CPR)**.

### Cardio-pulmonary resuscitation (CPR)

If someone has a cardiac arrest or heart attack, there are only a few minutes to act before it is too late. It is vital to know what to do beforehand. To perform CPR effectively, training and practice on resuscitation dummies are essential. Also, the procedure has been subject to change. You should check the current recommendations on the St John Ambulance website.

First aid courses are offered all over the country by voluntary organisations such as St John Ambulance or the Red Cross. Send someone else to fetch help by dialling 999, and ask for an ambulance.

- If the person is unconscious, check whether s/he is breathing. If s/he is breathing, the chest should rise and fall. Second, check whether there is a pulse. If you place two fingers on one side of the person's voice box, you should be able to feel the carotid pulse.
- If the patient has a pulse but is not breathing, check that the airway is clear (for example, clear any food from the mouth) and carry out **rescue breathing** (see spread 1.2.3.4).
- If the person is not breathing and has no pulse, he/she has suffered a cardiac arrest.
- Lay the person on his/her back. If possible, raise the legs up 30–40 cm to allow more blood to flow towards the heart.
- Place your hand at the centre of the chest. Place the other hand on top and interlock your fingers. Keeping your arms straight and your fingers off the chest, press down 4–5 cm then release the pressure. Your pressure should be applied through the heels of your hands only. You can see this in Figure 3.
- Repeat this 30 times at the rate of 100 compressions a minute, then give two rescue breaths. Continue this 30:2 procedure until help arrives.
- Check for any sign of breathing or a pulse occasionally. If a pulse starts again, continue with the rescue breathing. If both breathing and a pulse return, move the casualty into the **recovery position**, and monitor him/her carefully until help arrives. The person may be confused and anxious, so you will need to reassure him/her and help him/her to keep calm.

**Figure 3** CPR

## Questions

1 Distinguish between:
   (a) atheroma and angina
   (b) cardiac arrest and myocardial infarction.
2 What is the purpose of **(a)** the rescue breaths and **(b)** the 'pumping action' in CPR?
3 Diseases can be classified in several ways. Infectious and non-infectious are two classes. Suggest why CHD and cancers can also be classified as degenerative diseases.

## Emergency treatment for CHD

A person who has suffered cardiac arrest may be treated using a defibrillator. This will be done by a qualified medical practitioner or a trained first-aider. **Defibrillation** is when a powerful electric shock is delivered to the heart.

The defibrillator is turned on. The person's clothing and jewellery should be removed. The two pads are placed on the chest, one on the upper right side and the other on the lower left.

The pads are plugged into the connector. The defibrillator will determine whether a shock is needed. Meanwhile, nobody should touch the patient.

If the machine determines that a shock is needed, both visual and auditory prompts will be given. These will instruct the operator to press a button to deliver a shock. While this is happening, nobody should touch the patient. After this, the machine may instruct the operator to give a second shock.

The operator should check whether the patient's airway is open, and check for breathing and pulse. If the person has a pulse but is not breathing, rescue breaths should be given. If there is no pulse, defibrillation should be repeated.

You can see a defibrillator being used in Figure 1.

**Figure 1** Using a defibrillator

People who have suffered a **myocardial infarction** are usually given aspirin. Aspirin reduces the ability of **platelets** to cause blood clotting. Doctors often recommend that people who have had a myocardial infarction take a low dose of aspirin every day.

### Surgery to treat CHD

Angina is often treated by **angioplasty**. This involves inserting a tiny, flexible hollow tube called a *catheter* into an artery in the groin or arm. At the tip of the catheter is a tiny inflatable balloon. The catheter is pushed along the artery until it gets to a **coronary** artery. The doctor can see where the catheter is because it shows up on an X-ray screen.

When the catheter arrives at the narrowed or blocked section, the balloon is gently inflated. This squashes the fatty **atheroma** that has narrowed the artery, and so the artery is widened. Most balloons contain a *stent*. This is a short tube made of stainless steel mesh. The stent expands when the balloon is inflated, so that it stays behind when the balloon is removed, holding the narrowed blood vessel open. You can see this in Figure 2.

*Coronary bypass surgery* is used to treat blocked coronary arteries. In this operation, a piece of vein is taken from elsewhere in the body, often the leg. It is then used to bypass the blockage in the coronary artery. People may have a single, double, triple or even quadruple bypass, depending on the number of blocked arteries. Usually, the heart is stopped while the operation takes place and a machine is used to pump the blood round the body. However, more recently, some surgeons have carried out the operation without stopping the heart.

In a small number of cases, medication or the types of surgery mentioned above may not be effective at controlling the condition, or the heart may be severely damaged. In these cases, a *heart transplant* may be the only possible treatment. But some patients are not suitable for a heart transplant and there may be a wait of several months for a suitable donor. Inevitably, many people die waiting for a heart transplant as there are not enough donor organs available.

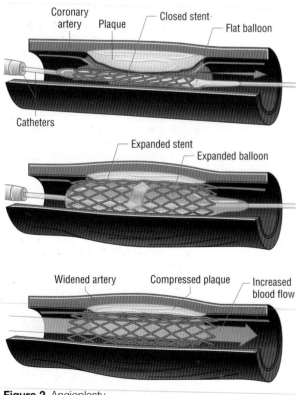

**Figure 2** Angioplasty

### The costs of treating coronary heart disease

The table shows the costs of treating cardiovascular and coronary heart disease in the UK in 2003 (figures from the British Heart Foundation).

| | Cardiovascular disease | | Coronary heart disease | |
|---|---|---|---|---|
| | £ million | % of total | £ million | % of total |
| Primary care | 639 | 4.3 | 101 | 2.9 |
| Outpatient care | 184 | 1.2 | 56 | 1.6 |
| Accident and emergency care | 51 | 0.3 | 19 | 0.5 |
| Inpatient care | 11 229 | 76.2 | 2773 | 78.6 |
| Medications | 2629 | 17.8 | 578 | 16.4 |
| Total health care costs | 14 732 | 100.0 | 3527 | 100.0 |
| Cost per person | £249.29 | | £59.49 | |

Treating diseases of the **cardiovascular system** is very expensive. One of the reasons for this is that we are living longer, and we are more likely to suffer from CHD, strokes, etc. as we get older. But we can do something about some of the factors that increase the incidence of these diseases (see spread 2.4.1.3).

NICE is the National Institute for Health and Clinical Excellence. It is an independent organisation responsible for providing national guidance on promoting good health and preventing and treating ill health.

NICE produces guidance in three areas of health:
- public health – guidance on the promotion of good health and the prevention of ill health for those working in the NHS, local authorities and the wider public and voluntary sector
- health technologies – guidance on the use of new and existing medicines, treatments and procedures within the NHS
- clinical practice – guidance on the appropriate treatment and care of people with specific diseases and conditions within the NHS.

NICE guidance is developed using the expertise of the NHS and the wider healthcare community, including NHS staff, healthcare professionals, patients and carers, industry and the academic world. For example, NICE produce guidance on the use of coronary artery stents. Their guidance is listed in the case study on the right.

NICE guidance is issued so that all health professionals can benefit from the latest research, and that best practice can be used in treating all patients. It also aims to make treatment as cost-effective as possible.

## Questions

1 Explain why it would be dangerous to use a defibrillator on a person whose heart was still beating.
2 Explain why the vein used in a coronary bypass operation comes from the patient and not a donor.
3 Primary care is care given in the community, for example at GP clinics and by district nurses. Explain why this is cheaper than hospital care.

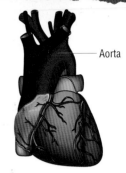

Aorta

Saphenous vein

Blocked coronary artery

**Figure 3**  Coronary bypass surgery

### NICE guidance on the use of coronary artery stents

A stent should normally be used during balloon angioplasty in a person who has angina or has had a heart attack.

The decision on which type of stent to use should depend on the person's symptoms, and on the size and shape of the narrowed part of the artery. A drug-eluting stent should be used if the person has angina, and the inside diameter of the artery is less than 3 mm across, or the narrowed area is more than 15 mm long. There are several different drug-eluting stents, which contain different drugs. NICE recommends stents that contain either a drug called sirolimus, or one called paclitaxel, because most of the research has been on these.

If more than one artery is narrowed, doctors should make the decision on which type of stent to use for each artery separately.

This guidance covers treatment for people who would normally be offered some form of balloon angioplasty. NICE has not made any recommendations on using stents to treat people who have had a heart attack in the previous 24 hours, or people who had a clot in the narrowed artery.

### The global distribution of CHD

Figure 1 shows the frequency of deaths from coronary heart disease (CHD) in many different countries. You will see that deaths from CHD are more frequent in some countries than in others. Epidemiologists study patterns of disease like this. This can help them to find out what causes CHD.

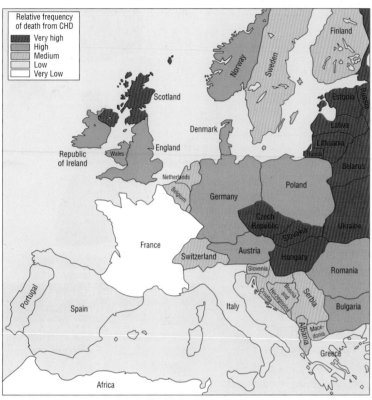

**Figure 1** Mortality from CHD in different countries

Epidemiologists also analyse data from individual populations. This can identify different factors that pre-dispose people to develop CHD. These factors are called **risk factors**.

### Risk factors for CHD

- Diet. A diet that is high in salt increases the risk of CHD, because it increases **blood pressure**. High levels of saturated fat in the diet increase blood **cholesterol** levels, which also increases the risk of CHD. But some foods contain **antioxidants** that reduce the chances of developing CHD. Vitamin C, found in many fresh fruits and vegetables, is an example of an antioxidant.
- Blood pressure. High blood pressure can cause damage to the **endothelium** in the arteries, and therefore increases the chances of **atheroma** developing.
- Exercise. Physical activity, especially **aerobic exercise**, lowers blood pressure and blood cholesterol which can help to prevent CHD. Also, physical activity reduces the chance of developing obesity, which is another factor that can lead to CHD.
- Smoking. Smoking is the greatest risk factor for CHD. Smoking raises blood pressure, increases the development of atheroma and increases the chances of a blood clot forming.
- Genetic influences. Some people have genes that predispose them to high blood pressure and high blood cholesterol levels. Also, men are more at risk of CHD than are women up to the age of menopause. After this, the risk for males and females is the same. There is also evidence that some races, for example Afro-Caribbeans in the UK, have a higher rate of CHD than others. But it is not clear whether this is because of genetic factors or whether social or environmental factors such as diet are important.

### Body mass index

As you learned above, being overweight is a risk factor for CHD. It is also a risk factor for developing **type II diabetes**, which is itself another risk factor for CHD. There are two ways to assess whether a person is overweight – by calculating a BMI or the waist-to-hip ratio.

Two people may have the same body mass, but one may be much taller than the other. The taller person may have a normal **body mass index** (**BMI**), while the shorter person could be obese. Height is an important factor in determining if the person's weight is healthy for him/her. To take this into account, we determine the relationship between weight and height by calculating the body mass index (BMI). To calculate your body mass index, take your weight (in kg), and divide by your height (in metres) squared.

Ideally, a person should have a BMI of between 20 and 24.9. If a person has a BMI of 25 or higher, they are considered to be overweight, and if they have a BMI over 30 they are considered to be obese.

### Examiner tip

Epidemiological studies are data collection studies carried out on large populations over long periods of time. The data can be used to look for correlation between the onset of CHD and lifestyle factors for example. One of the most famous – the Framingham Study – has been running for 50 years. You will find a visit to their website very useful.

However, the way that fat is distributed on the body is also important. People with a lot of fat around the abdomen are at most risk. These people are said to have 'central adiposity', and people with this fat distribution are said to be 'apple-shaped'. A person can determine whether they have central adiposity by measuring his/her waist in centimetres. A healthy man should have a waist circumference measurement of less than about 94 cm, and a healthy woman should have a waist circumference measurement below about 80 cm.

Another way to measure body size is the waist-to-hip ratio. You measure the waist circumference at the height of the umbilicus (navel), and the hips around the widest point. Then you divide the waist measurement by the hip measurement. Ideally the ratio should be 1.0 or less in men or 0.8 or less in women.

### Examiner tip

It is important that you define 'obesity' in terms of BMI and that you are aware how a BMI is calculated. There are several calculators available on the Internet. Some will require metric units (kg and metres) and some imperial (pounds and feet). The answer is a ratio so it is the value which is important, not the units.

### The risk of CHD

Figure 2 shows the deaths due to CHD in men and women of different ages

- Describe the trends shown in the data.
- More women aged 90 or over die of CHD than men of the same age. Suggest an explanation for this.

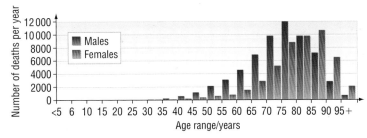

**Figure 2** The mortality from CHD in men and women of different ages

### Case study

Under recent government proposals, parents in England may be warned if their children are found to be overweight. Children in England are currently measured at the ages of five and ten, but parents are informed of the results only if they request them. The new plan may see parents getting results automatically.

Between one-quarter and one-third of children are thought to be overweight, and doctors fear there will be an epidemic of poor health related to obesity in coming decades. An obese person dies on average nine years earlier than somebody of normal weight, while a very obese person's life is cut short by an average of 13 years.

Under the proposals being considered, information obtained under the measurement programme could be given automatically to parents, and involvement may become compulsory unless people choose to opt out of the measurement scheme.

Dr Terry Dovey, a psychologist at the University of Staffordshire who specialises in childhood obesity, said 'stigmatising' children and parents for having an overweight child was not the way forward. He said: 'The most success you can possibly have is being vigilant in both what you're eating and how much of it you are eating.

'If you're wanting to target specific groups then you have to have the key stages in the process to help the child and parents.'

A recent report by the Foresight Programme argued that dramatic and comprehensive action was required to stop most of us becoming obese by 2050. Its authors predicted that if current trends continue, in that year 60% of men and 50% of women will be obese, and cases of type II diabetes will rise by 70%. The report also suggested that cases of stroke will rise by 30% by the middle of the century and cases of coronary heart disease will rise by 20%.
- Do you think it is necessary to tell parents that their child is overweight or obese?
- Some people worry that, if this proposal is implemented, it will lead to overweight children becoming stigmatised. Do you agree?
- To what extent are parents responsible for their child's body mass?
- Suggest some other ways of reducing childhood obesity.

Story adapted from BBC NEWS:
http://news.bbc.co.uk/go/pr/fr/-/1/hi/uk/7055735.stm

## Questions

1 Figure 1 shows that the number of deaths from CHD varies considerably in different countries. Suggest reasons for these differences.

2 List the risk factors for CHD under the following headings: environmental; behavioural; social; genetic factors.

### Case study

The ban on smoking in public places in Scotland is already beginning to have an impact on the nation's health. The number of non-smokers admitted to hospital after heart attacks fell by 20% in the 10 months after the ban came into force in March 2006, compared with the same 10 months in the year before.

Other studies have shown that children's exposure to second-hand smoke has fallen, except among children whose mothers smoke, or those with two parents who smoke.

One study covered nine hospitals, which between them account for two-thirds of all hospital admissions for heart attacks in Scotland. In the 10 months of the year leading up to the ban, there were 3235 admissions, while in the matching period after the ban, the figure was 2684. Patients were asked if they were smokers or non-smokers, and their answers double-checked through blood tests to detect levels of cotinine, the product into which nicotine is converted by the body. In non-smokers, the fall in heart attack admissions was higher, at 20%.

The scientists carrying out the study said that the reduction among non-smokers was biologically plausible, because smoke contained a lot of toxins that could trigger heart attacks in people with coronary heart disease. They say that the study is important because it has shown an effect in people who have never smoked. They believe that this can only be due to lower levels of passive smoke.

Rates of heart disease are falling everywhere, but not as fast as this. Over the same period of 10 months after the ban, admissions in England fell by 4%, and the reduction rate in Scotland over the decade before the ban was 3% per year.

Adapted from an article dated 11 September 2007 *Times online* http://www.timesonline.co.uk/tol/life_and_style/health/article2426743.ece

There are several different kinds of lung disease. Some are described as **acute**. This means that they only happen for a short time, and then the person gets better. An example of this might be acute bronchitis, an infection of the lungs caused by a virus. However, other diseases may be **chronic**. Chronic diseases often start slowly but they last for a long time. Chronic diseases are rarely cured. An example of a chronic disease would be diabetes, or chronic bronchitis caused by smoking cigarettes.

## The effects of smoking on the respiratory system

Tobacco smoke contains a large number of different chemicals. You learned about these on spread 2.1.1.4. The main chemicals in tobacco smoke that harm health are carbon monoxide, nicotine and tar.

Carbon monoxide is a gas that diffuses through the walls of the alveoli into the blood. It readily combines with haemoglobin in the red blood cells to form **carboxyhaemoglobin**. Carboxyhaemoglobin stays combined with carbon monoxide, making it unable to carry oxygen. Unfortunately, haemoglobin combines with carbon monoxide even more readily than it combines with oxygen, so even a small amount of carbon monoxide significantly reduces the amount of oxygen that a person's blood can carry. A smoker's blood may carry up to 15% less oxygen than a non-smoker's, which is why so many smokers suffer from breathlessness, especially when they exercise. Carbon monoxide also increases the heart rate, as the heart has to pump more often to get the same amount of oxygen to body tissues.

A large proportion of the tar in tobacco smoke is deposited in the bronchi, bronchioles and alveoli. Here they irritate the epithelium and cause inflammation. The cilia on the ciliated epithelium cells become paralysed, so they can no longer move mucus back upwards towards the throat. As a result, mucus containing dirt and microorgansisms builds up in the lungs. Tar also stimulates mucus production by the goblet cells. As a result, a smoker is prone to lung infections as microorganisms are not removed from the respiratory system. A persistent cough develops as the body tries to get rid of the accumulated mucus. This mucus has a number of effects on gas exchange in the lungs:

- Mucus reduces the diameter of the bronchi and bronchioles, reducing the rate at which air can reach the alveoli.
- Coughing can damage the bronchi, bronchioles and alveoli. Scar tissue is produced which lowers the rate of diffusion and narrows the diameter of the airways.
- Mucus builds up in the alveoli, reducing the volume of air that they can hold and increasing the diffusion pathway.
- Infections in the lungs lead to the development of emphysema, in which the walls of the alveoli break down. This reduces the surface area available for diffusion.
- The mucus building up in the lungs also contains allergens such as pollen. This can cause inflammation of the bronchi and bronchioles, reducing the rate at which air can enter the alveoli. It can also lead to asthma attacks.

### Chronic bronchitis

The build up of mucus in the lungs, and the effects on the respiratory system listed above, lead to a condition called **chronic bronchitis**. This condition develops slowly. The main symptoms are breathlessness and a persistent cough. Once the disease has established itself, it is very difficult to improve the symptoms.

### Emphysema

Smoking also leads to the lung disease, emphysema. In a healthy person, phagocytic white blood cells produce an enzyme called elastase, which helps them to digest tissues to reach the site of an infection. In smokers, the presence of irritants in the mucus that builds up in the lungs causes inflammation. As a result, large numbers of phagocytic white blood cells are attracted to the alveoli. The elastase produced by these phagocytic cells breaks down elastin and some other proteins in the walls of the alveoli. In healthy people, there is an inhibitor to stop tissue damage by elastase, but in smokers this inhibitor is inactivated. As the elastin and other proteins break down, the alveoli become enlarged and damaged. This reduces the surface area available for gas exchange.

This reduced surface area for gas exchange causes breathlessness in the smoker. They breathe in more deeply, stretching the elastin that remains in the walls of the alveoli. As a result, the elastin fibres become permanently stretched. In a healthy person, elastic recoil from these elastin fibres helps to force air out of the alveoli when breathing out. In a person with emphysema, some stale air remains in the alveoli, making gas exchange even less efficient.

As emphysema worsens, a person becomes increasingly disabled. Some become so breathless that they cannot get out of bed. The person can be given oxygen from a cylinder to breathe in, but there is no way to reverse the damage. Death from respiratory failure usually results.

### Chronic obstructive lung disease (COPD)
Chronic bronchitis and emphysema are both caused by smoking, so the two conditions usually occur together. People with both conditions are said to have chronic obstructive lung disease, or COPD.

Figure 1 shows the results of an investigation into the effect of smoking on lung function.
- Use the graph to describe the effect of smoking on lung function.
- Is it worthwhile for a 65-year-old smoker to give up? Explain your answer.
- Suggest one way in which this investigation could be made more reliable.

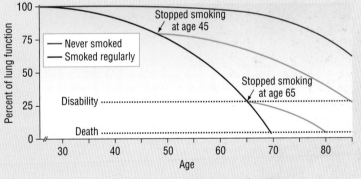

**Figure 1** Graph showing effect of smoking on lung function

### Lung cancer
Tobacco smoke contains a number of carcinogens, i.e. chemicals that cause cancer. These carcinogens cause mutations in the genes of epithelial cells that line the lungs. You will remember from spread 2.1.1.2 that some genes control cell division. If these genes mutate, uncontrolled cell division can take place. You learned on spread 2.1.1.3 that p53 is a tumour suppressor gene. Benzopyrene, a chemical found in the tar from cigarette smoke, binds to p53 and inactivates it. As a result, epithelial cells divide uncontrollably, leading to the formation of a tumour.

The symptoms of lung cancer include:
- breathlessness, as a result of the tumour blocking the airways and damaging the alveoli
- wheezing caused by air being forced down obstructed airways
- a persistent cough, resulting from the tumour obstructing the airways
- sputum stained with blood, as a result of the tumour causing damage to the lung tissue
- changes in the voice may occur if the tumour presses on the airways or larynx (voice box).

## Questions
1 Does this study prove that passive smoking can cause heart attacks?
2 Why was it important that the scientists compared the hospital admissions for heart attacks in Scotland since the smoking ban with those for England?
3 Some people think that the smoking ban will lead to a reduction in cases of lung cancer and COPD. Explain why it is too early to decide.

## Asthma

The term asthma comes from the Greek words meaning 'to breathe hard'. It is a disease that affects the trachea, bronchi and bronchioles of the lungs. Doctors sometimes refer to asthma as *reversible obstructive airway disease*, or ROAD. Asthma does not affect sufferers all the time. Many people who develop asthma have a family history of the disease, so there is a genetic component. But it is still not fully understood why some people develop asthma and others do not.

During an asthma attack, the membranes lining the airways release mucus and become inflamed. This inflammation causes the muscles of the airways to contract, narrowing the airways. This is called *bronchoconstriction*. Mucus also blocks the airways. This leads to wheezing, which is the sound heard as the airways constrict and make it difficult for air to escape. You can see this in Figure 1.

90% of all asthma is *allergic asthma*. This means that the asthma is triggered by **allergens**, which are substances capable of causing an allergic reaction. These include: fur from household pets; pollen; dust mites and their faeces; mould spores; pollutants; smoke; and many chemicals. An allergic reaction happens when the immune system over-reacts to something in the environment. This kind of asthma is more likely to develop in younger people, before the age of 40. Most childhood asthma is related to allergy. One cause of this is the mother smoking during pregnancy, or a parent smoking in the home. Cigarette smoke irritates the respiratory tract of children and increases the chances that they will develop allergic asthma.

*Intrinsic asthma* typically develops after the age of 40, and allergies do not play a part in this. It can be caused by respiratory irritants such as perfumes; cleaning agents; fumes; smoke; or infections of the upper respiratory tract. This kind of asthma is harder to treat than allergic asthma.

*Exercise-induced asthma* can affect anyone at any age. It is usually caused by the loss of heat and moisture from the lungs that happens during strenuous exercise. The symptoms are worse during cold, dry weather.

*Nocturnal asthma* (sleep-related asthma) may be triggered by allergens in the bedding or bedroom, or a decrease in room temperature, among other things. It is estimated that 75% of asthmatics suffer from nocturnal asthma.

*Occupational asthma* develops as a result of breathing chemical fumes, wood dust or other irritants over a long period of time.

### Treatment for asthma

**Steroids** are used to treat asthma. Treatment uses a group of steroids called *corticosteroids,* which are usually taken in an inhaler. The steroids are usually taken twice a day. It is best if they are taken by inhaling them, because this delivers the drug straight to the lungs, and causes the fewest side effects. But steroids may also be given in tablet or syrup form, or via an injection if the asthma is very severe.

Steroids work by reducing the inflammation in the airways. This means they work slowly, but have a long-term effect in reducing the severity of asthma attacks. However, they can also be used when a person is having an asthma attack.

**Beta-agonists** can also be used to treat asthma. They act as bronchodilators, relaxing the muscles that constrict the airways. Like steroids, they are usually inhaled so that they are delivered straight to the place where they are needed. However, they may also be taken in tablet or syrup form.

If a person is suffering from a very severe asthma attack, they may need to attend hospital. Steroids will be given, and they may be injected into the blood. *Beta-agonists* will also be given to relax the muscles, and the person may be given oxygen to help his/her breathing.

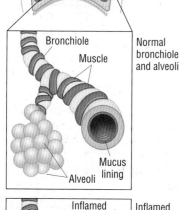

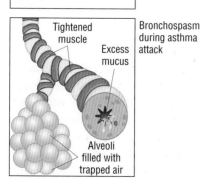

**Figure 1** How bronchoconstriction causes asthma

## Pets 'cut chance of children having asthma'

By Chris Bunting
Published: 28 August 2002 from Independent online

Babies who grow up in a house with pet dogs and cats have less than half the chance of developing asthma compared to those who do not, according to recent research. Scientists in the United States found that having two or more pets in the home during the first year of life also greatly reduced hay fever and other common allergies. The scientists who carried out the study were surprised, because they had expected to find that pets increased the rate of allergies.

The scientists say that the findings appeared to support the 'dirty hypothesis' – the theory that early exposure to bacteria can prime the immune system and help prevent allergies. They said it may be we have so many children with allergies and asthma because we live too clean a life. When children play with cats and dogs, the animals lick them. The lick is transferring a lot of bacteria and that may be changing the child's immune system response in a way that helps to protect against allergies. Recent studies have shown that children who live in cities have higher rates of allergy than children of farmers, perhaps because of their limited contact with animals.

In this research, doctors followed a group of 474 healthy babies in the Detroit area from birth to about the age of seven. Positive reactions to common allergens ranged from 33 per cent in children without exposure to dogs or cats, to 15 per cent in children who had been regularly exposed to two or more animals.

## Does pollution from cars cause asthma?

It is often claimed that asthma symptoms are made worse by pollution from traffic exhaust. The main pollutants are sulfur dioxide, carbon monoxide, nitrogen dioxide and particulate pollution (PM10s). The Association of British Drivers claims that this link is not supported by the evidence. They claim that the incidence of asthma has increased sharply at a time when traffic exhaust pollution has fallen. The results of their research are shown in Figure 2.

- How does this evidence support the claims made by the Association of British Drivers?
- Is there any other explanation for the results shown in the graph?
- What information would you need to know in order to evaluate the results of this study?

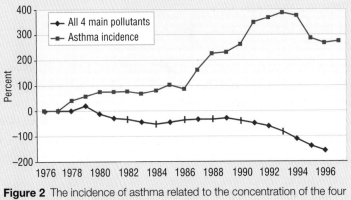

**Figure 2** The incidence of asthma related to the concentration of the four main pollutants in traffic exhaust

## Housework 'can cause asthma'

A study found using household cleaning sprays and air fresheners as little as once a week raised the risk of asthma. Heavy use of such products has already been linked with occupational asthma, but the latest work suggests occasional use in the home also poses a threat.

This research, conducted in Spain, involved the study of over 3500 people. The risk of developing asthma increased with frequency of cleaning and the number of different sprays used. Spray air fresheners, furniture cleaners and glass cleaners carried the highest risk. The researchers suggest that exposure to cleaning products could account for as much as 15%, or one in seven, adult asthma cases. On average, the risk was 30–50% higher in people who regularly used the sprays than in others.

The scientists who conducted this study said work was needed to determine the biological mechanism behind the increased risk. They think that sprays might contain irritants specific to asthma.

Story adapted from BBC NEWS: http://news.bbc.co.uk/go/pr/fr/-/1/hi/health/7041182.stm

- How could you carry out an investigation to find out whether cleaning sprays contain irritants that trigger asthma?
- What are the difficulties in carrying out such an investigation?

## Questions

1 Steroids are usually taken by asthmatics to prevent asthma, and beta-agonists to relieve the symptoms. Explain why.
2 Use your knowledge of the immune system to explain why the immune system will be more active in babies exposed to animals.
3 Use your knowledge of asthma to suggest an alternative reason why children living in cities have higher rates of asthma than the children living in the countryside.

# Diabetes

Diabetes is derived from a Greek word which means 'passing through of water'. The earliest record of diabetes has been traced to about 1600 BC in Ancient Egypt. An Egyptian doctor wrote on a papyrus that frequent urination was a symptom of this condition. In about the eleventh century AD, the word 'mellitus' was added to the term diabetes. Mellitus is the Latin word for honey, and it was used to refer to the sweet-tasting urine of the people with diabetes. Until recently, doctors could diagnose diabetes only by tasting the urine of people suspected of having diabetes. If the urine tasted sweet, diabetes mellitus could be diagnosed.

In normal sugar metabolism, the concentration of glucose in the blood is monitored by cells in the pancreas. If the glucose concentration becomes too high, the beta cells of the pancreas secrete insulin, which stimulates cells to take up glucose and respire it, and stimulate muscle and liver cells to take it up and convert it into glycogen for storage. When blood glucose levels fall, insulin secretion stops, and the hormone glucagon is secreted by the alpha cells of the pancreas. This stimulates the breakdown of glycogen to glucose and the release of glucose into the bloodstream.

There are two kinds of diabetes. **Type I diabetes** is usually diagnosed in children and young adults. It happens because the immune system has attacked the insulin-producing cells in the pancreas. As a result, the pancreas produces very little or no insulin. Insulin is a hormone that regulates blood glucose concentration. In a normal person, insulin is secreted when blood glucose levels are too high. The excess glucose is then converted to glycogen. In a person with type I diabetes, the blood glucose level can become far too high or far too low. If type I diabetes is untreated, a person may go into unconsciousness or coma. It is possible to treat this kind of diabetes with regular insulin injections.

It is not known what causes type I diabetes, although both genetic and environmental factors are involved.

**Type II diabetes** usually develops in adults over the age of 40, but there have been recent concerns that some children are developing this kind of diabetes. This is the commonest kind of diabetes. In this kind of diabetes, people may not be producing enough insulin, but they are also insensitive to the insulin that is produced. People who have regularly eaten a diet high in sugars are most likely to develop insensitivity, but there are other causes of this, too. This kind of diabetes is not usually treated using insulin, as the body does not respond properly to this. People with this kind of diabetes are advised to control their diabetes by exercise and diet. They are advised to eat more complex carbohydrates (such as starch) because these are digested slowly into glucose, avoiding a rapid rise in glucose levels in the blood. They are also advised to eat less sugar and salt, and to eat more fruit and vegetables.

A number of factors can trigger type II diabetes. These include old age, obesity, high blood pressure, physical inactivity and a family history of diabetes.

Some experts estimate that there could be hundreds of children with type II diabetes in the UK. Most of these cases are among the Asian population. Scientists think this is because of a combination of genetic factors, diet and lack of exercise. However, the scientists emphasise that the white population is also at risk. They suggest that clearer food labelling would help, and that schools should stop selling junk food.

### Health risks associated with diabetes

Uncontrolled diabetes leads to high blood glucose levels. Over a period of time, this can cause damage to nerve cells; damage to the retina of the eye that can lead to blindness; kidney failure; high blood pressure; heart disease; strokes; chronic infections: wounds that will not heal; and other conditions.

## Child diabetes time-bomb warning

Type II diabetes, more associated with adults, is much higher among children than thought, figures suggest. Research in east London has revealed 22 children under 16 have type II diabetes, which is largely caused by obesity. The findings suggest the latest audit of type II diabetes in under 16s two years ago, which identified 100 cases in the UK, is a gross underestimate.

Experts said there may be up to 1500 cases nationally now and warned the UK was sitting on a 'time-bomb'.

Ten years ago type II diabetes was unheard of among children, but it has begun to emerge as the obesity epidemic has exploded. All but one of the cases identified in east London by Barts and the London and Newham University Hospital NHS trusts were from the ethnic minority population, which generally has a higher prevalence of type II diabetes.

Doctors fear that, unless children are educated to eat a healthy diet and take up physical activity, the number of children with the condition will continue to increase.

People with type II diabetes, also known as non-insulin-dependent diabetes, do not make enough insulin or cannot make proper use of it. Type I diabetes, which is more common among children, is not linked to obesity.

Story adapted from BBC NEWS:
http://news.bbc.co.uk/go/pr/fr/-/1/hi/health/4462111.stm
Published: 2005/04/19

### Measuring blood glucose levels

A person with diabetes should try to keep their blood glucose concentration at a safe level. In a healthy person, blood glucose levels stay within narrow limits throughout the day: 4–8 mmol dm$^{-3}$. Blood glucose levels are higher after meals and usually lowest in the morning. A person with diabetes should try to make sure that their blood glucose concentration is:

- 4–7 mmol dm$^{-3}$ before meals
- less than 10 mmol dm$^{-3}$ 90 minutes after a meal
- around 8 mmol dm$^{-3}$ at bedtime.

A person with diabetes can use a biosensor to measure blood glucose levels. The biosensor contains a test strip with an enzyme on it. This enzyme is called glucose oxidase. This enzyme converts any glucose in the blood to gluconolactone. As it does this, a small electric current is produced. This is picked up by an electrode on the test strip, and a reading for blood glucose concentration is produced on a digital display screen.

You can see how to use a biosensor in Figure 1. First, you place a test strip in the glucose test meter. Next, you should disinfect your skin with alcohol. Then use a sterile lancet inside a device that looks like a pen to prick your skin. This produces a small drop of blood, which you should squeeze onto the test strip. After about 30 seconds, you will see a reading for blood glucose concentration on the digital display screen. The glucose test meter has a memory that can store measurements, so you will have a record of recent changes in blood glucose concentration.

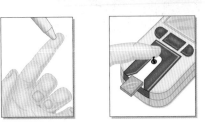

**Figure 1** Testing blood glucose concentration

## The link between obesity and type II diabetes

An investigation was carried out to find out whether obesity causes type II diabetes. Scientists calculated the body mass index of the people taking part in the study, and also their waist-to-hip ratios. The people were then studied for the next 13 years. The percentage of people in each group who developed type II diabetes was recorded. The results are shown in Figure 2.

- Describe the trends shown in this graph.
- What would you need to know to decide whether the results of this investigation are reliable?
- Does this investigation show that poor diet causes type II diabetes? Explain your answer.

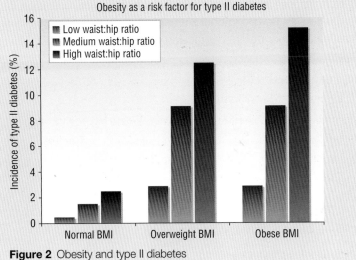

**Figure 2** Obesity and type II diabetes

## Questions

1  One symptom of diabetes mellitus is feeling thirsty. Use your knowledge of osmosis to explain why a person with diabetes feels thirsty.
2  Explain how diabetes mellitus can lead to high blood pressure.
3  A person with type II diabetes is advised to eat less salt. Suggest why.
4  Explain how exercise may benefit a person with diabetes.
5  Explain why a glucose test meter will detect the levels of glucose in the blood, but not any other substance.
6  How could scientists carry out a study to find out whether it is genetic factors or dietary factors that make Asian children more likely to develop type II diabetes?
7  How could clearer food labelling help to reduce the spread of type II diabetes?
8  How can junk food increase the risk of developing type II diabetes?

**Examiner tip**

You will meet diabetes again in A2 when you look at the role of hormones. There are also links here to water potential, carbohydrate structure and the role of blood plasma in transporting solutes. Watch out for synoptic questions which link these ideas.

## The fasting blood glucose test

The person being tested has to eat and drink nothing (except water) for 8 to 12 hours before taking this test. A blood sample is then taken. The results of the test are interpreted according to the table below:

| Blood glucose level 2 hours after 75 g glucose drink/mmol dm$^{-3}$ | Interpretation |
| --- | --- |
| 3.6–6.0 | Normal glucose tolerance |
| 6.1–6.9 | Impaired glucose tolerance |
| 7.0 and above | Probable diabetes |

**Table 1**

### Glucose tolerance test

The most common glucose tolerance test is the oral glucose tolerance test (OGTT). The patient is asked to eat normally for several days leading up to the test, but not to eat or drink anything after midnight on the evening before the test. When the person arrives for the test, a blood sample will be taken. The person will then be asked to drink a liquid containing a measured amount of glucose (usually 75 g). More blood samples will be taken every 30 minutes until 3 hours have passed. The results are interpreted according to the table below.

| Blood glucose level 2 hours after 75 g glucose drink/mmol dm$^{-3}$ | Interpretation |
| --- | --- |
| Less than 7.8 | Normal glucose tolerance |
| 7.8–11.0 | Impaired glucose tolerance |
| More than 11.1 | Probable diabetes |

**Table 2**

Figure 1 shows some typical results from a glucose tolerance test.

However, a high glucose level might indicate another medical condition, and some other drugs the patient might be taking could interfere with the test. A doctor might therefore need to investigate further before diagnosing diabetes mellitus.

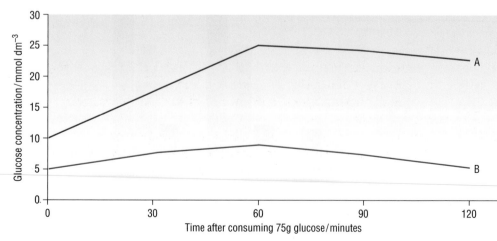

**Figure 1** Results from a glucose tolerance test on two patients

### The growing problem of diabetes

Diabetes is a growing problem in many countries. Look at Figure 2. You can see that the number of cases of diabetes has increased in all the countries shown on the graph.

This increase is blamed on growing wealth in many countries. This means that there is plenty of food available, and people are more likely to eat refined, processed foods. Also, more people have cars and so take less exercise.

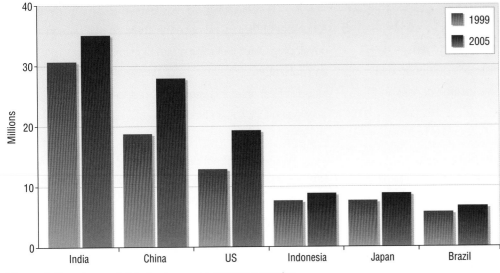

**Figure 2** Prevalence of diabetes in several different countries

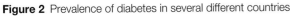

### Ethnicity and diabetes

The table shows the rates of diabetes among different ethnic groups in the UK.

| | Ethnic group | | | | | |
|---|---|---|---|---|---|---|
| | White | African Caribbean | South Asian | Indian or African Asian | Pakistani or Bangladeshi | Chinese |
| Diabetes prevalence (%) | 1.7 | 5.3 | 6.2 | 4.7 | 8.9 | 3 |

Do these data show that diabetes is likely to be controlled by genetic factors? Give reasons for your answer.

## Questions

1 Explain why it is important for people to have a period without eating or drinking anything except water before taking the glucose tolerance test.

2 Explain why blood samples are taken at 30-minute intervals after consuming glucose in the glucose tolerance test.

3 Figure 2 refers to 'prevalence' of diabetes. Explain what this term means.

4 Calculate the percentage increase in prevalence of diabetes in China and Brazil between 1999 and 2005. Suggest reasons for the difference.

## Self-check questions

### Fill the blanks

1 .................... diseases are caused by pathogens, e.g. TB or HIV/AIDS. .................... diseases do not spread from one person to another. Examples include cancer and .................... heart disease.

2 If the .................... lining of an artery becomes damaged, a fatty plaque called .................... can build up. Fatty deposits build up .................... the damaged endothelium. Atheroma builds up very slowly over many years. Eventually, the plaque breaks open. Blood cells enter the crack and a blood clot, or ...................., starts to form. A piece of the clot may break off and block an artery where it is narrowed by atheroma. If the coronary artery becomes blocked, a heart attack or .................... may occur. Some of the heart muscle cells may die, because they do not receive enough glucose or oxygen.

3 .................... occurs when the .................... arteries become narrowed by .................... . This causes .................... pain when a person starts to exercise, as the heart muscle cannot obtain enough glucose and oxygen for .................... through the narrowed .................... arteries.

4 Cardiac .................... occurs when the heart stops beating efficiently enough to pump blood round the body. Myocardial infarction is one cause of cardiac .................... When a person has suffered cardiac arrest, no .................... can be felt and the person loses .................... . The first aid treatment for a patient who has suffered cardiac arrest is cardio-pulmonary resuscitation (.................... ).

5 A number of factors increase the chances of a person developing CHD, including a diet high in .................... and saturated .................... . However, .................... found in fresh fruit and vegetables reduce the chances of developing CHD. High blood...................., lack of physical ...................., .................... influences and .................... are also causes of CHD. Obesity is another risk factor for CHD. A person can find out if they are obese by calculating their .................... (BMI). This is your weight (in kg), divided by your height (in metres) squared. Another way to measure body size is the .................... to hip ratio.

6 Tobacco smoke contains a number of ...................., i.e. chemicals that cause cancer. These .................... cause mutations in the genes of epithelial cells that line the lungs, leading to the formation of .................... .

7 During an asthma attack, the membranes lining the airways release .................... and become inflamed. Mucus blocks the airways and leads to .................... . This is called .................... Most asthma is triggered by ...................., which are substances capable of causing an allergic reaction.

8 .................... is a disease in which the person cannot control their blood glucose level. Type I diabetes, or .................... Dependent Diabetes, is usually diagnosed in children and young adults. The .................... system has attacked the insulin-producing cells in the .................... so that the pancreas produces very little or no insulin. Type II diabetes usually develops in .................... over the age of 40 and has been called Mature Onset Diabetes. However, there have been recent concerns that some children are developing this kind of diabetes

# Summary questions

**1** It is estimated that 3.4 million people in the United Kingdom have asthma.

Figure 1 shows the number of new cases of asthma diagnosed by doctors between the years 1976 and 2000, in two age groups.

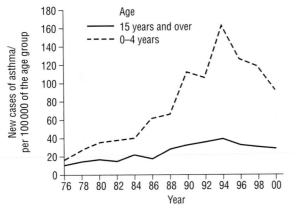

**Figure 1**

**(a)** Using information in the figure,

**(i)** Suggest why the figures are presented 'per 100 000 of the age group'. [1]

**(ii)** Describe the trends in the number of cases diagnosed between 1976 and 2000 in both age groups. Suggest reasons for these trends. [5]

**(b)** State two possible causes of an asthma attack. [2]

**(c)** Beta agonists are often used to relieve the symptoms of asthma. Describe how these drugs work. [3]

**2 (a)** Lung disease may be described as *chronic* or *acute*. Explain, using examples, the meaning of these terms. [3]

**(b)** Describe the long-term effects of smoking on the respiratory system. [6]

**3 (a)** Respiratory arrest can occur for a variety of reasons. State two possible causes. [2]

**(b)** The figure shows the use of the mouth-to-mouth method of resuscitation (ventilation).

Suggest how it is possible for exhaled air to resuscitate an unconscious person when this method is used. [2]

**4** Coronary heart disease (CHD) is one of the most common causes of premature death in the United Kingdom. Evidence has shown that a high level of saturated fat in the diet increases the risk of CHD.

**(a)** What causes coronary heart disease? [2]

**(b)** The table shows the number of deaths from CHD in four countries.

| Country | Deaths per 100 000 |
|---|---|
| Ukraine | 393.8 |
| Romania | 198.6 |
| United Kingdom | 150.4 |
| Japan | 35.7 |

**(i)** Suggest why the figures in Table 1 are quoted as deaths per 100 000. [1]

**(ii)** State two reasons, other than diet, for the differences in death rate from CHD in these four countries. [2]

**5 (a)** Describe how a glucose biosensor may be used to monitor blood glucose level. [3]

**(b)** People with untreated diabetes feel thirsty and need to drink a lot. Suggest and explain a reason for this. [2]

**(c)** A person with type 2 diabetes cannot normally be treated with insulin injections. Instead, they are advised to control their diabetes by exercise and diet. Suggest how they would do this. [3]

## Question 1

Dividing cells enter the cell cycle in G1. The cell cycle finishes when the cytoplasm splits during cytokinesis to form two new daughter cells.

(a) Complete the table by inserting the stages of the cell cycle in the correct order.

| 1 | G1 |
|---|---|
| 2 | |
| 3 | |
| 4 | |
| 5 | cytokinesis |

[3]

(b) As cells pass through the cell cycle, check points occur. At the end of G1, DNA in the cell is checked for damage before the cell progresses into the next phase of the cycle. If DNA is too damaged to repair, a protein called p53 will trigger apoptosis.

State what is meant by the term 'apoptosis'. [1]

(c) The gene which codes for p53 protein is called a tumour suppressor gene.

(i) Explain what is meant by the term 'tumour'. [2]

(ii) Distinguish between a malignant and a benign tumour. [3]

(d) Damage to the p53 gene can lead to DNA damage being undetected.

(i) State two possible causes of damage to the p53 gene. [2]

(ii) Explain why damage to the p53 gene could result in cancer. [4]

## Question 2

Figure 2.1 shows an ultrasound scan of a foetus taken during gestation (pregnancy). A scale has also been given

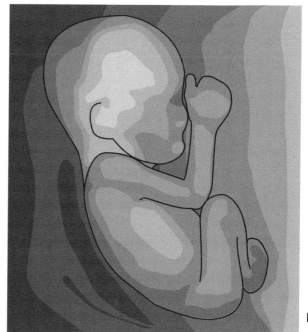

2 cm
Scale

**Figure 2.1**

(a) Measurements taken of biparietal diameter and crown–rump length taken during ultrasound scans can be used to determine the gestational age of the fetus.

(i) Explain what is meant by 'biparietal diameter'. [1]

(ii) Table 2.1 shows the relationship between the crown–rump length and the gestational age of the fetus

| Gestational age/weeks | Crown–rump length/mm |
|---|---|
| 16 | 112 |
| 20 | 160 |
| 24 | 208 |
| 28 | 242 |
| 32 | 277 |

Using the information in Figure 2.1, estimate the gestational age of the fetus in the ultrasound picture. [3]

(b) The Dietary Reference Values for energy, protein, calcium, iron and folic acid change during pregnancy.

Describe the changes in DRVs during pregnancy for the nutrients listed above and explain the reasons for the changes. In this question, one mark is available for the correct use of technical terms. [7]

(c) Pregnant mothers are also advised not to drink during pregnancy. Describe the possible effect of alcohol on the growth of the fetus. [4]

# Question 3

Antibiotics are only effective against bacterial cells. Antibiotics can be described as bacteriocidal when they kill bacterial cells, or they can be described as bacteriostatic. Bacteriostatic antibiotics do not necessarily kill bacterial cells but they do prevent them dividing. Figure 3.1 is a graph illustrating the effect of a bacteriostatic antibiotic. The antibiotic was added at the point marked X on the graph.

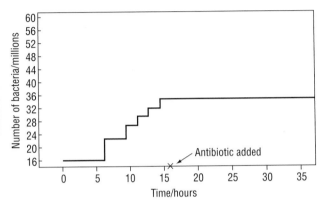

**Figure 3.1**

(a) With reference to Fig 3.1, describe the effect of adding the antibiotic on the bacterial population. [2]

(b) Bacterial cells are prokaryotic cells. Describe the differences between bacterial cells and a eukaryotic cell such as a neutrophil. One mark is available for the quality of written communication. [7]

(c) Suggest why bacteriostatic antibiotics are effective in treating bacterial infections even though they do not kill the bacteria. [3]

## Examiner tip

Questions which start with the command word 'suggest' tend to be targeted at high achieving students. This one requires time to reason why and builds on your knowledge of the immune system and the delayed response.

# Question 4

Coronary heart disease (CHD) is a major cause of death in developed countries. Several epidemiological studies have been carried out to determine the risk factors associated with coronary heart disease. One such study is the Framingham study.

This study was started in 1948. A total of 5209 between the ages of 30 and 62 were recruited to the study. A second generation of participants was recruited in 1971 with a further 5124 participants. These were the adult children of the first group along with their partners. The participants were questioned every two years to assess their lifestyle by questionnaire. In addition, they are given a detailed medical examination.

(a) Using the information given above, explain what you mean by an epidemiological study. [2]

(b) The Framingham study identified several lifestyle factors which increased the risk of CHD.

   (i) List four lifestyle factors which are known to increase the risk of CHD. [4]

   (ii) Suggest a reason for including the children of the original participants in the study. [2]

(c) CHD can be treated using a range of medical procedures including angioplasty.

   Describe how angioplasty is carried out and discuss the social and economic advantages of angioplasty compared with other forms of surgical treatment for CHD.

## What is energy used for in the body?

All living organisms use energy. The most common energy source is adenosine triphosphate (ATP) which is used in every cell to carry out processes such as active transport, movement, anabolic reactions and the maintenance of body temperature.

### The structure of ATP

ATP is a high energy molecule which is called a phosphorylated nucleotide. It is made up of an organic nitrogenous base, adenine, a 5-carbon ribose sugar and 3 phosphate groups. ATP is water soluble and easily transported within a cell. An important feature of an ATP molecule is the presence of free electrons that surround the phosphate groups. These electrons give the molecule its high energetic potential.

### Key definitions

**Phosphorylation** is the process that involves the addition of a phosphate group to a molecule. An example of this is the phosphorylation of ADP to ATP. There are three types of phosphorylation, substrate linked and oxidative in respiration and photophosphorylation in photosynthesis. So the synthesis of ATP is linked to three types of reactions.

**Hydrolysis** is the breaking down of molecules by the addition of water.

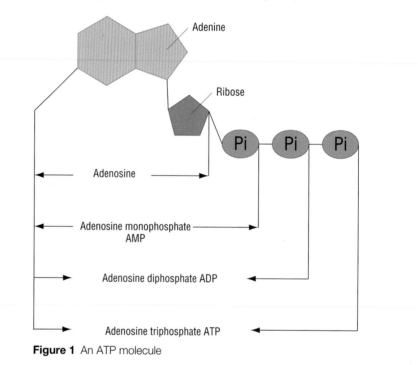

**Figure 1** An ATP molecule

### ATP as a source of energy

The phosphate 'tail' on the ATP molecule is the main source of energy. The removal of a phosphate group by the process of hydrolysis releases energy. ATP is hydrolysed to ADP, releasing 30.5 kJ of energy. This molecule can be further broken down to AMP, again releasing a small amount of energy. The energy released as a result of this hydrolysis can be channelled directly into other molecules and used by cells. A certain proportion of this is lost as heat.

This release of energy is just enough to drive chemical reactions in the body. ATP is continually broken down and reformed at a rate of anything up to 8000 cycles per day.

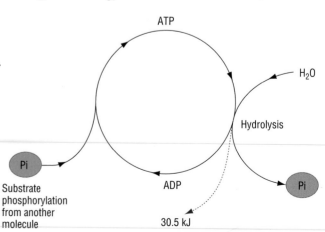

**Figure 2** ATP hydrolysis

## How is ATP synthesised?

ATP is used by every living organism and can be synthesised in many different ways. It can be made on the thylakoid membranes in chloroplasts by a process called photophosphorylation, using light to drive the reaction. It takes place in the cytoplasm of cells by a process called substrate-level phosphorylation and also on the inner membranes of a mitochondria during oxidative phosphorylation.

Different types of production produce different amounts of ATP and have different biochemical requirements and may or may not require oxygen. Photophosphorylation and oxidative phosphorylation both synthesise ATP using the enzyme ATP synthase, often referred to as ATPase.

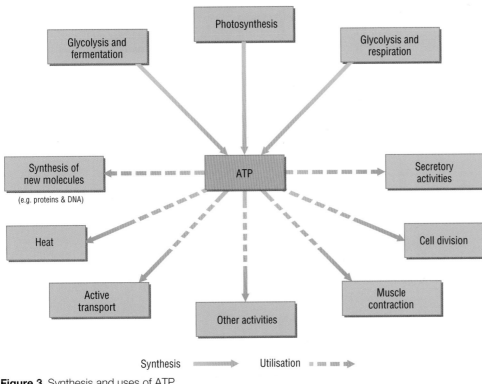

**Figure 3** Synthesis and uses of ATP

# Questions

1 Explain how an ATP molecule is similar to that of DNA or RNA.
2 Describe how the hydrolysis of ATP helps maintain the core body temperature.
3 Explain why it is an advantage for an organism to hydrolyse ATP to meet its energy requirements rather than hydrolyse glucose directly.

## Summary of cellular respiration

Aerobic respiration involves several key stages and occurs in the cell cytoplasm, the matrix and the cristae of the mitochondria. It is not possible to break down glucose and directly produce ATP, and so a series of metabolic pathways takes place that eventually leads to the production of many molecules of ATP. The stages include glycolysis, the link reaction, Krebs cycle and oxidative phosphorylation.

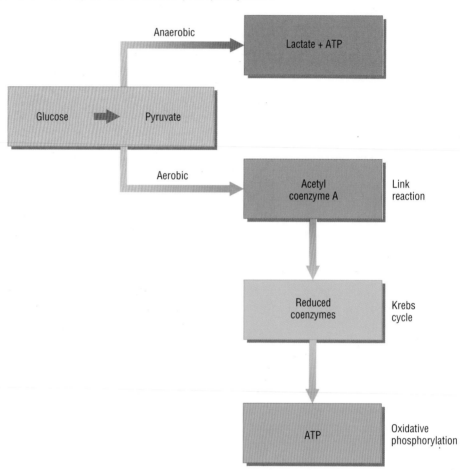

**Figure 1** Stages in cellular respiration

### The role of enzymes and coenzymes in respiration

The breakdown of glucose involves the action of many different enzymes and coenzymes. Decarboxylase and dehydrogenase enzymes are involved throughout the process of respiration. Dehydrogenase enzymes remove hydrogen from other molecules and make this available and able to be passed on to molecules such as coenzymes. Decarboxylase enzymes hydrolyse the carboxyl group (COOH) from a molecule, usually producing carbon dioxide. Coenzymes such as NAD and FAD act as hydrogen acceptors for the dehydrogenase enzymes.

NAD and FAD enable potential energy to be transferred from one molecule to another. Coenzyme molecules are important because they can be both oxidised and then reduced. They exist in a finite quantity in the cell and many are synthesised from vitamins – particularly the B group vitamins.

## Glycolysis

This process takes place in the liquid part, the cytosol, of all cells and involves the splitting of glucose to produce ATP, pyruvate and reduced NAD.

Glucose enters the cytoplasm by either facilitated diffusion or active transport and to ensure that it does not leave the cell it must first be chemically altered. The glucose is initially altered by becoming phosphorylated using up two molecules of ATP. This phosphorylated molecule is then gradually broken down through a series of enzyme steps to release four molecules of ATP by the process of substrate-level phosphorylation, two molecules of reduced NAD and two molecules of the 3-carbon compound pyruvate.

Pyruvate which, unlike glucose, can enter the mitochondria, is able to be further broken down to produce ATP, as we shall see in the next section.

The energy stored in the reduced NAD has the potential to make ATP in the later stages of the respiratory pathway. In summary, the role of glycolysis is not to produce lots of energy but to activate the glucose molecule and so prepare it for further breakdown and subsequent ATP production.

## Link reaction

Pyruvate, the product of glycolysis, is a molecule with lots of potential energy that can be channelled into ATP synthesis. The fate of the pyruvate however depends on the presence or absence of oxygen.

With oxygen present, the pyruvate is actively transported into the mitochondria whereby it undergoes the link reaction and combines with a substance called coenzyme A (co-A).

Pyruvate is dehydrogenated with the aid of NAD and dehydrogenase enzymes, and decarboxylated using decarboxylase enzymes to release hydrogen atoms and $CO_2$ gas. The result of these reactions is to produce a 2-carbon molecule called acetyl coenzyme A, which is fixed in the matrix of the mitochondria and can then enter the next stage of respiration, called the Krebs cycle. The hydrogen is accepted by NAD, forming reduced NAD.

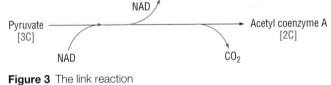

**Figure 3** The link reaction

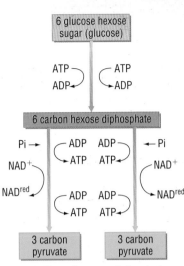

**Figure 2** Flow diagram of glycolysis

# Questions

1  Explain why glycolysis can be described as a metabolic pathway.
2  What does the term glycolysis mean? (Hint: split the word up into glyco and lysis.)
3  What is the importance of substrate-level phosphorylation to a cell?
4  Copy and complete the table summarising the processes involved in glycolysis and the link reaction.

| | Starting molecules | Products of reactions | Site of reactions | Oxygen required |
|---|---|---|---|---|
| Glycolysis | Glucose, ATP, NAD | | | No |
| Link reaction | | Acetyl co-A, $CO_2$, reduced NAD | | |

## The Krebs cycle

This is a series of chemical reactions that take place in the matrix of the mitochondria. These result in the complete breakdown of the acetyl group into carbon dioxide, the removal of the hydrogen atoms to form more reduced coenzymes, and the synthesis of more ATP directly.

The first step in this pathway involves the combination of the 2-carbon acetyl coenzyme A with a 4-carbon acceptor molecule called oxaloacetate. This reaction produces a 6-carbon intermediate called citrate. This is rapidly decarboxylated in a series of enzyme-linked steps to regenerate the 4-carbon compound. It is also dehydrogenated during these stages, releasing hydrogen atoms and forming reduced NAD and reduced FAD.

The regenerated 4-carbon compound can combine again with further acetyl coenzyme molecules, starting the cycle over again.

### Examiner tip

The Krebs cycle sequence was identified and explained by the British scientist Hans Krebs in 1937. This process involves many steps, each driven by a specific enzyme. These enzymes can be inhibited by certain compounds, and this gave scientists tools to investigate the process. You are not expected to understand each step of the cycle but to appreciate the overall importance. That is, to reduce the 2-carbon molecule, produce a little ATP and form more reduced coenzyme as a potential energy source – a source of high energy electrons which will be used to form ATP in oxidative phosphorylation.

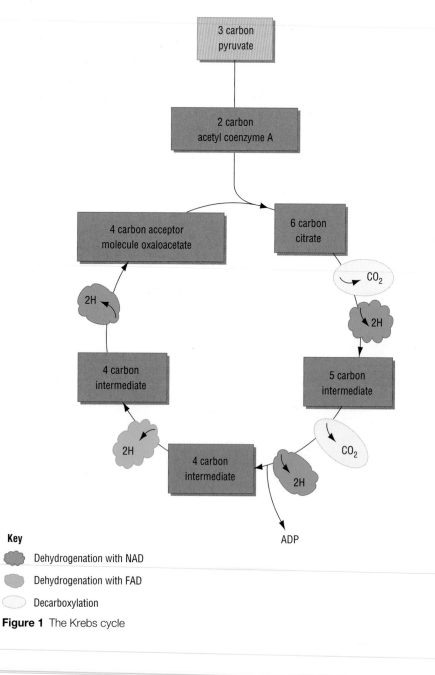

**Key**

- Dehydrogenation with NAD
- Dehydrogenation with FAD
- Decarboxylation

**Figure 1** The Krebs cycle

# Importance of Krebs cycle

The Krebs cycle breaks down the product of the link reaction, acetyl co-A to $CO_2$ and also releases many hydrogen atoms by the action of decarboxylase and dehydrogenase enzymes. This process of dehydrogenation with the help of coenzyme hydrogen carriers is important for the latter stages of ATP synthesis. Acetyl co-A can be produced from sources other than the link reaction, such as amino acids and fatty acids. The outcome of the Krebs cycle, irrespective of the original source of acetyl co-A, is:

- three molecules of reduced NAD
- one molecule of reduced FAD
- one molecule of ATP produced by substrate-level phosphorylation
- two molecules of $CO_2$
- one molecule of regenerated oxaloacetate.

## Control of the Krebs cycle

A similar process occurs in the enzymes involved in the Krebs cycle, in particular the dehydrogenase enzymes that are responsible for reducing the NAD and FAD as stores of potential energy for synthesis of more ATP.

The first three steps in the cycle are inhibited by high levels of ATP, then the following enzymes are inhibited by high levels of reduced coenzyme. The result is to stop the cycle and so allow the breakdown of substrates to occur only when needed. The presence of large quantities of citrate also inhibits the continual glycolysis of glucose and so regulates the amount of substrate flowing through the pathways.

This type of inhibition that occurs on these enzymes is called allosteric inhibition – the molecule acting as an inhibitor binds temporarily somewhere other than the active site, but the active site shape is changed.

**Examiner tip**

Enzymes in the Krebs cycle – you should be able to describe the action of enzymes on substrates and also the effect of inhibitors from AS work. The action of enzymes is crucial to the workings of this cycle and you should be prepared for synoptic-linked questions in your exams.

# Questions

1 How many molecules of $CO_2$ are produced from glucose during the link reaction and the Krebs cycle?
2 Explain the importance of coenzymes and dehydrogenase enzymes in the link reaction and Krebs cycle.
3 Describe precisely the location of the reactions of the Krebs cycle and explain why this does not occur in the cytoplasm of the cell.
4 Explain how allosteric inhibition differs from competitive inhibition. (Hint: recap AS material.)

Using the potential energy in reduced coenzymes to make ATP is called oxidative phosphorylation. This process occurs on the cristae, the inner membrane of the mitochondria and involves a series of interlinked enzyme reactions.

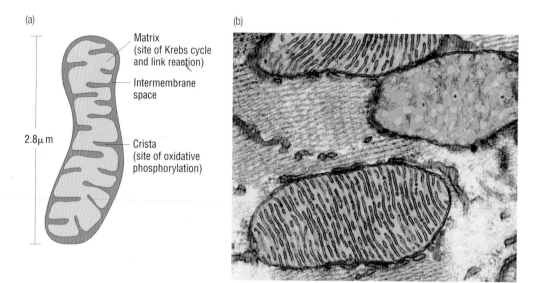

**Figure 1 a** Mitochondria and site of oxidative phosphorylation; **b** electronmicrograph of mitchondria

Oxidative phosphorylation involves the transfer of electrons along membrane-bound carriers called cytochromes. The energy released is captured by phosphorylating ADP to ATP.

This process can be described simply as follows (see numbering in Figure 2):
* Reduced NAD and FAD lose their hydrogen atoms and become oxidised when they come into contact with the cytochrome carriers situated on the cristae. **(1)**
* The hydrogen atoms split up into electrons and protons.
* Electrons pass along the cytochrome carriers and the energy released is used to pump protons (H+) into the intermembrane space. **(2)**
* Protons create an electrochemical gradient as they cannot pass back through the membrane.

* Protons which have built up in the intermembranous space can diffuse through specialised protein channels back into the matrix. **(3)**
* These channels contain a structure on the matrix side which consists of the enzyme ATP synthase and is where ADP is phosphorylated to form ATP. **(4)**
* As the protons flow through these channels they 'turn' this enzyme and fuel the process of ATP production. The 'used' protons then combine with electrons and oxygen atoms to form water.
* The electrons move along the membrane carriers until they reach this final carrier molecule oxygen. **(5)**
* This process is referred to as chemiosmosis.

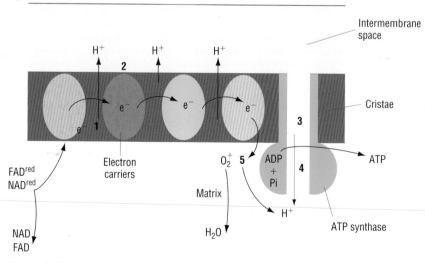

**Figure 2** Diagram of oxidative phosphorylation

# Evidence for chemiosmosis

The Nobel Prize winner Peter Mitchell carried out two experiments to demonstrate the process of ATP production by oxidative phosphorylation in mitochondria. These experiments proved that a gradient of hydrogen ions or protons was needed to synthesise ATP and that the process required the enzyme ATP synthase.

In the first experiment, mitochondra without a source of potential energy (pyruvate or acetyl co-A) were fooled into making ATP when researchers raised the hydrogen ion concentration in mitochondria, which had previously been in a low hydrogen ion concentration. The outer membrane is permeable to hydrogen ions, so these moved into the intramembranous space, creating a gradient between here and the matrix, and ATP was made.

For the second experiment a 'proton pump' was isolated from a bacterium and added to membrane vesicles. This 'pumped' hydrogen ions into the vesicles. When the enzyme ATP synthase was inserted into the vesicles as well, ATP was made even though the electron carriers were absent.

For chemiosmosis to work, ATP synthesis must be 'coupled' with the formation of a hydrogen ion gradient. If the ATP synthase is replaced by another molecule, the energy stored in the gradient is released as heat. This happens in newborn infants, in the so called 'brown adipose tissue' where the mitochondria release heat rather than generate ATP.

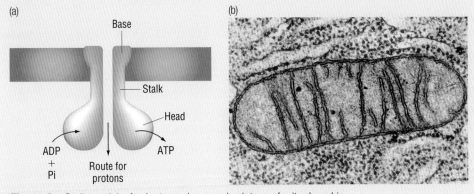

Figure 3a Stalk particle; b electronmicrograph picture of mitochondria

# Questions

1 How is a mitochondria adapted to synthesise ATP?
2 Describe the role of oxygen in oxidative phosphorylation.
3 Summarise the process of oxidative phosphorylation.
4 Explain why the protons which build up in the intermembrane space cannot pass back through the inner membrane unless they go through specific protein channels. (Hint: think back to polar molecules from AS and membrane structure.)
5 Calculate the net gain in ATP molecules from the electron transport chain in oxidative phosphorylation, assuming the total gain from glycolysis, the Krebs cycle and all other processes is 35 molecules of ATP.

## Continual production of ATP

In aerobic respiration, most ATP is produced by oxidative phosphorylation using the electron transport chain coupled to ATP synthesis. Oxygen is needed as the terminal acceptor for the electrons and protons. The presence of oxygen means that ATP can be replaced as it is used. It is important that oxygen is available to continually accept the electrons and protons from the stalked particles and electron carrier systems.

A summary of where ATP comes from in aerobic respiration is given below in the table.

In aerobic respiration, the ATP produced at each of the metabolic stages in the breakdown of one molecule of glucose is as follows:

| Glycolysis | 2 (by substrate level phosphorylation) |
|---|---|
| Krebs cycle | 2 (by substrate level phosphorylation) |
| Electron transport chain | 28 (by oxidative phosphorylation) |

The number of ATP molecules produced during aerobic respiration varies depending upon the physiological and biochemical conditions of the cells releases the energy. Up to 34 molecules of ATP can be produced. This is based on conditions of pH7 with excess substrate – optimum conditions which are rarely found in cells.

### Using other substrates to make ATP

Lipids and proteins can also be used to generate ATP and Figure 1 shows where these substrates enter the respiratory pathway.

Lipids in the form of triglycerides are broken down into glycerol and fatty acids. These fatty acids are then further broken down, producing 2-carbon acetyl fragments. These can then combine with coenzyme A. Glycerol can also be used as a fuel in glycolysis. The potential to generate ATP molecules from lipids is high, far greater than that of glucose. For example, one molecule of stearic acid can yield 146 molecules of ATP.

Proteins cannot be stored in the body and are broken down in the liver by a process called deamination. This splits the amino acid molecules, releasing an amino group that enters the ornithine cycle and is incorporated into urea and an organic acid molecule that can enter the Krebs cycle, and so generates energy. The number of ATP molecules produced by these organic acids is fewer than both glucose and lipids.

### Metabolic poisons

Scientists have searched for safe and effective diet pills for many years and early in the 1930s a breakthrough was close. A compound called 2,4-dinitrophenol (DNP) was seen to have dramatic effects on overweight patients. Weight loss was dramatic, with a huge decrease in body fat. Unfortunately the patients became very listless, had very little energy and many died.

DNP acted to prevent a buildup of protons in the intermembrane space, so effectively stops oxidative phosphorylation. The electron transport chain was working and oxygen consumption increases to try and provide ATP. Carbohydrates, fats and protein were still broken down and the patients lost weight initially, however the energy released was not being captured as ATP but lost as heat.

### Examiners tip

It is important for you to appreciate that the process of respiration is a continuous metabolic pathway. The product of one reaction becomes the substrate for the next reaction step and so on. As a result of this, if the system was to stop or be inhibited in some way, it will result in ATP synthesis stopping, hence the reactions that use ATP will stop. This is a key principle to the workings of many metabolic poisons such as cyanide. This blocks the electron transport chain so protons are not pumped and ATP is not made.

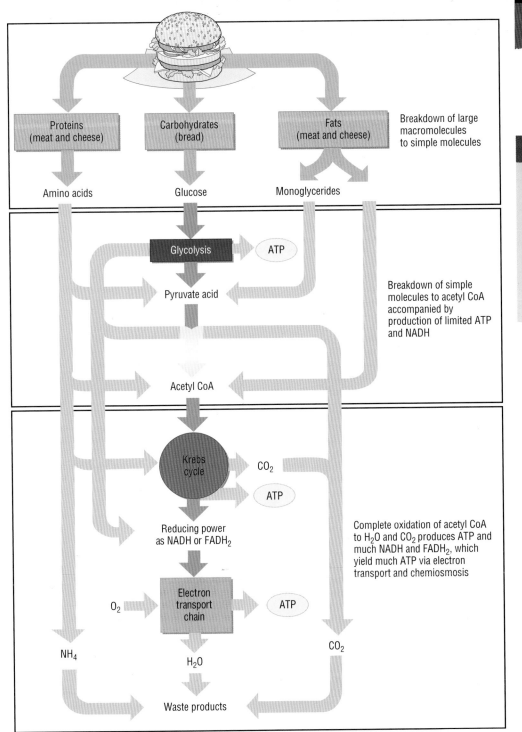

Breakdown of large macromolecules to simple molecules

Breakdown of simple molecules to acetyl CoA accompanied by production of limited ATP and NADH

Complete oxidation of acetyl CoA to $H_2O$ and $CO_2$ produces ATP and much NADH and $FADH_2$, which yield much ATP via electron transport and chemiosmosis

**Figure 1** Summary diagram of aerobic respiration

# Questions

1 Which steps in the aerobic pathway cannot continue without the involvement of oxygen?

2 Describe the importance of enzyme inhibitors in explaining the theory of chemiosmosis.

3 Cells are described as having a finite quantity of NAD and FAD. What is the importance of this to the cell?

4 Describe four uses of ATP to a cell.

5 FAD and NAD are derived from vitamin B. What symptoms would someone with a vitamin B deficiency have?

## Anaerobic production of ATP

In the absence of oxygen, there is no final acceptor for hydrogen to be passed to and so NAD and FAD will not be regenerated. The result is that oxidation will be blocked and the link reaction, Krebs cycle and the electron transport chain will all stop.

Anaerobic conditions occur in any cell deprived of oxygen. At the start of physical exercise the circulatory system cannot work quickly enough to supply the muscle cells with sufficient oxygen. These cells still need to generate some ATP, but do not have sufficient oxygen to do so using aerobic respiration. Therefore, in the absence of oxygen, the only energy available to the cell is from glycolysis and through substrate-level phosphorylation. For this to continue, it is important to remove the pyruvate and to recycle the NAD molecule in order to make more ATP.

The NAD is recycled by passing on the hydrogen to a different hydrogen acceptor – pyruvate. The pyruvate becomes reduced to lactate and the NAD is made available to pick up another hydrogen so glycolysis can continue.

Lactate build-up in muscle cells could inhibit glycolysis so that eventually even this supply of ATP stops. To prevent this, the lactate can be oxidised back to pyruvate by the enzyme lactate dehydrogenase. This is present in liver and muscle cells.

### The fate of lactate

Most of the lactate produced will leave the muscle cells and enter the blood. The build-up of lactate in the blood can lead to 'lactic acidosis' but lactate is also a valuable source of potential energy. To 'rescue' the lactate for further use it is taken to the liver. Liver cells have up to 50 times the levels of the enzymes necessary to convert the lactate firstly back to pyruvate, and from this back to glucose. The glucose can be returned to the muscles via the blood or stored as glycogen. This process is referred to as the Cori cycle and is shown in Figure 2.

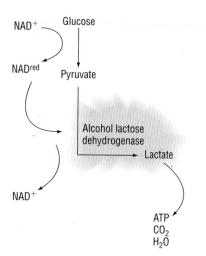

**Figure 1** Flow diagram of anaerobic respiration

### Key definition

**Oxygen debt** is the amount of oxygen taken in by the body above that required for resting metabolism after exercise. It is also known as 'excess post-exercise oxygen consumption', or EPOC.

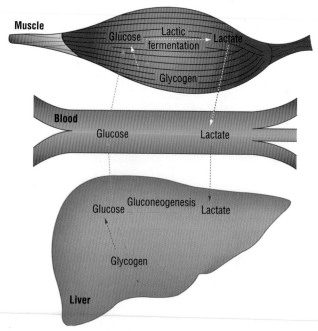

**Figure 2** Diagram of Cori cycle

## Other anaerobic pathways

In other organisms such as yeast, it is not pyruvate that acts as the hydrogen acceptor in anaerobic respiration, but a compound called ethanal. Yeast is a eukaryotic cell and produces pyruvate from glycolyis as human cells do. Ethanal is made by decarboxylation of pyruvate. The ethanal then picks up the hydrogen from the reduced NAD and ethanol is formed. A few grams of ethanol are produced by microorganisms in the large intestine in humans every day! The enzyme responsible for this reaction is alcohol dehydrogenase since, in the reverse direction, it is responsible for removing the hydrogen from ethanol. This is an important enzyme in human liver cells.

Yeast is a useful organism to use in investigating factors that affect the rate of respiration.

### Experiment using redox indicators and yeast

Using an indicator such as triphenyl tetrazolium oxide (TTC) that act as artificial hydrogen acceptors, respiration in organisms such as yeast can be investigated. During both aerobic and anaerobic respiration, hydrogen atoms are transferred to molecules such as NAD and FAD that act as hydrogen acceptors. TTC can act in a similar way to NAD and FAD and will change colour when reduced by hydrogen atoms.

Hydrogen atoms from substrate + TTC (colourless) $\rightarrow$ reduced TTC (pink).

If a yeast suspension containing a known concentration of glucose is mixed with TTC, you can record how long it takes for the colour change to occur. This experiment can be carried out over a range of temperatures and used to calculate the temperature coefficient (Q10) for the reaction.

The formula for the Q10 is:

$$\frac{\text{rate of reaction at } t + 10\,°C}{\text{rate of reaction at } t\,°C}$$

where $t$ is the chosen temperature for the experiment. For a value of 2 the rate of reaction would double with every 10 °C temperature rise.

Using this information, suggest the following:
- a suitable temperature range for an experiment investigating the effect of temperature on respiration in yeast
- a prediction of what the effect of temperature will have on the rate of TTC reduction
- a reason why it is important to repeat the experiment at least three times for each temperature chosen
- a qualitative method for assessment of the reduced TTC.

**Examiner tip**

Look at the experiment described on this page. You should be able to identify the dependent and independent variables. What factors would you need to control in order to make this a valid investigation?

## Questions

1 How is lactate biochemically different to pyruvate?
2 Describe the physiological conditions under which lactate is produced.
3 How is it possible to survive on a protein-only diet?
4 Complete the following table:

| Method of respiration | Site of reaction(s) | Reactants | Products | Relative contribution to cells ATP total |
|---|---|---|---|---|
| Aerobic | | | | |
| Anaerobic | | | | |

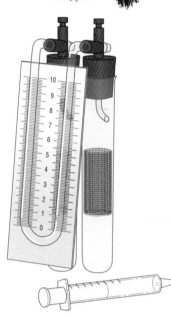

**Figure 1** A respirometer

### Examiner tip

You are likely to be asked at some point to calculate RQ values and to comment on the significance of this figure. An RQ of 1 or below indicates aerobic respiration has occurred and anything above 1 generally suggests that anaerobic respiration has occurred or that carbohydrates are being turned into fat.

## The respiratory quotient

When substrates such as glucose, proteins or lipids are used to produce ATP they are oxidised, using up oxygen and producing carbon dioxide. The respiratory quotient, RQ, is the volume of $CO_2$ produced by organisms divided by the volume of $O_2$ consumed in any given time period.

$$RQ = \frac{\text{volume of } CO_2 \text{ produced}}{\text{volume of } O_2 \text{ consumed}}$$

RQ values allow us to deduce which substrates are being used in respiration and to deduce whether aerobic or anaerobic respiration is taking place. An RQ value for the respiration of glucose is 1, lipids have a value of approximately 0.7 and proteins 0.9.

Measuring oxygen consumption and carbon dioxide production in humans gives an average RQ during rest of 0.85 due to the respiration of both lipids and glucose. During prolonged periods of exercise this value decreases as carbohydrate stores deplete and the body begins to use up lipid stores that give a lower overall RQ value.

### Calculating the RQ value

Stearic acid can be completely oxidised in aerobic respiration. The equation for the reaction is as follows:

$$C_{18}H_{36}O_2 + 26O_2 \rightarrow 18CO_2 + 18H_2O$$

To calculate the RQ we need to work out how many $CO_2$ molecules have been produced and then divide this number by the number of $O_2$ molecules consumed.

From the equation therefore, the $RQ = \dfrac{18}{26} = 0.7$

### Measuring the RQ of an organism

A piece of equipment called a respirometer is used to measure the gases exhaled and inhaled by organisms and to calculate the RQ value. From this value you can deduce what is being respired. A simple respirometer is shown in Figure 1. This involves using small animals or yeast cells to calculate the RQ value. For human subjects, a Douglas bag can be used to collect and measure expired air over a period of time and so used to calculate the RQ value.

## Respirometer experiment

This simple respirometer consists of two test tubes linked by a manometer. One of the test tubes contains the living specimens and the other acts as a pressure control. The syringe is used to reset the level of manometer fluid between the experiments and allows for repeats to be taken easily. The role of the potassium hydroxide is to absorb any $CO_2$ given off by the organisms during respiration.

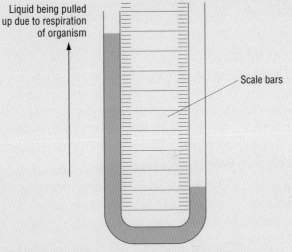

Liquid being pulled up due to respiration of organism

Scale bars

**Figure 2** Diagram of a respirometer being used to calculate RQs

**Examiner tip**

You should be able to both set up a simple experiment to calculate RQs and critically evaluate the procedure, and suggest precautions and improvements and highlight sources of error. You should be clear how you could investigate variables such as temperature. Remember, respiration relies on the activity of enzymes. Remind yourself from spread 4.1.1.6 what the effect of temperature is on enzyme-controlled reactions.

To calculate the RQ value, the oxygen consumed is measured in mm per minute (X), as shown in Figure 2. The experiment is then repeated using water instead of potassium hydroxide. Any change in the manometer fluid is recorded as mm per minute (Y). $CO_2$ given off will be (X–Y) mm per minute. To calculate the RQ value you would use the following equation:

$$RQ = \frac{X - Y}{X}$$

# Questions

1  Calculate the RQ value for a lipid with the formula $C_{17}H_{35}COOH$, assuming it was completely oxidised in respiration.
2  Suggest reasons why the RQ values obtained in a respirometer experiment might vary during the day for a particular organism.
3  Describe how you would set up an experiment to investigate the effect of temperature on the substrates used by organisms to produce ATP.
4  Why should the RQ value for anaerobic respiration be greater than 1? (Hint: how much oxygen is used? How much carbon dioxide is produced?)

Exercise is important for the overall health and wellbeing of body systems. **Aerobic exercise** in particular has many short-term and long-term benefits. Long-term exercise involves repeated short-term bouts of exercise and is normally referred to as 'training'. Over time, the body adapts and changes.

## The short-term effects of exercise

Short-term aerobic activity benefits both the circulatory and the gas exchange systems. An immediate response to exercise is to increase the heart rate, initially due to an increase in adrenaline secretion and the stimulation of the sympathetic nervous system.

As exercise continues **vasodilation** of the arterioles is brought about by muscle cells, which respond to a drop in oxygen levels by secreting *nitric oxide*. Nitric oxide has the effect of dilating blood vessels, so it increases the amount of blood flowing to the muscles and returning to the heart. This in turn raises **cardiac output** due to *Starling's Law* – the more the ventricle is filled with blood during **diastole**, the greater the volume of the blood pumped out in **systole**.

Blood is also redirected to areas of the body demanding more oxygen, such as the muscles. Less blood flows to areas such as the digestive system. The arterioles of the skin also receive more blood to enable them to lose heat by radiation. Again this is brought about by adrenaline.

Another immediate effect is to increase the rate and depth of breathing, allowing more oxygen to enter the blood to support aerobic respiration. This increase in breathing is also essential to help remove the increasing level of blood **carbonic acid**. This is formed from an increase in the amount of $CO_2$ released from cells as a result of more aerobic respiration. The carbonic acid lowers the blood pH and stimulates **chemoreceptors** in the brain and the aortic arch, triggering an increase in the overall breathing rate.

## The long-term effects of exercise

The body adapts to long-term, regular aerobic exercise. Changes occur to the overall muscle structure, and in the circulatory and gas exchange systems. Your ability to take in, transport and use oxygen improves and there will also be positive benefits to the immune system and general wellbeing of the body. The maximum amount of oxygen that can be taken in, transported and utilised is known as the **$VO_2$ max**. Government health guidelines now recommend that adults should carry out at least three 30-minute sessions of vigorous exercise a week to avoid an increased risk of obesity, coronary heart disease and strokes. But to really improve fitness, a programme of regular exercise is needed.

Muscles develop specific adaptations to increasing levels of aerobic exercise. There is an increase in the cross-sectional area of slow-twitch fibres and also in the number of mitochondria in the cells. These slow-twitch fibres rely on the increase in ATP production brought about by the presence of more mitochondria. The mitochondria also increase in size, store more enzymes and have more available glycogen as an increase in energy stores. An increased number of capillaries provide the mitochondria with more oxygen. Muscle cells also adapt by using fats more efficiently and conserving glucose. Remember, fats can only be broken down aerobically (see spread 4.1.1.6), but as oxygen supplies to muscle cells improve with training, fats can be used to fuel exercise and this contributes to an improvement in body mass index (BMI) (see spread 2.4.1.3). One way in which oxygen supplies are maintained is by increasing the levels of myoglobin – an oxygen store in muscle cells.

### Key definition

Arteriole walls contain smooth muscle that, as body temperature rises, will be made to relax by the action of chemicals such as nitric oxide or by stimulation from centres such as the hypothalamus. The result is that more blood will flow through the vessels so that heat can be lost from the skin through radiation. **Vasodilation** also increases the blood supply to exercising muscles.

### Examiner tip

It is important when you design a training programme that you choose a combination of aerobic activities such as jogging and swimming and aim to exercise so that your heart rate is approximately 70% of theoretical max (220 – age). It is important to increase levels of exercise gradually over time, to include rest periods in the programme and to warm up and down for each bout of exercise. It is also recommended that everyone should have a medical health check-up before starting a training programme.

The effects of long-term exercise on the heart are much more noticeable. Overall **resting heart rate** is lowered as the **stroke volume** is increased. Since each beat can pump out more blood, fewer beats are required at rest to deliver the same cardiac output. Exercise increases the thickness of the muscular ventricular wall – particularly the left ventricle. Recovery time from exercise (the time it takes for the heart to return to its resting rate) is shorter, and there is a decrease in both systolic and diastolic blood pressure. The shortening of recovery time is partly a response to an increase in the $VO_2$ **max**, which means that the body experiences anaerobic conditions for a shorter time while exercising. Less lactate will have built up, so less 'extra' oxygen is required to break it down.

In response to the longer periods of prolonged exercise, the lungs increase their maximum breathing rate. During exercise there is also an increase in the **tidal volume** and an increase in **vital capacity**, so there is a greater potential to get more oxygen into the body quicker. Interestingly, the tidal volume at rest may actually decrease. This is due to an increase in the density of capillaries in the lungs – more oxygen is able to diffuse so at rest ventilation can be less! In other words, there is a reduction in the physiological 'dead space'.

## Summary of the short- and long-term effects of exercise on the body

| Short-term effects | Long-term effects |
|---|---|
| Increases heart rate | Increases size and number of blood vessels, and structure of muscles |
| Increases cardiac output | Increases glycogen stores |
| Redirects blood to muscles away from areas in the body such as the digestive system | Increases $VO_2$ max |
| Dilates arterioles under the skin increasing heat loss | Increases heart muscle, stroke volume but decreases overall pressure and resting heart rate |
| Increases breathing rate and tidal volume | Increases maximum breathing rate, tidal volume and vital capacity |
| Increases acidity of blood | |

### Genetic potential of athletes

There is a fixed limit to the athletic potential of individuals as determined by their genetic code. It is obviously impossible to carry out assessments of whether athletes have reached this limit, but we know that with a sustained training schedule improvements in personal performance can be made.

Each athlete will have a different athletic potential that can be reached. In theory, the speed at which an athlete can potentially run 100 m should steadily decrease as training methods and technology improves. Many scientists speculate that certain populations of people share specific genetic groups that provide advantages in certain sports. It has often been assumed that because long-distance running events tend to be dominated by athletes of East African descent then there must be a set of genes present that are involved in athletic performance and that exceptional athletes may express these. All evidence so far suggests that this is not the case, that these athletes are actually hugely genetically diverse and that training at altitude is more important in improving athletic performance than genetic predisposition.

**Figure 1** Trimesh Dibaba, the Ethiopian long distance runner

## Questions

1 List the main short-term and long-term benefits of exercise on the body.
2 Nitric oxide is often used by angina sufferers. Angina attacks involve a narrowing of blood vessels supplying blood to the heart, causing extreme pain. Use the information in the text above to explain how nitric acid might help such sufferers.
3 Suggest reasons why muscles of the upper leg might contain lots of slow twitch muscle fibres and what a cross-section of this muscle might look like.
4 Refer back to your AS notes and describe how the breathing rate is monitored in the body and increased as a result of increased exercise.

# Changing diets to enhance sporting performance

## Carbohydrate loading

Muscles and liver cells store carbohydrate in the form of glycogen. This can be easily broken down to form glucose and so can be used to generate ATP. Some endurance athletes may load up on carbohydrates a few days prior to a race so as to maximise the glycogen stores in the muscle cells. Many athletes will deplete stores of glycogen prior to a race and then aim to eat approximately 10% of their body mass in carbohydrates. The idea is that this will provide plentiful supplies of available energy through the conversion to glycogen and then to ATP.

### How to carboload?

Most sports physiologists suggest you should follow the following regime prior to a race:
- 10 days before the event you should eat only protein and restrict carbohydrate intake.
- Continue this regime until three days before the event.
- Three days prior to the event switch to carbohydrates and consume as much as possible to build up glycogen stores in the body – this will cause a gain in weight due to the storage of both the glycogen and water.
- After the event it is important to recover by consuming carbohydrates and proteins.

## Enhancing athletic performance

Some athletes might choose alternative methods to training to enhance athletic performance. These methods such as blood doping, steroid treatment or the use of recombinant erythropoietin all carry a medical risk, are illegal in competitive sports activities, and work by changing normal body function.

### RhEPO and blood doping

When blood levels fall due to bleeding or oxygen levels fall at high altitude, the body responds naturally by secreting a hormone called erythropoietin from cells surrounding the glomerulus in the kidney. This has the effect of stimulating red blood cell production in the bone marrow. This tightly regulated mechanism of blood production can be manipulated by athletes. This can involve the use of bacterially produced hormones, RhEPO, and blood infusions called doping. The following table describes these two methods:

| Blood doping | Enhanced EPO (RhEPO) |
|---|---|
| Athletes remove a quantity of blood a few months prior to the competition | Athletes inject genetically engineered EPO hormone into blood |
| Blood cells are extracted and stored at cold temperatures | The EPO targets the bone marrow stimulating further increases in red blood cell count |
| Cells are then warmed prior to the competition and injected back into the athletes blood | Red blood cell numbers are raised within a few days and the athlete will also benefit from a greater oxygen carriage and ability to carry out aerobic respiration |
| The increase in red blood cells enhances the athletes' ability to carry oxygen and carry out aerobic respiration | |

**Table 1** Blood doping and RhEPO

> **Key definition**
>
> **Glycogenolysis** is the breakdown of glycogen to glucose by the hormones glucagons and adrenaline. The glucose is released from liver cells called hepatocytes into the blood. The many branches in the glycogen molecule provide many 'access points' for the enzymes that will hydrolyse it so breakdown is rapid.

## Steroid enhancement

Some athletes enhance performance by using steroids that act very differently on the body compared to EPO, or through the addition of extra red blood cells. Steroids are complex lipids made from cholesterol and include certain sex hormones such as progesterone, oestrogen and testosterone. The overall effect of these steroids is to stimulate anabolic reactions in the body such as protein synthesis and promote growth.

Anabolic steroids are artificially produced molecules, structurally similar and chemically based around naturally occurring steroids. An example is nandrolone, which acts like a sex hormone in that it easily diffuses through the cell membrane and promotes protein synthesis in target cells. These anabolic steroids are used to increase muscle mass, and make athletes more competitive, aggressive and able to train for longer periods of time.

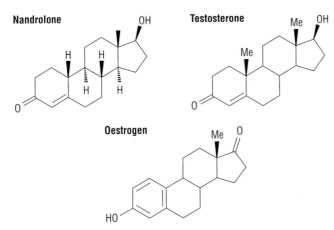

**Figure 1** Diagram of nandrolone with comparisons to testosterone and oestrogen

### Health problems with blood doping, RhEPO and anabolic steroids

Blood doping introduces extra blood cells into the cardiovascular system, but not necessarily more plasma! This can make blood more viscous, slow down the cardiac output and so actually decrease blood oxygen content. It can lead to heart failure, strokes and also introduces the possibility of infections from such agents as HIV and hepatitis through unclean needles.

RhEPO can create severe cardiovascular complications and kidney failure. There are also the risks and ethical implications of using hormones from genetically engineered bacteria.

Steroids mimic the action of sex hormones, and increasing these levels can cause rages, acne, hair loss and large rises in blood cholesterol level. Men may also develop additional breast tissue and are at risk of becoming infertile.

## Questions

1 Carboloading increases the glycogen stores in muscle cells. Recap your AS knowledge to describe the structure of glycogen and explain why it is an excellent storage molecule for muscle cells.
2 Why is it important for athletes who are involved in blood doping to use their own blood cells to improve their performance? (Hint: review blood types in AS.)
3 Describe why steroids are able to pass through cell membranes easily and influence cell synthesis.
4 Explain why athletes must remove blood for doping a few months before an event.

Humans use two respiratory pigments, both of which bind reversibly to the oxygen needed for ATP synthesis. These are haemoglobin found in red blood cells that transports oxygen, and myoglobin located in muscle cells that acts as an oxygen store. Haemoglobin is more abundant and carries approximately 98% of the body's total oxygen. The remainder is carried in solution.

## Structure of haemoglobin

Haemoglobin is a 3D quaternary protein consisting of four polypeptide chains, two alpha and two beta polypeptides. Each of these is associated with an iron-containing haem group. The whole molecule consists of 574 amino acids ordered in a precise way. Haemoglobin is a good example of a protein with a quaternary structure and a prosthetic group.

### Constructing a haemoglobin molecule

The genetic information to construct the haemoglobin molecule on the ribosomes of a cell is found on two genes on chromosomes 11 and 16 in the nucleus of every cell. This genetic code consists of an array of triplets made up from the bases adenine, guanine, cytosine and thymine in the DNA nucleotides. These genes are lengths of DNA and are the blueprints for constructing the haemoglobin molecule. Each triplet of three bases codes for one amino acid. With four possible bases for each position in the triplet, there are 64 possible combinations. Three are 'stop' signals and the remainder code for the 20 amino acids, with some amino acids having more than one triplet. This makes the genetic code a 'degenerate' code.

When haemoglobin is made in the erythrocyte precursor cell, the genes coding for the polypeptides are copied onto RNA intermediate molecules. These RNA molecules are transferred to the ribosomes in the cytoplasm, and then converted from a sequence of codes to a sequence of amino acids forming the polypeptides. These polypeptides are then combined with the haem group and the haemoglobin is assembled. When the haemoglobin content of the cell reaches about 30%, the nucleus, endoplasmic reticulum, golgi and mitochondria begin to break down and the process is complete by the time the red blood cell enters the circulation.

### RNA

Copying DNA and producing haemoglobin in the cell requires the use of RNA nucleotides. There are three types of RNA used, messenger RNA (mRNA), transfer RNA (tRNA) and ribosomal RNA (rRNA). RNA is a very different molecule to DNA and Table 1 outlines some of the key differences:

### Transcription

Transcription takes place in the nucleus and involves a short strand of mRNA copying the genes for haemoglobin, then passing out through a nuclear pore and binding to a ribosome in the cytoplasm. The sequence of events in transcription is as follows:

- the length of DNA that contains the code for making a haemoglobin polypeptide unzips under enzyme action – the hydrogen bonds holding the double helix are broken and this then exposes the nucleotide bases
- the coding strand of DNA (see Figure 2) is copied by complementary mRNA nucleotides under the control of the enzyme RNA polymerase. There is complementary base pairing with the thymine base on the DNA strand being equivalent to uracil on the RNA strand
- mRNA nucleotides copy the entire DNA sequence from the first triplet-base start codon to the final triplet-base stop signal. The DNA triplets have become RNA codons
- mRNA strand then detaches and leaves the nucleus through a nuclear pore, allowing the bonds on the DNA molecule to reform and the chromosome to return to its original state.

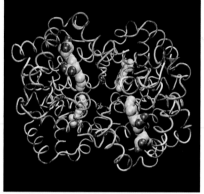

**Figure 1** Diagram of haemoglobin molecule

| DNA | RNA |
|---|---|
| It is a large, long-lived molecule of heredity | It is a short, single-stranded molecule involved in protein synthesis |
| It is composed of a double helix | It contains the bases adenine, uracil, cytosine and guanine |
| It contains the bases adenine, thymine, cytosine and guanine | It contains the 5-carbon sugar ribose |
| It contains the 5-carbon sugar deoxyribose | It can be found in one of three different forms, mRNA, tRNA and rRNA, and each form plays a different role in protein synthesis |
| It is confined to the nucleus attached to chromatin | |

**Table 1** Key differences between DNA and RNA

## Translation

Translation then occurs on the ribosomes, and involves the conversion of the sequence of codons held on the mRNA into a sequence of amino acids and subsequently polypeptide strands. The process involves the following:

- mRNA attaches to a ribosome either on the rough endoplasmic reticulum or free-floating in the cytoplasm
- specific tRNA molecules pick up specific amino acids in the cytoplasm (activation)
- each codon on the mRNA is linked to a complementary anticodon on a tRNA molecule (see Figure 2)
- as the tRNA bring in the amino acids in the order laid down on the mRNA, adjacent amino acids form peptide bonds and a polypeptide grows. Once a 'link' is made, the tRNA detaches from the amino acid and the mRNA, and returns into the cytoplasm to be activated again
- translation continues until the final stop codon is reached for which there is no tRNA
- the polypeptide chains are then combined with the haem, modified in the endoplasmic reticulum and Golgi apparatus, and become recognisable as a haemoglobin molecule.

**Examiner tip**

You should recap cell structure and familiarise yourself with the arrangement and function of the organelles in a typical animal cell. In particular you should revise the functions and roles of endoplasmic reticulum, Golgi apparatus and ribosomes.

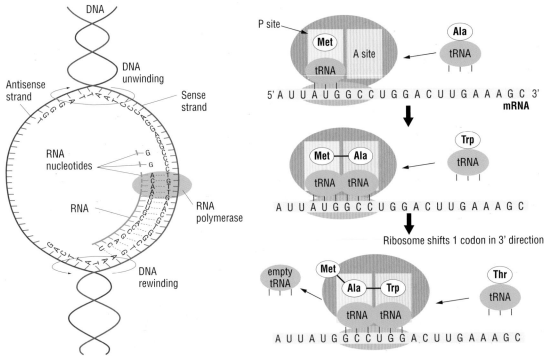

**Figure 2** Summary diagram of transcription and translation

## Questions

1 Complete the following table to summarise the roles of RNA.

| mRNA | tRNA | rRNA |
|------|------|------|
|      |      |      |

2 Compare the structure of haemoglobin to that of an enzyme that you studied at AS.

3 Use the information in the text and diagrams to construct a flowchart to describe the process of transcription and translation.

4 Describe the bonding that would be found in a haemoglobin molecule. (Hint: recap on AS material.)

## Haemoglobin and myoglobin

98% of all the blood's oxygen is carried combined with haemoglobin as *oxyhaemoglobin*. Each molecule of haemoglobin can bind to four molecules of oxygen. The oxygen molecule binds tightly to the **haem** group, and if all the haem groups are occupied by oxygen, the molecule is said to be *saturated*.

**Myoglobin** differs from haemoglobin in that it contains only one haem group and consists of only one polypeptide chain. Its main role is to act as an oxygen reserve in the muscles. Oxygen is released by myoglobin pigments only when oxygen levels in the muscle cells are very low, for example, during intense exercise. Myoglobin has a higher affinity for oxygen than haemoglobin (see Figure 2).

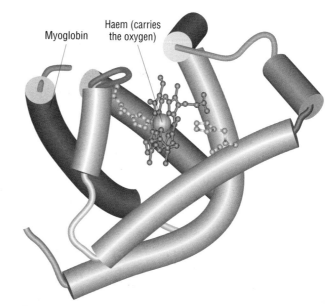

**Figure 1** Myoglobin

**Examiner tip**

When discussing gas concentration in relation to haemoglobin we use the term partial **pressure**. If you consider all the dissolved gases in the blood, the total pressure they exert is the sum of the partial pressures of each component gas. So the partial pressure will depend on the concentration of one gas relative to the other. The partial pressure of a gas is given in kPa units (kilopascals). You often see a comparison of the partial pressures of $CO_2$ and $O_2$. In the lungs, the partial pressure of oxygen in the alveoli is high (about 14 kPa). Since the partial pressure of oxygen in blood returning to the lungs is much lower, oxygen will diffuse into the blood until it is in equilibrium. With a high partial pressure of oxygen, haemoglobin will be fully saturated.

## Dissociation curves for haemoglobin

Haemoglobin binds to and releases oxygen. The dissociation curve describes the relationship between partial pressure of oxygen and the saturation of haemoglobin. The graph is called 'S-shaped' or 'sigmoid' because of the way in which the haemoglobin molecule binds and releases its four oxygen molecules.

### How $CO_2$ is transported in blood

$CO_2$ can be transported in the blood in three ways. A small amount is dissolved in **plasma** and some is carried in combination with haemoglobin as *carbamino* compounds. Mostly it is carried as the hydrogen carbonate ion ($HCO_3^-$). The $CO_2$ reacts with water to form **carbonic acid**. This happens in red blood cells and the reaction is catalysed by the enzyme **carbonic anhydrase**.

$$CO_2 + H_2O \xleftrightarrow{\text{carbonic anhydrase}} H_2CO_3$$

The carbonic acid then dissociates to form hydrogen ions and hydrogen carbonate ions. It is in this form that most of the carbon dioxide is transported in plasma.

$$H_2CO_3 \longleftrightarrow H^+ + HCO_3^-$$

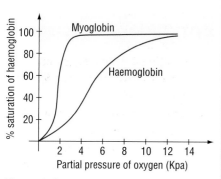

**Figure 2** Oxygen dissociation curve for haemoglobin and myoglobin

At very low $pO_2$ (partial pressure of oxygen) it is difficult for $O_2$ to combine with haemoglobin and so the curve is shallow. When the first haem group binds to $O_2$, the molecule of haemoglobin is distorted making it easier for successive oxygen molecules to bind, hence a steeper curve. Likewise, at higher $pO_2$ it is difficult to fully saturate the haemoglobin and so the graph plateaus.

The S-shaped dissociation curve for haemoglobin is physiologically important for the body. Haemoglobin becomes saturated in the lungs where there is high $pO_2$. In respiring muscles, the $pO_2$ will be low. When the red blood cells reach the capillaries in these muscles, the haemoglobin can quickly unload $O_2$. Conditions in muscles correspond to the steep part of the curve.

### The Bohr shift and effect of $CO_2$ on the oxygen dissociation curve

When the carbonic acid dissociates, the hydrogen ions are picked up by the haemoglobin to form **haemoglobinic acid**. The haemoglobin is acting as a **buffer** to prevent a drop in pH. The effect of this is to lower the affinity of haemoglobin for oxygen – it will release oxygen more easily, for example, at higher $pO_2$ levels. The effect of the $CO_2$ on the oxygen saturation of haemoglobin is called the **Bohr effect** and the subsequent effect on the dissociation curve is called the Bohr shift – the curve is shifted to the right. This is important because as more exercise is done, more $CO_2$ is produced and there is a greater unloading of oxygen from the oxyhaemoglobin. The change in partial pressure of $CO_2$ changes the dissociation of haemoglobin and results in greater oxygen being delivered to the tissues when needed.

### The oxygen debt and excessive post-exercise oxygen consumption

Following a period of exercise, the body's stores of oxyhaemoglobin and oxymyoglobin need to be replenished. This occurs during the period referred to as the *oxygen debt*. Figure 4 explains the pattern of oxygen uptake during exercise and recovery.

Oxygen uptake rises over the first few minutes of exercise, but the demand for ATP is such that, until oxygen transport can increase, not all the ATP can be produced aerobically. **Creatine phosphate** will be used and oxygen will be unloaded from myoglobin, but inevitably some lactate will be produced as ATP is made anaerobically. The graph then plateaus, during which time the oxygen usage in the tissues is equal to oxygen delivery and uptake. The Bohr effect will be occurring, causing a rapid release of oxygen to respiring tissues.

As exercise stops the oxygen uptake decreases, but it still remains higher than the level taken in at rest. This is known as the *oxygen debt* and the purpose of this 'additional' oxygen is to:

- reload stocks of oxyhaemoglobin and oxymyoglobin
- replace stocks of ATP, creatine phosphate and glycogen
- convert lactate, made during anaerobic respiration, to glucose or glycogen
- meet the demands of increased metabolic rate, heart and respiratory functions as a result of the temperature rise during exercise.

The *excessive post-exercise oxygen consumption* (EPOC) is actually the total oxygen consumed after exercise minus the pre-exercise level of oxygen consumption.

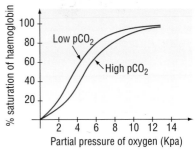

**Figure 3** The Bohr shift

## Questions

1. Compare the structure of haemoglobin and myoglobin.
2. Suggest a reason why haemoglobin is used to transport most of the oxygen in the body.
3. Explain why the dissociation curve for oxygen is described as a S-shaped curve.
4. Explain why most exercise will result in an oxygen debt being formed even if the person does not push themself so hard that they go above their aerobic threshold.

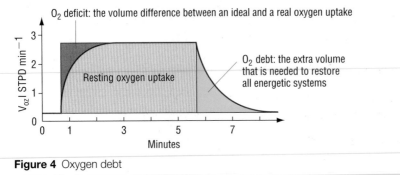

**Figure 4** Oxygen debt

## Sickle cell anaemia and the effect on haemoglobin

Sickle cell anaemia is a genetic disease that leads to a fault in the production of the beta polypeptide chain in haemoglobin. It is caused by a mutation that has a considerable effect on the properties of haemoglobin, the shape of red blood cells and the carriage of oxygen.

At low $pO_2$, haemoglobin forms long, sharp crystals that make red blood cells become sickle-shaped and then rupture. The red blood cells carry less oxygen, block capillaries, and can lead to **hypoxia** in the body.

### How does sickle cell anaemia occur?

Sickle cell anaemia is a result of a single base **substitution** in the RNA sequence that codes for the beta chain of haemoglobin. The single change, uracil replacing adenine, alters the coding of one of its amino acids from glutamate to valine. This is a result of a **point mutation** in the DNA.

The first seven **codons** for the normal base sequence in haemoglobin are:

GUA-CAU-UUA-ACU-CCU-GAA-GAG

The sequences in the mutated sickle cell sequence are:

GUA-CAU-UUA-ACU-CCU-GUA-GAG

The inheritance of sickle cell anaemia is discussed in spread 1.1.2.2.

Remember protein structure from spread 1.1.2.2.

### How the mutation of haemoglobin was discovered

The discovery that haemoglobin from sufferers contains more amino acid residues of valine than glutamic acid was the culmination of many years of research.

It involved several scientists working in different laboratories in different countries, investigating different physiological aspects of sickle cell anaemia. This involved analysis of the transmission of light through blood samples and electrophoresis of normal and mutated haemoglobin. The data from the scientists was collated.

An overall conclusion into the effect of sickle cell anaemia on haemoglobin was pieced together by the British scientist Dr Linus Pauling.

During this discovery it is interesting to see how the scientists working on the problem acted towards each other and used the scientific discoveries presented to them to further their knowledge.

The sociologist, Robert Merton, described the four norms or standard behaviours that scientists follow during such discoveries.

Scientists can do the following:
- universally judge and test any results presented to them objectively, or
- cooperate and collaborate with other colleagues for the greater benefit of science, or
- regard their work unselfishly and so seek only scientific recognition and not money, or
- wait until all the facts are presented before making their final judgement or conclusion.

In the case of the discovery of the change in haemoglobin by the sickle cell mutation, this is a good example where all these four norms were seen in the behaviour of the scientists working on the problem.

### How DNA is repaired

In sickle cell anaemia, the change that occurs to DNA cannot be repaired. Most mutations occur during **DNA replication** and it is estimated that approximately one mistake occurs for every 10 **genes** copied. Changes to DNA structure can be caused by a number of factors, as shown in Figure 2. If these mistakes were not corrected then most humans would die!

### Examiner tip

Remember protein structure from spread 1.1.2.2. If the **primary structure** of the polypeptide is changed, the **R groups** and/or their positions will change. This leads to a change in the **tertiary structure**. The haemoglobin will have different properties – in this case it becomes less soluble at low oxygen concentrations.

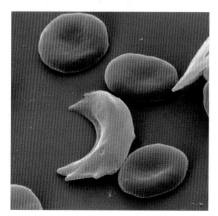

**Figure 1** Sickle cell

### Key definition

**Hypoxia** is where the oxygen level of the blood falls to very low levels. This results in a reduction in cellular oxidation and a reduction in ATP synthesis and, as a result, depressed aerobic activity.

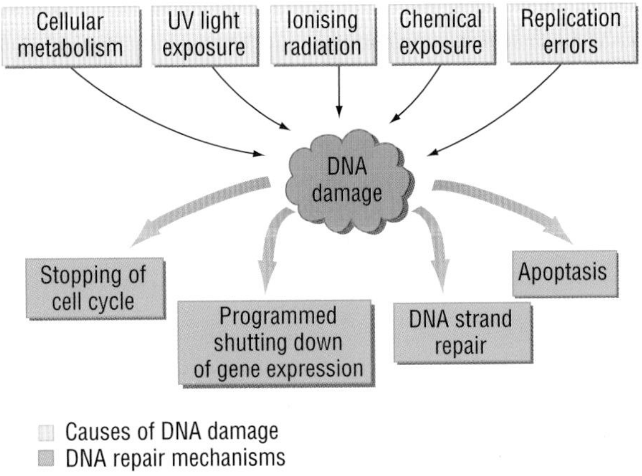

**Figure 2** DNA repair

A proof-reading mechanism carried out by DNA polymerase enzymes corrects these mistakes. After DNA has replicated, a second set of proteins 'surveys' the DNA looking for mismatches in the base pairs. One form of colon cancer is due, in part, to a failure in a 'mismatch' repair. When DNA is damaged during the life of the cell, further enzymes cut out or *excise* the mispairing nucleotides, and DNA polymerases and *DNA ligases* repair the strands. In the disease *Xeroderma pigmentosum*, the mechanisms that repair DNA damage due to ultraviolet light are missing. The result is that even low exposure to sunlight triggers the development of skin cancers.

### DNA and cellular ageing

At the ends of chromosomes are small structures called **telomeres**. These act almost like bookends, and act to protect the genes on the chromosomes. They also help to regulate the division of the cell. Every time the cell divides, these telomeres are not copied, so they become slightly shorter during each division. When they become too short, essential parts of the DNA can be destroyed. With the telomere removed, the genes are exposed and are easily damaged and mutated.

Telomere length governs the number of times a cell can replicate. In humans, most cells replicate between 20 and 30 times before the telomeres become too short. This explains why cells do not last for an entire lifetime. Constantly dividing cells such as bone marrow and **embryonic stem cells** can maintain their telomeres using the enzyme **telomerase**, but most **somatic** cells do not express this enzyme. As you would suspect, telomerase is also expressed in 90% of human cancer cells. It is interesting to speculate what research on this enzyme could lead to in terms of understanding and combating ageing.

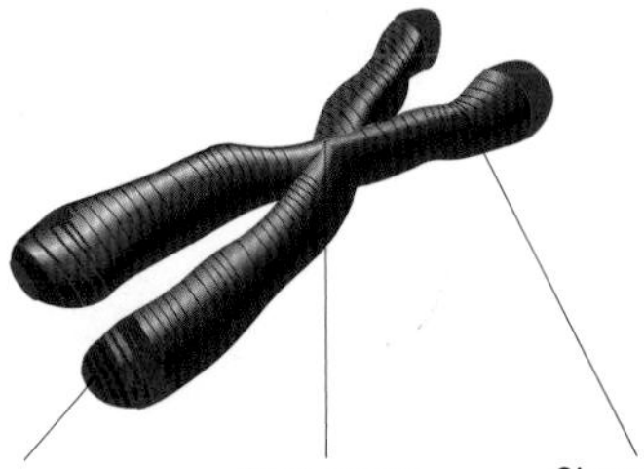

Telomere          Centromere          Chromatid

**Figure 3** Chromosome and telomere structure

## Questions

1 Describe the importance of polymerase enzymes during cell division and replication.
2 Describe why hypoxia might lead to less ATP synthesis in a cell.
3 Describe other forms of gene mutation.
4 During apoptosis the cell commits suicide and destroys itself in an ordered way. What might be present in the cell to carry out such programmed cell death?

## Types of muscle in the body

Muscle tissue is capable of contracting and allowing movement to occur. There are three types of muscle tissue – cardiac, skeletal and smooth. Each of these tissues contains elongated cells containing large quantities of the proteins **actin** and **myosin**. These proteins form myofilaments in the cell which allow contraction to occur. The three types of muscle carry out different roles in the body and their key features are highlighted below. You will have met all three types of muscle in the AS units. It is important when you are writing about 'muscle' to say which type you mean.

| Type of muscle | Examples of distribution and key features in the body |
|---|---|
| Skeletal | Attached to bones, abdomen wall, diaphragm, facial muscles and extrinsic muscles of eye. Primarily voluntary controlled by neurones of the somatic system |
| Cardiac | Heart control of the autonomic nervous system. Cells are myogenic |
| Smooth | Iris and ciliary body of the eye, artery, vein and arteriole walls, walls of the respiratory tract, reproductive tract and alimentary canal. Involuntary muscle and contraction is controlled by the motor neurones of the autonomic nervous system. |

**Table 1** Muscle type

### Skeletal muscle

Skeletal muscle is able to receive stimuli and respond to them by contracting. The contraction of the muscle uses ATP to produce movement. In doing so, much of the heat that warms the body is also released. The specialised cells or muscle fibres that make up the muscle's structure enable it to carry out these important functions.

Muscle fibres are surrounded by a membrane called the **sarcolemma**, which encases a specialised cytoplasm called the **sarcoplasm**. This membrane has special finger-like invaginations called **T tubules** which link up to the **sarcoplasmic reticulum**, creating a complete cellular network. The cytoplasm contains many mitochondria, plus stores of glycogen and myoglobin. The cytoplasm also contains the muscle proteins or filaments, actin and myosin, arranged in bundles as **myofibrils**.

The sarcoplasmic reticulum, another membrane system, contains calcium ions that are essential in the contraction process of muscles.

### The structure of a sarcomere

Within a myofibril, the myofilaments are arranged in a very ordered and regular way forming a **sarcomere**. The thin filaments are actin, and the thicker filaments are myosin. The actin filament is made up of three proteins bound together, actin, **tropomyosin** and **troponin**. The myosin filament consists of bundles of polypeptide molecules with a globular head for binding to ATP and to the actin filament.

Under the microscope the myofibril is seen as being made up of a thick filament of myosin surrounded by six thin filaments of actin, which in turn are surrounded by three thick myosin filaments. This arrangement is illustrated in Figure 1.

### Examiner tip

You should be able to use the photomicrographs and identify and label the key structures in a muscle fibre and be prepared to make simple annotated sketches of these to illustrate and explain questions in exams. You need to be able to describe which proteins you would find if you 'sliced through' the different bands that are seen in a sarcomere.

### Sliding filament theory

This description of muscle contraction is often referred to as the *sliding filament theory*. It dates back to the work of scientists in the 1950s. They observed micrograph pictures of muscle fibres during and after contraction and sought to explain their observations.

They noticed that when a muscle contracted and sarcomeres reduced in length, I bands became narrower but the A bands did not. They also saw that if a resting muscle was stretched, the H zones increased in size. These observations can be explained by the overlap of thick and thin filaments. During contraction, the thin filaments slide in towards the M band. If the muscle is stretched, the thin filaments slide away, so increasing the size of the H zone.

The movement of the cross-bridges between the filaments of actin and myosin is like the rowing of oars in a boat. It is sometimes described as a *ratchet mechanism*. Just remember that the ATP is needed to *break* the cross-link between the proteins, not to form it.

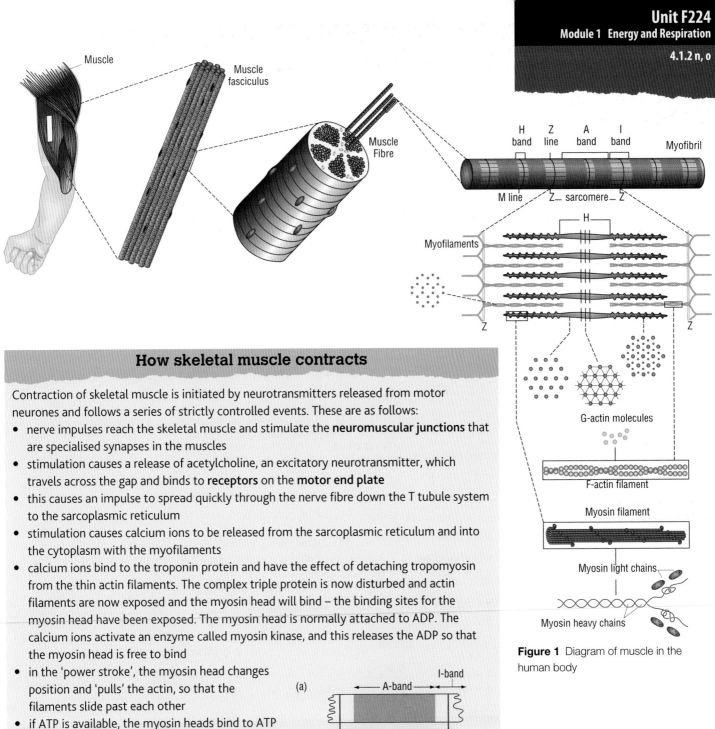

**Figure 1** Diagram of muscle in the human body

## How skeletal muscle contracts

Contraction of skeletal muscle is initiated by neurotransmitters released from motor neurones and follows a series of strictly controlled events. These are as follows:

- nerve impulses reach the skeletal muscle and stimulate the **neuromuscular junctions** that are specialised synapses in the muscles
- stimulation causes a release of acetylcholine, an excitatory neurotransmitter, which travels across the gap and binds to **receptors** on the **motor end plate**
- this causes an impulse to spread quickly through the nerve fibre down the T tubule system to the sarcoplasmic reticulum
- stimulation causes calcium ions to be released from the sarcoplasmic reticulum and into the cytoplasm with the myofilaments
- calcium ions bind to the troponin protein and have the effect of detaching tropomyosin from the thin actin filaments. The complex triple protein is now disturbed and actin filaments are now exposed and the myosin head will bind – the binding sites for the myosin head have been exposed. The myosin head is normally attached to ADP. The calcium ions activate an enzyme called myosin kinase, and this releases the ADP so that the myosin head is free to bind
- in the 'power stroke', the myosin head changes position and 'pulls' the actin, so that the filaments slide past each other
- if ATP is available, the myosin heads bind to ATP and hydrolyses it, releasing energy that causes it to release the actin and returns the myosin back to its original position. The cycle will repeat for as long as ATP and calcium ions are available
- as the filaments slide past each other the attachment breaks and new attachments form further along the molecule. Over 100 attachments occur every second
- the contractions will continue to shorten the fibre until nervous stimulation ceases and the calcium ions return to the sarcoplasmic reticulum
- with a fall in calcium ions, the actin filaments re-attach to troponin and tropomyosin, preventing cross-bridges from forming and so returning the muscle fibres to their original shape.

(a)
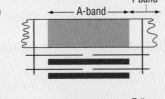

(b)
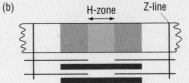

(c)
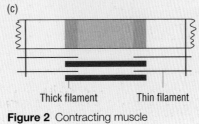

**Figure 2** Contracting muscle

## Questions

1 Suggest why the properties of cardiac muscle need to be different from those of skeletal muscle.
2 Describe the structure of a muscle fibre.
3 Explain the importance of the T tubule system in the sarcomere.

## Summary and practice questions

**1** Outline three roles of ATP in organisms.

**2** ATP is made during the process of ..................... and is composed of a molecule called adenosine, made up of ..................... and ....................., and three ..................... ions. The addition of a ..................... ion to ADP involves a condensation reaction and occurs in two structurally different organelles, the ..................... and the ..................... . It can also occur in the cytoplasm during ..................... .

**3** Explain the importance of the coenzymes NAD and FAD in the production of ATP.

**4** Glycolysis involves the splitting of ....................., a ..................... carbon sugar molecule into two molecules of ..................... . In the process, four molecules of ..................... and two molecules of ..................... are also made. The ..................... then enters the link reaction if ..................... is present. A 2 carbon molecule, ..................... is formed by dehydrogenation and ..................... . The ..................... then binds to a 4 carbon acceptor molecule called ..................... and passes through a series of steps in the Krebs cycle. The products of the Krebs cycle are ....................., ....................., ....................., ..................... and ..................... and the 4 carbon molecule is regenerated.

**5** Describe the fate of pyruvate in the absence of oxygen in human cells.

**6** Explain, using annotated diagrams, the process of oxidative phosphorylation and describe how the structure of mitochondria is related to its function?

**7** Complete the following table:

| Substrate | Where this is broken down in the respiratory pathway and what is its energy contribution? |
|---|---|
| Fatty acids | |
| Amino acids | |

**8** Describe what is meant by the RQ of a substrate molecule.

**9** Explain the difference between dehydrogenation and decarboxylation.

**10** Explain why the energy efficiency of aerobic respiration differs to that of anaerobic respiration.

**11** Explain what is meant by carbo loading.

**12** DNA is a code for making ..................... . It is created from repeated ..................... joining together using the enzyme DNA ..................... . Each of these ..................... is composed of a deoxyribose sugar, a ..................... group and an organic base. The bases in DNA are ....................., ....................., ..................... and ..................... .

**13** Describe how the information of DNA is transcribed and translated.

**14** How is mRNA different to that of tRNA?

**15** Complete a list of the short and long term effects of exercise upon the body.

**16** Haemoglobin is a complex .................... with a
.................... structure. Each molecule can carry up to
.................... molecules of oxygen. This oxygen is
loaded onto the haemoglobin in the lungs where the partial
pressures of oxygen are .................... .
Oxyhaemoglobin will lose this oxygen to respiring tissues
such as in .................... and this unloading is
accelerated with increasing levels of .................... and
lowering blood .................... .

**17** Describe what is meant by the Bohr shift.

**18** Explain the importance of ATP in contracting muscles.

**19** Complete the following table:

| Muscle component | Description at rest | Description during contraction |
|------------------|--------------------|--------------------------------|
| z lines | | |
| I band | | |
| A band | | |

## The female reproductive system

The female reproductive system has a number of roles. It produces **haploid gametes** and the oestrogen sex hormones. The **oviduct** is the site of fertilisation, which normally leads to the **implantation** and development of an embryo in the lining of the **uterus**.

From **puberty** until the **menopause** the ovaries release ova, which are the female gametes. Ova develop inside **follicles**, and at **ovulation** they are released into the oviducts (also called the **Fallopian tubes**).

Figure 1 shows the key structures in the female reproductive system, together with the associated organs of the urinary system.

### The ovary

Each ovary is approximately 40 mm in length and consists of over 200 000 specialised cells called primary follicles. Of these follicles, only about 1% will ever mature and develop into secondary oocytes, the female gamete.

A cross-sectional view through the ovary (Figure 2) shows the follicle cells at different stages of development. During this development of the follicle into an **oocyte**, **oogenesis**, the follicles will go through a sequence of staged events. These follicles change and mature at different times during these stages. These individual stages will be discussed in the next section.

### The development of a follicle

The production of egg cells is called oogenesis. It begins during the development of the female **fetus** prior to birth. **Germinal cells** divide first by **mitosis** to form **oogonia**. These then begin to divide by **meiosis** but only reach prophase 1. These are now called *primary oocytes.* When a girl is born she has over one million primary oocytes. By puberty, only about 200 000 primary oocytes remain. Each primary oocyte is surrounded by a layer of ovarian cells and this will become the 'functional unit' of the ovary – the follicle – at the onset of puberty.

From puberty, some of the primary oocytes progress from prophase 1 to the end of the first division of meiosis. The two daughter cells are both haploid, but only one becomes a *secondary oocyte*. The remaining cell is smaller and forms the **polar body**. Between six and 12 follicles begin to mature each month, but after about a week one follicle becomes dominant and the rest stop developing. This is the ovarian or **Graafian follicle**. The secondary oocyte continues to develop, but stops at metaphase 2 of meiosis. It is this secondary oocyte that is released at ovulation.

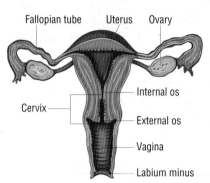

**Figure 1** The female reproductive system

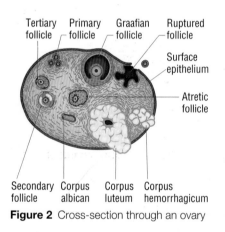

**Figure 2** Cross-section through an ovary

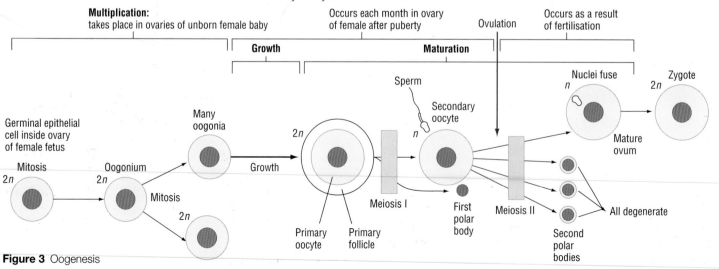

**Figure 3** Oogenesis

Figure 3 describes the process of oogenesis and highlights the key times for each stage and the stage in meiosis that is involved. It is important to notice that oogenesis occurs over many years, and the staged processes occurring in meiosis are not continuous.

As seen in Figure 3, the original oocyte produces four daughter cells during its division. Three of these are very small and are called polar bodies. These eventually disintegrate. The fourth daughter cell is much larger and develops into the secondary oocyte. One of these will be released once a month during a female's reproductive cycle.

## The male reproductive system

The male reproductive system has several roles. It produces haploid gametes and the male sex hormone, **testosterone**.

The male reproductive structures are found on the outside of the body. This is important for sperm production because the male gametes mature at a temperature slightly lower than that of the rest of the body. Figure 4 shows the key structures in the male reproductive system, together with the associated organs of the urinary system.

### The testes

The reproductive structures of the male include the penis and the testes. The testes produce sperm cells and also secrete the male sex hormone, testosterone. Unlike the development of oocytes, sperm production begins at puberty and continues into old age. Looking at a cross-section through a testis you can see that the structure is composed of a series of lobules each containing many **seminiferous tubules**. These tubules are the site of **spermatogenesis** – the production, growth and maturation of sperm – and are also where testosterone is produced and secreted. Figure 5 shows the key features of a seminiferous tubule.

### Spermatogenesis

Each seminiferous tubule contains many diploid cells called **spermatogonia**, which have developed from cells in a tissue called the *germinal epithelium*. These will later develop to become four haploid and mature sperm, each capable of fertilising an ovum. In the initial stage of development the spermatogonia multiply by mitosis, then grow into **primary spermatocytes**. After this first division, these cells develop into **secondary spermatocytes**. After the second meiotic division they become haploid **spermatids**. In the final stage of permatogenesis these cells differentiate and mature into sperm cells. Figure 6 shows the development of sperm from spermatogonia.

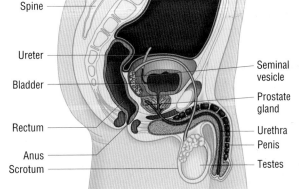

**Figure 4** The male reproductive system and urinary system

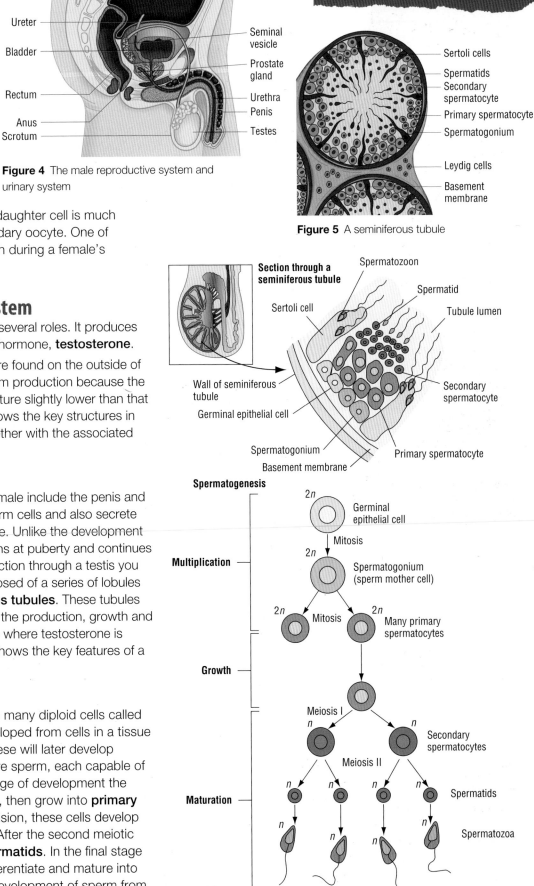

**Figure 5** A seminiferous tubule

**Figure 6** Spermatogenesis

## Secondary oocytes

The secondary oocyte is a large structure clearly visible in the ovary through a microscope. It is released from a mature (Graafian) follicle during ovulation. The structure of these follicles and the associated secondary oocyte is shown in Figure 1. The cells of the follicle aid in the development of the mature oocyte.

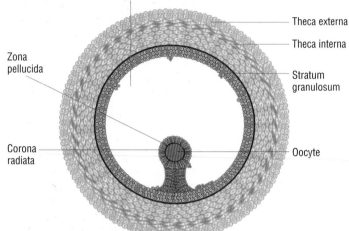
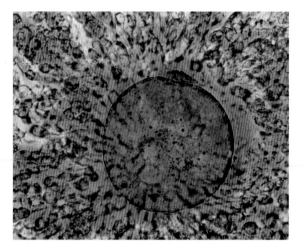

**Figure 1** A mature follicle and secondary oocyte

When the oocyte leaves the follicle during ovulation, you can see that it is surrounded by a layer of cells called the **zona pellucida** and a collection of **granulosa cells** called the **corona radiata**. The zona pellucida helps to protect the oocyte, as it is made up of a tough layer of glycoproteins that will play a crucial role in any fertilisation by sperm. The corona radiata are cells from the original follicle, but they do offer some protection to the oocyte as it travels through the oviduct.

## Sperm

Sperm cells are much smaller than secondary oocytes and their internal structure is not clear when observed under a light microscope. Sperm develop and mature in the testes. Through the action of specialised nurse cells called **Sertoli cells**, the sperm cells differentiate so that each is made up of a head, a midpiece and a tail section. This is shown in Figure 2. The importance of these sections is shown in Table 1.

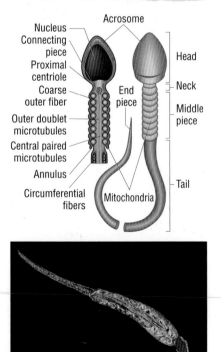

| Region of sperm | Role in fertilisation |
|---|---|
| Head | Has an acrosome cap containing hydrolytic enzymes needed to digest through the zona pellucida |
| | Contains the haploid nucleus, which will fuse with the haploid nucleus of the secondary oocyte to form a zygote |
| Midpiece | Contains many mitochondria that provide the energy needed to move the microtubules in the tail which form the contractile filaments |
| Tail | Contains contractile filaments that create a whip-like movement, allowing the sperm to swim at speeds of 1–4 mm/min |

**Table 1** Roles of key regions in structure of sperm

**Figure 2** Sperm

# Hormonal control of gametogenesis

Both oogenesis and spermatogenesis are controlled by hormones. In both sexes these processes are controlled by the release of hormones from the hypothalamus and the **anterior pituitary gland**.

The hypothalamus secretes a hormone called **gonadotrophin releasing hormone** (GnRH) into the blood. This hormone stimulates the anterior pituitary gland to secrete two hormones, follicle stimulating hormone (FSH) and luteinising hormone (LH). Both these hormones act on the ovaries and testes to trigger the development of the follicles and sperm.

| Female | Male |
| --- | --- |
| FSH and LH levels rise in bloodstream. Hormones bind to follicle cells. Follicle cells mature and produce oestrogen. One follicle becomes 'dominant' and secretes increasing levels of oestrogen | LH binds to receptors on Leydig cells in testes leading to secretion of testosterone |
| Rising oestrogen levels inhibit the release of LH and FSH initially but stimulate a 'surge' of LH (there is a smaller response for FSH) | Testosterone then affects Sertoli cells in testes, which stimulate spermatogenesis |
| | FSH binds to surface of Sertoli cells, making them more receptive to testosterone |
| LH surge causes dominant follicle to release the oocyte | Rising blood testosterone levels act to inhibit LH and an excessive increase in activity of Sertoli cells inhibits FSH by secreting the hormone inhibin |

**Table 2** Action of hormones on the production of gametes in males and females

### The menstrual cycle

Reproduction in females really involves two linked cycles, both regulated by the same hormones – the ovarian cycle and the uterine cycle. By relying on the same hormones, oocyte release can be synchronised with the development of the uterine lining. Together they are generally called the menstrual cycle. The cycle involves four phases, and each of these is tightly controlled by the action of the hormones, oestrogen, progesterone, LH and FSH. These phases in the cycle are:

- the *proliferative phase* whereby the lining of the uterus, the **endometrium**, regenerates
- the *ovulation phase* whereby the oocyte is released from the mature follicle into the oviduct
- the *secretory phase* where the endometrium secretes nutrients in preparation for **implantation**
- the *menstrual phase* where the lining of the uterus is shed.

Figure 3 shows how the actions of the hormones affect the lining of the uterus and also the changes that occur within the maturing follicle. The flowchart also summarises the action of the four hormones.

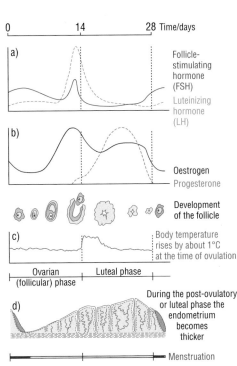

**Figure 3** Flowchart showing control of the menstrual cycle

## Menarche, menopause and breast cancer?

Studies have shown that the earlier the age of the first period – *menarche*, and the later the *menopause* (the cessation of the menstrual cycle), the greater the risk of developing breast cancer. This is probably due to the number of periods a woman will experience in her lifetime – early menarche and late menopause increases the number of periods and so increases her exposure to oestrogen and progesterone. This in turn increases the risk of breast cancer.

But it is important to understand the level of risk – starting periods early probably means that an additional one woman in every 100 may develop breast cancer. Studies have shown that the average age of menarche has not changed much in the UK over the last 20 to 30 years. So this seems to suggest that any increase in the incidence of breast cancer is unlikely to be due to changes in menarche. But what about the menopause? We can delay the onset of menopause with treatments such as HRT – hormone replacement therapy. What issues might this raise in the light of studies into the effect of menarche and menopause on breast cancer?

## Questions

1 Compare the ultrastructure of an egg and a sperm cell.
2 Describe the role of hormones in the development of gametes.
3 Explain the principle of negative feedback using the menstrual cycle as an example of this control mechanism. (Hint: look at 5.3.2.1)

## The passage of sperm

Mature sperm are stored in the **epididymis**, where they mature and become motile. During the process of ejaculation these sperm are released into a tube called the **vas deferens**. The sperm are pushed through this tube in a series of waves generated by the action of smooth muscle in and around the penis. The sperm pass through a collection of accessory glands that add various fluids, forming a liquid called **semen**. About two-thirds of the volume of the semen is *seminal fluid* from the **seminal vesicles**. This fluid contains mucus and proteins to help with swimming, and fructose to provide energy. About one-third of the fluid is provided by the **prostate gland** and this fluid helps neutralise the acidic vagina.

During copulation the semen is released into the vagina of the female. The sperm swim through the cervix and into the uterus with increasingly stronger action of their tails as time goes on. The sperm will continue to swim further up into the **oviducts** and, if an **ovum** is present in the oviduct, fertilisation can occur.

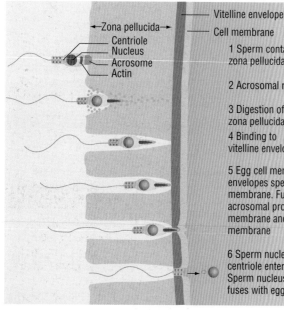

Vitelline envelope
Cell membrane

←Zona pellucida→
Centriole
Nucleus
Acrosome
Actin

1 Sperm contacts zona pellucida

2 Acrosomal reaction

3 Digestion of zona pellucida

4 Binding to vitelline envelope

5 Egg cell membrane envelopes sperm cell membrane. Fusion of acrosomal process membrane and egg membrane

6 Sperm nucleus and centriole enter egg. Sperm nucleus then fuses with egg nucleus

**Figure 1** An oocyte being fertilised by sperm

## How fertilisation occurs

After it has been released, the secondary oocyte is swept into the **Fallopian tube** by the movement of feathery **fimbriae** and then moved along by peristalsis in the tube and by the **cilia** of the cells that line it. For successful fertilisation and implantation, the sperm must reach the oocyte before it travels too far down the oviduct.

There is a mechanism that makes sure that only one sperm cell penetrates through the outer layers of the oocyte. The events are as follows:

- when sperm reach the oocyte, the capacitated sperm binds to a glycoprotein **receptor** on the **zona pellucida**. This triggers the **acrosome reaction**
- one sperm penetrates the outer layer of cells surrounding the egg and passes through the zona pellucida. The sperm head reaches the **plasma membrane** of the oocyte and binds to another receptor
- **lysosomes** present in the oocyte called *cortical granules* are immediately stimulated to fuse with the zona pellucida and change the proteins there. This is the **cortical reaction** and forms a **fertilisation membrane** stopping the entry of any further sperm
- lastly, the presence of the haploid sperm causes the oocyte to finally complete meiosis. Of the two potential 'cells', one becomes a second **polar body**. The sperm nucleus now enters the oocyte enabling the two parental nuclei to fuse. The cell is now diploid and is called a **zygote**
- rapid mitosis takes place, turning the zygote into a bundle of cells called a **blastocyst**. The blastocyst will continue to travel down the oviduct towards the uterus, where it will implant into the lining of the uterus about seven days later.

# Implantation of the blastocyst and its future development

As the blastocyst (bundle of cells) makes its way to the uterus, it can implant only if the lining of the uterine wall, the **endometrium**, is at the correct stage in the **menstrual cycle**, usually around the twentieth day.

Implantation starts when the blastocyst makes contact with the endometrium and stimulates an inflammation-like reaction. The endometrium grows out around the blastocyst, forming a structure called a **trophoblast**. This structure will eventually go on to form the placenta. The blastocyst now develops into an **embryo**.

The blastocyst starts to secrete the hormone **human chorionic gonadotrophin**, or HCG. This continues for up to eight weeks. This has the effect of stimulating the **corpus luteum** so that, rather than declining, it carries on producing **progesterone**. The progesterone maintains the lining of the uterus and no menstruation occurs. The hormone is similar to LH and it also inhibits the release of FSH so for the duration of the pregnancy, no further **follicles** will ripen.

Further development of the trophoblast will continue and one of the membranes surrounding the embryo, the **chorion**, will develop projections that will grow deep into the endometrium. These **villi** will then develop into **chorionic villi** and will form the functional unit of exchange in the placenta. This will allow the blood systems of the mother and the developing fetus to come into very close contact but never mix.

# Hormones involved in pregnancy, birth and lactation

As soon as the blastocyst becomes implanted, new hormones are being synthesised and secreted into the bloodstream. During the development of the fetus, its birth and also during the period that it receives breast milk, considerable hormonal changes will take place in the woman's body. The action and importance of these hormones during these three stages are outlined in the following table.

| Hormones during pregnancy | Key roles in the body |
| --- | --- |
| Human chorionic gonadotrophin (HCG) | Stimulates the corpus luteum to release oestrogen and progesterone to maintain the lining of the uterus and ensure implantation |
| Human placental lactogen (HPL) | Ensures that breast tissue is receptive to oestrogen and progesterone and is thought to help regulate maternal blood glucose levels |
| Oestrogen and progesterone | During pregnancy these hormones cause the breast tissue to develop in readiness for milk production |
| Oxytocin | Secreted towards the end of pregnancy to start the process of birth by stimulating uterine muscles |
| **Hormones during birth** | |
| Oxytocin | Secreted in increasing quantities towards the fortieth week of pregnancy and stimulates the contractions that will allow the birth of the child. This coincides with declining levels of progesterone that had acted to inhibit oxytocin |
| **Hormones during lactation** | |
| Prolactin | Maintains milk production in the breast tissue and also inhibits any further ovulation |
| Oxytocin | Released as the infant sucks the nipple and stimulates muscular contractions around the milk glands, causing a release of milk |

**Table 1** The importance of hormones during pregnancy

### Examiner tip

It would be good practice to produce a flowchart explaining the journey that a sperm and an oocyte have to make from the point of production to the point where they combine to form a zygote. There are many terms and sequences that you need to get in the correct order and most of these are highlighted in the text. Use these key words in your description to achieve maximum credit for your answer.

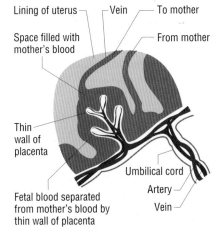

Lining of uterus — Vein — To mother

Space filled with mother's blood

From mother

Thin wall of placenta

Umbilical cord

Artery

Fetal blood separated from mother's blood by thin wall of placenta

Vein

**Figure 2** The placenta

### Key definition

The **placenta** is an organ formed from membranes from the foetus and the lining of the uterus, and is the organ of nutrient exchange. It has a large surface area and uses steep concentration gradients to ensure efficient exchange of gases, nutrients and waste.

# Questions

1  Explain why only one sperm can enter the egg and complete fertilisation.
2  Describe the importance of the uterine wall in implantation.
3  When the blastocyst is implanted into the uterine wall there is a reaction to these cells almost like an inflammatory immune response. Suggest a reason for this. (Hint: think back to work on antigens and antibodies.)
4  Produce a spider diagram to show how hormones regulate and control the processes in pregnancy, birth and lactation.

## Family planning

**Contraception** is one method of birth control or family planning. It is a strategy that couples can use to prevent fertilisation. Another method is to prevent implantation and the start of a pregnancy. Both strategies give couples choices regarding whether or when to have a child. The most commonly used methods are:

*Contraceptive devices*
- Birth control pill
- Condoms
- Diaphragm
- Implants and injections
- Sterilisation

*Anti-implantation devices*
- Intra-uterine device – IUD or coil
- Morning-after pill

### The birth control pill

The pill contains a combination of synthetic female hormones, **oestrogen** and **progesterone**, and works by 'fooling' the body into thinking it is pregnant. The pill is taken orally. The hormones it contains take advantage of the **negative feedback** mechanism that controls hormone release from the hypothalamus and the **pituitary gland**. As a result, gonadotrophin release is inhibited. This in turn stops follicles maturing and so stops **ovulation**. The monthly menstrual bleed – the uterine cycle – can be allowed to continue by stopping taking the pills every 21 to 23 days. The woman often takes a placebo tablet or spacer that does not contain the active hormone.

### Condoms and diaphragms

This is one of the most widely use methods of contraception. Condoms act as barriers, preventing the meeting and fusion of sperm and egg. Male condoms can be used with spermicidal cream, which makes them more reliable. The female condom, also known as a diaphragm or femidom, can also be used. This is a cap-shaped device that must fit tightly over the cervix, and women need advice on fitting this. It also needs to be left in place for at least six hours after intercourse. Both male and female condoms have the advantage of not only offering excellent prevention from pregnancy, but also reducing the risk of spreading sexually transmitted infections such as HIV or **chlamydia**. Like a condom, the diaphragm or cap prevents sperm reaching the egg. The diaphragm has to be inserted several hours before sexual intercourse and  must also be used with some spermicidal cream over the cervix to prevent any entry of sperm into the uterus. Because of the inconvenience of fitting and the fact that women cannot leave a diaphragm over the cervix for long periods of time, it is no surprise that this is not a very popular contraceptive method.

### Implants and injections

Women can receive contraceptive hormones via implants under the skin and also through a one-off injected dose. Implants such as the drug Implanon are injected under the skin, and slowly release a progesterone-like hormone that inhibits ovulation. The action of the implants can last for up to three years. The drug Depo-Provera, containing synthetic progesterone, is injected every 12 weeks. Its effects are similar to those of the implants. The advantage of these methods is that they offer nearly 100% protection from pregnancy and you do not run the risk of forgetting to take your pill! Like the contraceptive pill, these methods offer no protection from sexually transmitted diseases.

## Negative effects of the birth control pill

There are still concerns about the negative side effects associated with oral contraceptive pills. These include an increased risk of **thrombosis**, leading to heart attacks and strokes, and an increased risk of breast cancer. But much of the supporting data was generated from women who were using high dosages of hormones as opposed to the low doses in modern versions of the pill. For today's pills the risk of side effects is low, except for women over the age of 35 who smoke. For this group, the risk is considerably higher. For most other women the risk of death from using the pill is less than the risk associated with full-term pregnancy. Interestingly, the 'mini-pill' works in a very different way altogether. The low levels of progestins alter the mucus composition around the cervix, keeping it thick and sticky so it blocks the passage of sperm.

## How does male and female sterilisation work?

Sterilisation for both sexes involves cutting or tying (**ligating**) the tubes that allow either the sperm to travel from the testes (via the **vas deferens**) or the **oocytes** to travel from the ovaries (via the oviducts or **Fallopian tubes**).

Both processes are 100% effective and are normally irreversible. Sperm production continues, but since the sperm cannot migrate from the testes, they are destroyed by **macrophages**.

Figure 1 shows how sterilisation occurs in males and females.

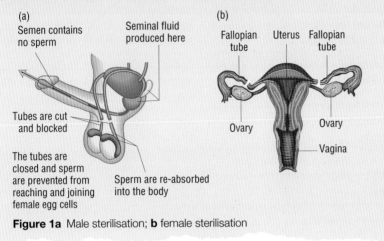

**Figure 1a** Male sterilisation; **b** female sterilisation

### Intra-uterine device (IUD)

This device is a small piece of copper or plastic that is specially fitted into the uterus and prevents the implantation of embryos. Once inside the uterus, the device actually stimulates an inflammation response, which prevents implantation. Also, the copper present in the IUD itself is toxic to the sperm and the embryo. The IUD has a failure rate of 1–7%.

### The morning-after pill

If a contraceptive method fails or unprotected intercourse has taken place, then a woman can use the morning-after pill as a method of emergency contraception. This can be used only within 72 hours of sexual intercourse and involves taking a large dose of synthetic **steroids**. The result is to prevent any implantation of the embryo. The short-term effects of the pill are extreme abdominal pains and sickness. There are no known long-term side effects of using this method of emergency contraception, but women are advised not to take this pill more than three times in a year.

## Questions

1 Explain how the oral contraceptive pill affects the natural production of oestrogen, FSH and LH.
2 Suggest two advantages and disadvantages of male and female sterilisation in controlling fertility.
3 Describe some of the ethical arguments that surround the use of contraception.
4 Copy and complete the following table:

| Contraceptive device | Condoms | IUD | Implanon |
|---|---|---|---|
| Advantage of use | | | |
| Disdavantage of use | | | |

## Natural family planning

Humans have always had ways of controlling their fertility and preventing unwanted pregnancy. More than 2000 years ago women used to make a drink containing active natural ingredients from the silphium plant. The use of natural family planning can reflect both religious beliefs and a reluctance to introduce artificial chemicals or objects into the body. Natural family planning relies on a woman keeping daily records of 'fertility indicators' such as temperature and mucus composition or levels of **LH** in urine. Success varies, and depends on how well a woman has been taught the technique. It can be as high as 98% effective. But lifestyle changes, illness, stress or travel can all alter the indicators, resulting in pregnancy.

### Examiner tip

For all of the fertility treatments available to infertile couples, you should prepare ethical and moral arguments for and against each treatment. It is important that you try to be impartial and view the treatments from the viewpoint of the rights of the parents and also of the unborn child. Each treatment brings with it specific moral, ethical and religious arguments that you should investigate further. Some of the treatments also carry an element of risk – particularly the hormone treatments.

Simon and Megan now have a family of three sons, but at one time it seemed as if they would have to remain childless or adopt. In 1997, after two pregnancies had ended in miscarriage and after an ectopic pregnancy, Simon and Megan had their first cycle of IVF. Megan became pregnant eventually after a frozen embryo transplant, but unfortunately miscarried. They adopted a 5-month-old baby boy – thinking that this was their only chance of being parents, but their local Primary Care Trust offered them another cycle of IVF. This was successful, and Simon and Megan now have twin boys. They also agreed to donate their remaining frozen embryos to assist other childless couples.

### Asherman's syndrome

Suzanne gave birth to her first daughter, but after a week she developed a uterine infection. After unsuccessful antibiotics, a medical scan and a D&C she had no periods for over a year following the procedure. After a series of failed diagnoses, Asherman's Syndrome was suspected. This is scarring in the uterus, which had resulted from the D&C. Suzanne underwent a hysteroscopy to try to remove the scarring, but although her periods returned, they were very light. Further surgery followed, alongside a course of HRT.

The surgery had been quite severe. The result meant that her uterus was now mostly scar-free but her **endometrium** was very badly damaged and thin. Suzanne has had two unsuccessful attempts at IVF, and, even with no pregnancies, the IVF drugs helped thicken up the uterine lining.

### Examiner tip

You should be able to understand what treatments to assist with fertility are trying to do. For example, you should recap hormonal action and control of the menstrual cycle when considering ovulation induction.

## Causes of infertility

Currently one in seven couples in the UK has problems conceiving.

The problem of infertility could be a result of problems with the female reproductive system, problems with the male reproductive system or a result of some other factor. The following table outlines some of the potential causes of infertility in both males and females.

| Females | Males |
|---|---|
| Failure to ovulate due to abnormal menstrual cycle or insufficient hormone levels | Sperm may be abnormal and fail to develop correctly in the testes, for example they might not contain tails |
| Blockage of the oviducts, which could be a result of a bacterial infection | Semen may contain such a low number of sperm cells that there is not enough to be successful at fertilising an egg |
| Endometriosis may occur which causes the uterine lining to develop outside of the uterus and may block the oviduct | Some men also produce antibodies into their own semen, which actually destroy the sperm at source |
| Anti-sperm antibodies may develop, which are secreted into the uterus, destroying sperm before it reaches the egg | |

**Table 1** Causes of infertility

## Fertility treatments

There are treatments available for infertility in both males and females. Ovulation can be induced with hormone treatment, surgery can unblock oviducts and sperm ducts, and couples can also be offered methods of assisted fertilisation. Some of the most commonly used treatments available to men and women include:

- ovulation induction
- artificial insemination
- *in vitro* fertilisation
- frozen embryo replacement
- gamete intrafallopian transfer
- intracytoplasmic sperm injection
- use of sperm banks and donor sperm.

### Ovulation induction

Many infertile women produce follicles that do not fully develop to form viable eggs. This is a condition called *polycystic ovarian syndrome*. It is treated by giving these women drugs that act to inhibit oestrogen. The effect is to promote the release of **GnRH** from the hypothalamus and so stimulate the release of FSH and LH from the anterior pituitary gland. This will induce ovulation and also help to develop immature follicles. LH and FSH can also be given directly into the bloodstream if other treatments fail. One possible result of the treatment is that by stimulating the development of follicles there is a high risk of multiple pregnancies.

### Artificial insemination

This involves injecting semen into the top of the vagina or the uterus through a small plastic tube. The process can involve either intra-cervical insemination (ICI), or intra-uterine insemination (IUI). With ICI, the semen is collected and insemination occurs within two hours at the top of the vagina. A plastic cap is often placed in the vagina for several hours to give the sperm the chance to enter the uterus through the cervix. IUI places sperm near the oviducts and this treatment is more successful. Semen rather than sperm is inserted into the uterus using a thin plastic tube that is passed through the cervix.

## How does IVF work?

*In vitro* fertilisation aims to fuse oocytes and sperm outside of the body. These cells fuse and the zygote goes on to divide, forming a blastocyst. This is then artificially implanted back into the endometrium. The first successful procedure took place in 1978, and although the technology has been refined, the basic process still involves the following steps:

- a woman is 'super-ovulated' using synthetic hormones so that several follicles ripen at the same time
- an ultrasound probe of the vagina is used to locate ripe follicles in the ovaries. This is done under local or general anaesthetic
- these follicles are aspirated (sucked out using a special instrument) a few hours before ovulation
- the oocytes are removed using a suction device and placed in a test tube containing a special medium
- these oocytes are then maintained in separate test tubes at body temperature
- sperm are prepared and at least 100 000 are added to each oocyte in a small Petri dish
- after 16–20 hours, the oocytes are checked to see if they have been fertilised
- the resulting embryos are then left to develop for two to three days in the incubator
- they can then be transplanted back into the uterus.

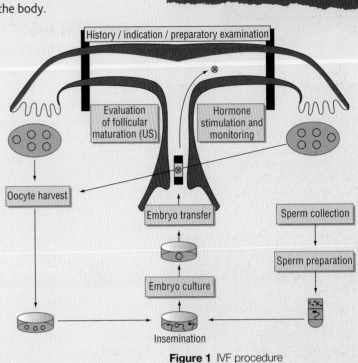

**Figure 1** IVF procedure

By providing many possible embryos in the process of IVF, there is a possibility that these can be frozen and stored for later use. These embryos can be used by women who, possibly through a medical procedure such as chemotherapy, cannot produce their own eggs. Some may be donated to another infertile woman.

### Alternatives and modifications to IVF

There are three possible alternatives for infertile couples that use the same principles as IVF, and also others that slightly modify it. These are explained below:

| GIFT – gamete intrafallopian transfer | ZIFT – zygote intrafallopian transfer | ICSI – intracytoplasmic sperm injection |
|---|---|---|
| Sperm and oocytes are passed directly into the oviduct and allowed to fertilise naturally – these oocytes may have been donated | This involves a zygote being created by the process of IVF, but the zygote is implanted into the oviduct and allowed to implant naturally | IVF is given a helping hand as the sperm is injected directly into the egg, forcing fertilisation |

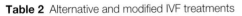

**Table 2** Alternative and modified IVF treatments

### Sperm bank

Sperm has been frozen in sperm banks and used to fertilise oocytes for over 50 years. This treatment has had a good success rate and has helped infertile couples and single woman to conceive.

Since the 1950s there have been many changes to the law regarding the use of donor sperm. Currently in the UK, the identity of the sperm donor cannot be revealed to the recipients, though his physical characteristics, blood group and other information can be told in order to help in matching.

## Questions

1 Explain how the use of reproductive hormones can help some infertile women.
2 Using the hormones used in fertility treatment, discuss their roles in the ovarian and menstrual cycle.
3 Explain how female antibodies may act to destroy male sperm and so prevent fertilisation.
4 Describe why it is important to screen donated sperm for diseases such as HIV.
5 Between 97% and 98% of men with cystic fibrosis may be infertile due to development problem with the sperm duct (vas deferens). Which procedure would you recommend? What other advice might you also offer?

**Examiner tip**

You need to read questions carefully and watch the terms that you use in the answers. Also, oocytes, sperm and embryos can all be stored frozen.

**Figure 1** Storage of embryos

**Vanishing twin syndrome**

Occasionally a twin observed during ultrasound in the early stages of development will disappear during later pregnancy. This is called a vanishing twin. The developing fetus has died during the pregnancy and has been reabsorbed by the woman's body. This can occur as a result of chromosomal abnormality, fetal development problem or a fault with the placenta.

If the fetus dies early on in the pregnancy then there are usually no health worries for the mother. But if the death occurs during the later stages of development it can result in infection, premature labour and death of the other fetus.

**Examiner tip**

You should re-read the section you studied on antibody structure. Remember that antibodies have variable and constant regions. The variable region is specific to the antigen and here, the hCG is the antigen.

## Embryo storage

Human embryos have been successfully frozen and thawed since 1983. Embryo storage initially involves screening for HIV and hepatitis B and C. The embryos can be frozen at any time between day one and day six inclusive. They may then be stored unharmed, frozen in liquid nitrogen at temperatures of −300 °C, for 50 years or more. But the current law restricts the time of freezing to between five and 10 years. With oocytes, the survival rate is around 50% but the fertilisation rate using conventional methods is low. But, using ICSI (intracytoplasmic sperm injection), the fertilisation rate is much improved. The first baby born using frozen oocytes was in 1986, but the success rate in 1997 resulted in only 13 healthy children from 212 transfers!

In the UK there is a legal requirement to state how long you would like your embryos frozen and what you would like to happen to them. If the couple separate and one partner withdraws his or her consent to the ongoing storage of the embryos, then the current law dictates that the embryos are removed from storage.

When freezing embryos, couples have the choice of using their own embryos, donating embryos to another couple undergoing fertility treatment, or donating the embryos to a research project. The frozen embryo cycle is relatively non-invasive compared to an egg collection cycle. The embryos can be replaced in a natural cycle if the time of ovulation can be monitored easily. Failing this, a 'controlled cycle' is used and hormonal tablets or nasal sprays are used to prepare the endometrium for implantation. The development of the endometrium is monitored by ultrasound. When the endometrium is the correct thickness, and providing both partners give consent, embryos can be thawed for transplantation.

## Multiple pregnancy

Multiple pregnancies occur when more than one fetus develop simultaneously in the womb.

These pregnancies occur naturally in about 1 in every 100 births. This natural incidence has been upset by current advances in fertility treatments. When women are given hormone treatment that acts to increase follicle development, this has the effect of creating a state of *super-ovulation*, which leads to the release of many **secondary oocytes** into the oviducts and the possibility of a multiple pregnancy.

Most multiple pregnancies produce twins – identical or fraternal. Identical twins occur as a result of the splitting of a single fertilised zygote into two separate individuals. Fraternal twins occur when two eggs are fertilised by separate sperm.

When twins fail to separate completely, the result is Siamese (or conjoined) twins.

### Risks involved in multiple pregnancies

Women who are pregnant with more than one fetus run a higher risk to their health than those who carry a single fetus. Their unborn children are also at a higher risk of problems such as low birth weight (less than 2500 g) and premature birth. They are also at a higher risk of being stillborn. For mothers, there is an increased risk of high blood pressure and **pre-eclampsia**, anaemia, haemorrhage, and early labour, and more chance of needing a Caesarean section. Although the risk of mortality in pregnancy is small, it is double for women expecting twins than for those expecting just one child.

If multiple pregnancies are a result of fertility treatment, a choice can be made to selectively reduce the number of fetuses. This is known as *fetal reduction*. The abortion of one or more fetuses can be carried out to increase the chances of survival for the other developing fetuses. This procedure does risk all of the fetuses and some women may miscarry or enter premature labour. It is a difficult and emotional decision for all involved.

Clinics will transfer a maximum of two embryos per IVF cycle if the woman is aged 39 or under, or three if the woman is over 40 and using her own eggs.

# Premature birth

Premature birth is the birth of a baby before the standard period of the 40-week pregnancy is completed. Babies born any time before the thirty-seventh week of pregnancy are classed as premature. By being born premature the baby has a risk of developing certain health problems. These are outlined in Table 1.

Advances in medical care and technology have meant that the age when a fetus can survive prematurely is actually quite low. The age above which prematurely born fetuses have a good chance of surviving is considered to be about 24 weeks. However, babies born this early are at considerable risk of brain damage because the lungs are not sufficiently developed to supply the brain with adequate oxygen.

# Pregnancy testing

There are tests that can be used very early on in pregnancy which can detect whether or not an oocyte has been fertilised by a sperm. If the woman has been receiving fertility treatment or has a history of multiple pregnancies, the outcome of this test will be vital in preparing the couple for the potentially difficult time to come. Most pregnancy tests use **monoclonal antibodies** or MABS, to detect the presence of the hormone **human chorionic gonadotrophin** (HCG) in the mother's urine. The test is described in Table 2.

### Biological, ethical and economic arguments in fertility treatment

In Table 3 there are some arguments for and against fertility treatments from three very different perspectives. It is important to be objective when considering reasons why people might seek help through fertility treatment and the effects such treatment might have on the couple or on the individuals born to the treatment.

- poor neurological development
- congenital heart defects due to a failure of the child's ductus arteriosus to close after birth
- *respiratory distress syndrome* or chronic lung disease
- gastrointestinal and metabolic problems such as hypoglycaemia or feeding problems
- blood problems such as anaemia or jaundice
- infections of the urinary tract.

**Table 1** Possible problems of premature birth

- the pregnancy stick is dipped into the mother's urine – an early morning sample is normally used as this will have the highest concentration of the hormone
- any HCG in the urine will bind to specific antibodies held on the stick
- these antibodies are bound to a colour bead and form a HCG–antibody–colour complex
- the urine then seeps up the stick until it reaches a region of fixed immobilised antibodies
- these antibodies bind to the HCG–antibody complex if it is present, giving a coloured band – this is the mark of a positive result for pregnancy
- a further band is used as a control, and will give colour with the antibody alone. This is a control and shows the test is working.

**Table 2** Pregnancy testing

| Biological | Ethical | Economic |
|---|---|---|
| There have been huge scientific advances in assisted fertilisation | Assisting fertilisation raises issues regarding the question of embryo research. What time limits should be placed on this research? Furthermore, when does life begin and what are the rights of an embryo? | Infertility treatment is expensive and because success is not guaranteed there can often be no specific end point to treatment |
| This is partly due to a decline in the fertility of couples in modern times | There are issues around sperm donor banks and the future child's right to access to information about his or her genetic background and mode of conception | Assisted reproductive techniques, such as IVF, have made treatment even more expensive because so much expertise and technology is needed for these procedures |
| Reasons for this include the increasing age of women at the time of marriage and childbearing. Also there has been an increase in the incidence of sexually transmitted diseases that damage the reproductive tract in both men and women | World religions also differ on the ethical implications of such treatment. For example, to the Catholic Church, artificial insemination by IVF is not acceptable. Procreation without sexual union in considered unnatural. In Judaism, however, IVF is accepted as necessary to heal the illness of infertility | Couples need to consider the cost-effectiveness of each treatment option. For example, it is true that an IVF cycle is four times as expensive as an IUI cycle, but the chance of a pregnancy is also four times as great |
| There have been overall decreasing sperm counts in men worldwide in the last few decades. The reasons for this are unclear but could be due to environmental pollution or to the stresses of modern life | | What is more, most insurance companies take the view that infertility is not a medical problem, so many couples are reluctant to claim for medical expenses for treatment. There is also evidence of a 'postcode' lottery with conditions for offering IVF and the numbers of cycles offered on the NHS varying between authorities |

**Table 3** Arguments for and against fertility treatments

# Questions

1 Discuss the ethical implications around the storage of frozen embryos.

2 Suggest why embryos are stored only for a limited time period such as 5 or 10 years.

3 Explain how antibodies might detect the presence of chemicals such as the hormone HCG.

4 What might be an advantage of a vanishing twin to the mother and remaining fetus in early pregnancy?

## Photosynthesis and respiration

Plants use photosynthesis to trap sunlight and use the energy to convert water and carbon dioxide into glucose, ATP and oxygen. In this way, they convert light energy into chemical potential energy. The sugars produced are then used in cellular respiration in the plant and converted into ATP to fuel active processes in plants such as protein synthesis and active transport. Converting and using the Sun's energy into usable chemical energy involves the actions of light-harvesting pigments such as chlorophyll.

Globally, photosynthesis is important in that it produces oxygen for human respiration and, as autotrophs, plants form the basis of every food chain.

### The site of photosynthesis

Photosynthesis occurs in chloroplasts present in some plant cells. These membrane-rich organelles are particularly abundant towards the top of the leaf in specialised cells that make up the palisade layer.

The biochemical processes occurring in photosynthesis involve two clearly distinct stages and involve the action of many enzymes. The reaction sites for these processes occur in the stroma and on the thylakoid membranes of the chloroplast. Thylakoid membranes span the whole chloroplast. Photosynthetic pigments are housed in the membranes, and these are the sites of the **light dependent** reaction. The stroma is the enzyme-rich matrix of the chloroplast where the **light independent** reactions occur.

### Light-dependent reaction

The first stage in photosynthesis involves the pigments, held on the thylakoid membranes. These harvest the energy from photons of sunlight. The pigments are grouped together in specialised structures called 'photosystems' and it is here that the Sun's energy is transformed from light to chemical energy. The photosystems include many different pigments that can trap light wavelengths of differing energy values. There are two photosystems, each with a molecule of chlorophyll as the primary pigment and a network of associated accessory pigments.

The energy from the sunlight energises electrons in the chlorophyll molecules. These electrons are boosted to a higher energy level in the photosystems. These electrons cannot remain energised and so return to their lower energy level through a series of electron carriers in the thylakoid membranes. As the electrons do this they release energy.

Electrons expelled from a chlorophyll molecule go through one or two electron transport systems, resulting in formation of ATP and reduced NADP. NADP is a coenzyme molecule that acts as a carrier of potential energy. It picks up hydrogen atoms, becoming reduced. The reduced NADP will be used in the light-independent stage to produce sugars. The energy for this process will be provided by the ATP.

ATP production in chloroplasts is similar to ATP production in mitochondria during aerobic respiration. The formation of ATP during photosynthesis is called photophosphorylation. In addition to producing ATP, light energy is used to split water molecules to provide a source of electrons and hydrogen ions. This process is called photolysis. It releases oxygen as a waste product and also supplies hydrogen to the coenzyme NADP.

### Key definition

Plants are said to be **autotrophic** and form the basis of all food chains. They use inorganic molecules and an external energy source to produce their own organic molecules. These organic molecules include glucose, starch, amino acids and fatty acids. These are then available to higher trophic levels that will feed on the plants directly (primary consumers), or feed on the primary consumers (secondary consumers). Organisms further up the food chain are termed **heterotrophs**.

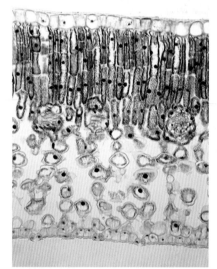

**Figure 1** A leaf section showing the pallisade layer in purple (see Figure 3 spread 1.1.1.5)

## The light-independent reactions

The light-independent reactions do not need the energy from light to proceed but do require the ATP and reduced NADP made in the light-dependent stage. This process is an enzyme-driven metabolic pathway known as the Calvin cycle. All the reactions take place in the stroma of chloroplasts and involve the reduction of atmospheric $CO_2$ to sugars. The sequence of events that occurs during this reaction is as follows:

- $CO_2$ diffuses into a leaf, across the plasma membrane of a photosynthetic cell, and into the stroma of the chloroplasts.
- An enzyme called ribulose bisphosphate carboxylase (Rubisco) combines a $CO_2$ molecule to a 5-carbon sugar called ribulose biphosphate (RuBP).
- This produces an unstable 6-carbon intermediate.
- This then splits to form two 3-carbon molecules of glycerate 3-phosphate (GP).
- The GP molecules are then reduced to triose phosphate using the hydrogen atoms from reduced NADP and the ATP from photophosphorylation (this is why the light-dependent stage is important).
- Most of the TP molecules continue in the cycle, and more ATP is used to regenerate the RuBP and so continue to fix more carbon dioxide.
- TP molecules will join to form sugar-phosphate, which can then be modified to form hexoses such as glucose, sucrose, starch, cellulose or fatty acids. Provided the plant has access to nitrates and other minerals, amino acids and nucleic acids can also be synthesised.

### Examiner tip

It is important that you understand that for both photosynthesis and respiration, the conversion of energy occurs. Energy is not lost or gained but simply transformed from one form to another. Key molecules in both processes act as energy transducers converting, for example, light energy to chemical energy. Make sure you don't confuse the NAD in respiration with the NADP of photosynthesis. Think 'P' for photosynthesis. But remember – plant cells will be carrying out both photosynthesis and respiration.

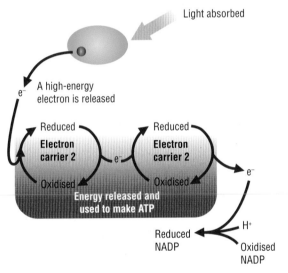

**Figure 2** Summary diagrams of light dependent stage

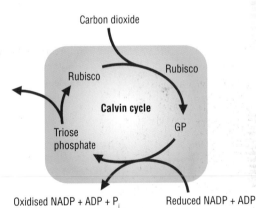

**Figure 3** Summary diagram of light-independent stage

## Questions

1. Outline the similarities between the structure and function of mitochondria and chloroplasts.
2. Explain why humans are totally dependent on photosynthesis.
3. Plants are autotrophic and humans are heterotrophic feeders. What is meant by these terms?
4. Complete the following summary table of photosynthesis:

| | Light-dependent stage | Light-independent stage |
|---|---|---|
| Site of reaction | Thylakoid membranes on chloroplast | |
| Reaction molecules involved | | RuBP, Rubisco |
| Reactants | Light and water | |
| Products | | TP, ADP, NADP, RuBP |

## The role of $CO_2$ and the uses of TP

$CO_2$ is a key molecule in photosynthesis as it forms the building blocks of all organic molecules made in the plant. $CO_2$ is fixed, reduced and transformed by the action of enzymes and reactive molecules in the stroma. The carbon atoms form the backbone for all organic molecules, to which other atoms will attach, creating sugars, lipids and proteins.

When the $CO_2$ is reduced during the Calvin cycle it goes through a series of metabolic steps, finally producing molecules of triose phosphate, TP. The molecules of TP and GP can be used by the plant to synthesise other organic molecules. These are outlined in the following table:

### Key definition

An **organic molecule** is any molecule that is constructed about a central carbon backbone. These molecules are made autotrophically by plants during photosynthesis.

### Examiner tip

Table 1 shows how interlinked the processes of photosynthesis and respiration are. The products of one process in photosynthesis can act as a reactant in a process in respiration. You should be able to produce a list of similarities and differences between these two important metabolic processes.

| Organic molecule | Production mechanisms |
|---|---|
| Carbohydrates | TP undergoes a series of enzyme reactions and can form simple monosacchrides such as glucose and fructose, the transport disaccharide sucrose, or complex polysaccharides such as starch and cellulose |
| Lipids | TP can be converted to glycerol, a component of triglycerides. GP produced during the light-independent stage of photosynthesis can enter the glycolytic pathway, can react with acetyl co-A and then be converted to fatty acids |
| Proteins | GP and TP can both be turned into amino acids via enzyme-driven pathways linked to the Krebs cycle. Of all the $CO_2$ taken in during photosynthesis it is estimated that over 30% of this is used to make proteins |

**Table 1** Use of TP and GP in plants

## Nitrogen, amino acids and photosynthesis

Many plants are able to synthesise all their amino acids using a nitrogen source such as ammonia or nitrate and GP/TP from photosynthesis. It is important that nitrogen, in an available form, is found in the soil so that plants can produce amino acids. Soil microorganisms can help recycle nitrogen within an ecosystem, making it available for plants to use. Some plants also have microorganisms associated with their roots which act to provide available nitrogen to the plants. Figure 1 shows how three different types of microorganism recycle nitrogen within an ecosystem, and allow plants to use this and produce proteins as a result of photosynthesis.

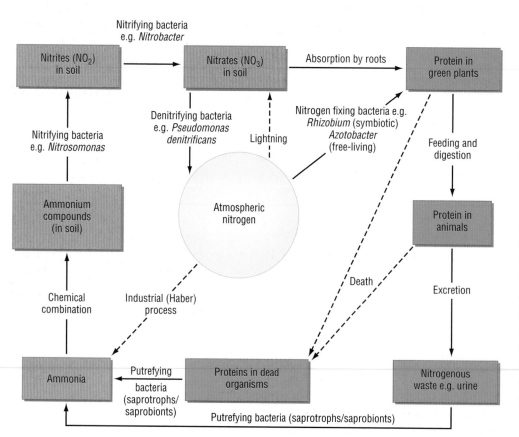

**Figure 1** Nitrogen recycling

## Photosynthesis and energy flows

The process of photosynthesis creates organic molecules and transfers the energy from the Sun, through plants and into animals (the food chain). As the flow of energy passes through each stage or trophic level of the food chain, successive organisms use this energy and so pass less on to the next organism in the chain. The result of this energy flow is a decrease in energy efficiency as you progress from producer to consumer. This can be demonstrated with the simple food chain:

maize  →  cattle  →  human

When the cattle consume the maize, not all of the available energy trapped in the maize plants by photosynthesis will pass into the tissues of the cattle. Some of the carbohydrates will be digested by the cattle and used to generate ATP in respiration. This ATP might be used for muscle movement and so heat energy will be lost to the surroundings and not incorporated into the tissues of the cattle.

Likewise, when humans consume the cattle only a small fraction of the animal will be eaten and an even smaller proportion of this energy will be incorporated into the tissues of a human. This loss of energy can be seen of as inefficiency in the system.

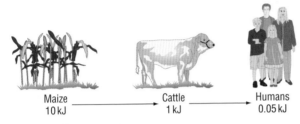

Maize 10 kJ — Cattle 1 kJ — Humans 0.05 kJ

**Figure 2** Energy passing on

Figure 2 shows how energy transfer occurs from maize to cow to human. At each trophic level there is a loss of approximately 90% of all of the available energy. The efficiency of energy transfer is therefore about 10%. The losses are due primarily to the following:

- parts of the plant and animal being inedible
- losses due to excretion and egestion
- losses due to respiration.

In calculating the efficiency of ecosystems it is often apparent that many energy transfers from one trophic level to the next are much more inefficient than 10%. This is especially the case in farm animals and where humans are the consumers. Using the example shown in Figure 3, the human would only use 4.8% of the total energy taken in by the cow.

> Total energy taken in by cattle = 2500 kJ
> Losses = 800 kJ due to respiration + 1580 kJ due to waste (urine and faeces)
> Energy passed on = 120 kJ
> % efficiency = 120/2500 x 100 = 4.8%

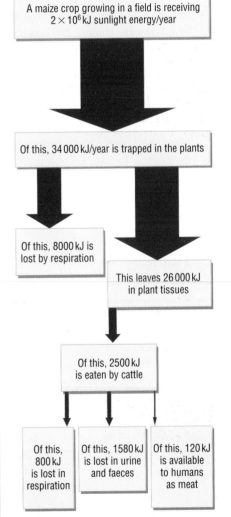

A maize crop growing in a field is receiving $2 \times 10^6$ kJ sunlight energy/year

Of this, 34 000 kJ/year is trapped in the plants

Of this, 8000 kJ is lost by respiration

This leaves 26 000 kJ in plant tissues

Of this, 2500 kJ is eaten by cattle

Of this, 800 kJ is lost in respiration

Of this, 1580 kJ is lost in urine and faeces

Of this, 120 kJ is available to humans as meat

**Figure 3** Efficiency in a food chain

## Questions

1 Suggest reasons for the plant producing lipids and amino acids as well as carbohydrates.
2 Explain what the result would be of limiting the amount of nitrate available to plants.
3 Suggest reasons why the number of trophic levels in a food chain is limited to a maximum of normally six links.

## Food production

Farmers produce the food that we eat, and the farming systems that they use vary considerable both within a country and between different countries. In the UK, farming systems tend to be either extensive or intensive. Both these systems provide food for a steadily growing population, and each has advantages and disadvantages.

### The 'green revolution'

In 1950, 692 million tons of grain came from 1.7 billion acres of arable (cropped) land to feed 2.2 billion people. By 1992, 1.9 billion tons was being produced from just 1.73 billion acres to feed 5.6 billion people. Until just over 50 years ago, each person 'took' more land to feed than his parents did! These amazing increases in the efficiency of crop production were brought about by the 'green revolution', and one of the chief architects of this revolution was Norman Borlaug. Borlaug was awarded the Nobel Prize in 1970 for developing, through plant breeding, short-straw varieties of cereal. Nitrate fertilisers used on these varieties goes into grain production rather than straw extension. However, the short plants would not compete well with weeds and hence herbicides also have to be used to control weeds. This means that global cereal production has been tied into an intensive farming system, and this has supported a huge increase in the global population.

Clearly this production system is heavily dependent on chemical inputs and is energy demanding. But can we turn the clock back? It is generally recognised that current production has probably peaked but global populations are still growing. In order to avoid Malthus's warning of demand outstripping supply, we will need to consider both new technologies and a change to the foods we use as staples.

### Extensive farming

This involves producing crops and livestock on large areas of land with little or no added fertiliser. The farmer relies on the natural fertility of the land and nutrient recycling within the system. Examples of this are the farming of sheep on upland areas and the conifer industry.

### Intensive farming

This system of farming involves produces large amounts of food in relatively smaller areas of land. These are said to be open systems of agriculture because there is a net loss of nutrients such as nitrates from the system. These have to be replenished with the repeated application of fertilisers. Pesticides also need to be used, since modern varieties of crop plants and domestic animals do not compete well with parasites, pathogens and weeds. Examples of intensive farming in the UK include poultry and pig farming and intensive cereal production.

### Sustainable farming

Sustainable agriculture involves successfully producing a harvestable resource over a long period of time with little or no negative effect upon the natural environment. At the same time, it has to provide a reasonable income for farmers. With appropriate and sensitive management, both intensive and extensive systems can be classed as sustainable.

| | Extensive agriculture | Intensive agriculture |
|---|---|---|
| Advantages | Low maintenance costs to the farmer due to reduced expenditure on fertilisers and pesticides. There is also very little environmental damage and it can help maintain upland areas by the prevention of the land's natural succession | High yields produced in smaller areas of agricultural land. This keeps the price of food down and supports ever increasing populations |
| Disadvantages | Very low yields produced and, because the grazing is low quality, the farmer has a limit to the number of animals they can farm. Product cost need to be high to make this type of farming viable | Excessive use of fertilisers and pesticides can result in pollution. Energy demanding in terms of machinery and chemical production. Also creates environments that increase the risk of diseases spreading due to overcrowding and reduces biodiversity |

Extensive farming may become unsustainable if overstocking occurs or if energy costs continue to rise for cereal farming. Overstocking will lead to environmental damage due to overgrazing, increased trampling and soil erosion. High energy costs for fuel and for chemical production make large-scale cereal production less viable. A solution to this in the UK has been to pay subsidies to farmers not to overstock and so put a limit on the number of animals they can keep.

Intensive farming requires farmers to provide increased quantities of fertilisers to replace those lost to the system. The effect of this can lead to the over-application of inorganic or organic fertilisers. A result of this can be that inorganic ions such as nitrates and phosphates run off into surrounding watercourses such as streams and lakes. This can lead to **eutrophication**. This pollution can be avoided and intensive farming can be carried out in a sustainable way. By adding rotting vegetable and animal matter, the humus content of the soil increases and there will be a reduced need for inorganic fertilisers. The soil will hold onto more moisture and run-off and erosion will be reduced. Release of nitrates will also be more gradual as it depends on the activity of the nitrifying bacteria in the soil.

### Agricultural conflict

Whichever farming practice is used to produce our food, there is always likely to be a conflict between the benefits of agriculture and the conservation of the environment. This is much more of an issue with intensive farming, the most important farming system in the UK.

Intensive farming uses large quantities of agrichemicals and can produce lots of animal waste that has the potential to damage the environment.

Pesticides can spread via spray droplets into other areas. They can enter water systems, kill beneficial insects and leave residues on food – something that many consumers have concerns over. Antibiotics are used in farming to reduce infection in intensively reared animals. The vast quantities used result in antibiotic-resistant bacteria through the processes of mutation and natural selection. This is a problem for humans who rely on these antibiotics to fight infection.

Animal wastes such as manure and slurry pose very great environmental effects due to the high BOD value (biological oxygen demand) of these wastes. If they are released directly into the environment without first being treated, eutrophication can occur.

## Eutrophication and the effects of animal waste released into the environment

When an accidental spillage of animal waste or silage effluent enters a water system such as a river or a lake, there is a sequence of events that can take place.

- Nitrates, phosphates and potassium ions enter the water from the effluent or waste.
- Small microscopic algal plants grow rapidly using these additional nutrients. An algal bloom is likely to cover the water surface.
- The algae quickly die, as light is now limited in the cloudy water, and are decomposed by aerobic bacteria in the water.
- The bacteria remove the oxygen from the water during decomposition due to their aerobic respiration.
- Oxygen levels in the water are reduced.
- Other aquatic life cannot survive due to the reduced oxygen content of the water.
- The aquatic ecosystem becomes devoid of some organisms, particularly fish.

These problems can also be caused by the overuse of fertilisers and can be alleviated by adding oxygen to the water via a series of pumps, restricting the use of fertilisers or closely monitoring the holding tanks for agricultural waste.

## Hedgerow removal

Hedgerows are man-made, narrow belts of vegetation that are used to mark boundaries between fields and act as boundaries for livestock. They contain dense vegetation of shrubs, trees and grasses and have high levels of biodiversity.

Increasingly the mechanisation of agriculture and the shift towards more intensive and larger farms has resulted in the loss and destruction of hedgerows.

During the 1990s the loss of hedgerows in the UK was running at 5000–6000 km a year. When hedgerows are lost there is a loss of species and the destruction of natural wildlife corridors that act to link fields together for animals such as voles, mice and stoats.

Recently there has been a realisation of the importance of hedgerows to agriculture and the role they play in preventing pest attacks on crops, as well as acting as homes for pollinators and other wildlife. Subsequently, removal of hedgerows occurs at a much-reduced rate, many are being replanted and most farms now are using hedgerows as an integral part of their farm.

## Questions

1 Describe the features of extensive and intensive agriculture.
2 Explain what is meant by sustainable agriculture.
3 Suggest some advantages and disadvantages of hedgerows in farming.
4 Explain how the use of antibiotics in farming might affect the use of these medicines to fight infections in the human population.

## What is succession?

Succession is the process where different species of organisms that are found in a community change over time. Succession takes place in stages, termed seral stages or seres. Each of these has a distinctive community of organisms. The organisms present change the physical characteristics of the habitat. They may increase the depth of soil or the water-holding capacity. As these change, new organisms can become established and these may grow better and out-compete the earlier species. Succession can be seen if you walk over a sand dune from the seashore, through the dune and into the grassland or woodland beyond. The differences you see will have occurred as a result of many changes taking place over time. The steps in such a community change are as follows:

* Bare sand is first colonised by pioneer plants such as marram grasses that can tolerate the dry, salty and unsheltered conditions.
* Marram grass begins to hold sand together with its dense root system, stabilising the sand and adding nutrients through the rotting of roots.
* As the marram grass makes the environment less hostile and improves the water-holding capacity, allowing soil to build up, plant species such as ragwort, rest harrow and grasses can become established. Rest harrow is a legume which, along with other species, grows faster and out-competes the marram grass.
* Marram grass decays and further fertilises the soil. Humus content increases, as does the soil depth, allowing taller plants to become established, such as willow.
* More species grow, further adding to the increasing level of biodiversity by providing food and breeding sites for insects and the birds which feed on them. Biodiversity increases.
* This process will continue to develop until a mature and stable vegetation type is found such as a woodland.
* This end point of the succession is termed the 'climatic climax community'.

Succession is a continuous process and in ever-changing systems such as sand dunes, it is often not possible to state where an end point occurs within the ecosystem. Figure 1 shows some of the stages in the development and change of a sand dune and how the system is liable to unforeseen change.

## Effect of agriculture on the development of climatic climax communities

All farming practices will have an effect on the surrounding environment. This may be through the use of inorganic fertilisers or by the selective grazing of sheep and cattle. Preventing a climatic climax community from developing naturally results in a **plagioclimax**, which forms as a result of **deflected succession**.

Grassland areas in Britain, if not used as grazing areas for livestock, would change over time and go through the stages of succession eventually leading to a climax vegetation of woodland. This could be oak, beech or natural conifer woodland, depending on abiotic factors such as underlying soil pH, rainfall and the maximum or minimum temperature. This is shown in Figure 2.

Much of the grassland found in Britain does not develop into woodland because of the agricultural management practices of burning, mowing or controlled grazing by cattle and sheep. These practices have been taking place for hundreds of years and have resulted in the rural landscape we are familiar with.

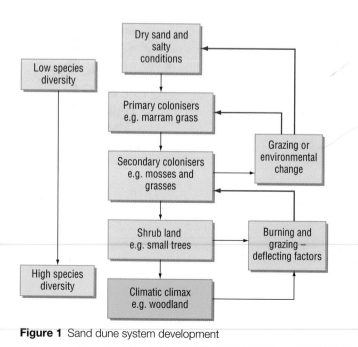

**Figure 1** Sand dune system development

| Boulders and rocks with mosses and lichens | Grassland with small flowering plants, like daisies | Taller herbaceous plants, like willowherb and foxgloves, grow and cut off light to small ones. Tree seedlings can become established | Bushes and shrubs, such as hawthorn and bramble, grow. Most non-woody plants die out | Fast-growing trees, such as birch, grow up, forming dense, low forest | Larger, slower growing, but stronger oak trees grow above the birch and establish the climax community |

**Figure 2** Woodland succession

Burning, mowing and grazing keep the grass short and act to prevent any tree seedlings from growing. The growing points of grasses are below the leaf level, so grazed grass can re-grow. However, tree seedlings have growing points higher up the plant. Grazing these means the plant cannot regrow.

| Burning and mowing – direct human intervention | Selective grazing – indirect human intervention |
|---|---|
| Controlled burning or mowing are practices used to stop the dominant growth of certain plant species | Grazing animals such as cattle have specific behaviours of grazing and will select certain species to eat |
| The effect of burning or mowing is to encourage the growth of new shoots. In effect, the development of a climatic climax community is being taken back a few seral stages and you are restarting the process | Sheep, for example, select short grass and early-flowering shoots. An effect of this grazing is to remove any tree seedlings, decreasing the overall level of biodiversity and the ecosystem becomes dominated by grasses |
| Burning vegetation releases nutrients back into the soil as the remaining organic material decays. Often, following a grassland fire, there is an increase in plant biodiversity. These species are likely to be opportunistic and fast growing pioneer species. Some of these species require the presence of fire to germinate and produce seedlings | The grazers will also trample the soil, increasing its level of compaction, so preventing colonisation by certain other plants. Animal waste such as cowpats can also affect the quality of the soil and provide nutrients that allow certain species to become dominant, out-competing others |

**Table 1** The effects of burning and mowing and selective grazing

### Key definition

**Biodiversity** refers to the variety of living organisms in a particular area. It is comprised of species diversity, genetic diversity and ecological diversity. A biodiversity index can be calculated. This is based on the richness or number of species and the number of organisms of each species that are present. A high value represents a rich and diverse community.

## Questions

1 Explain the differences between succession and deflected succession.
2 Other than grassland, give two further examples of a plagioclimax in the UK.
3 Are intensive or extensive farming practices more likely to prevent an ecosystem reaching its climatic climax? Give reasons for you answer.
4 Why would biologists study soil properties when investigating succession in a specific community?

The human population growth in the last 100 years has resulted in more human beings in existence than in the whole history of the human race; a period of hundreds of thousands of years. From 1950 to 1990, the total human population doubled from 2.5 billion to 5 billion. In the spring of 2007 it was announced that the 7 billionth human had been born. Every 3 seconds a new human is born and present estimates suggest that the population could reach 12 billion before the end of the twenty-first century if this growth rate is maintained.

Factors that affect the growth in human populations can be grouped together as either density-dependent or density-independent factors.

## Population size

There are four factors that affect population size. These include:

- Birth rate (natality), which can be calculated by the number of offspring produced, divided by the total number of adults in the population over a year.
- Death rate (mortality), which can be calculated by the number of deaths in the population in a year.
- Immigration, which is the number of individuals joining the population from surrounding populations.
- Emigration, which is the number of individuals leaving the population to join other surrounding populations.

The global human population has a birth rate that currently exceeds the death rate. The result of this is global population growth in the exponential phase of a typical growth curve. However, in some countries death rate is higher than or equal to birth rate, so these individual countries are not experiencing exponential growth. Figure 1 shows the growth curve for a population of humans.

### Factors affecting birth and death rates

There can be no doubt that the huge changes seen in the human population in the last 100 years have been as a result of technological advances in the production and distribution of food and medical advances in disease prevention and control. These have resulted in changes both in total population number and in the age ranges within a population. A population number can stabilise if fewer births occur and people live longer, but, over time, the average age of the population would rise. The factors affecting birth and death rates are described in more detail in Table 2.

| Density-dependent |
| --- |
| These factors depend on the actual size of the population and are usually biotic factors. For example, a large population is likely to be more prone to diseases and parasites than a smaller population, since these can be easily transmitted. Food supply and waste build-up would also be in this category |

| Density-independent |
| --- |
| These factors have similar effects regardless of the actual size of the population and are generally abiotic factors. For example, the effects of a forest fire or climate change would affect all members of a population regardless of its size |

**Table 1** Factors that affect growth in human populations

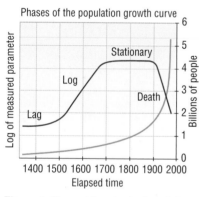

Phases of the population growth curve

**Figure 1** The red line shows the stages in a population growth curve. The yellow line shows population growth

## Population growth in Ethiopia

Ethiopia is an African country with problems similar to many developing countries. Birth rate is high because there is a need for large families to work the agricultural land. Death rate is also high because of the poor distribution of food, lack of medical services, poor hygiene, sanitation, and a high incidence of HIV in the population. The challenges to wealthier nations wishing to support and aid Ethiopia are complex. What is more, density-independent factors such as drought, leading to failed crops and subsequently famine, are common occurrences.

**Figure 3** Drought and famine in Africa

# Infant mortality and health

Infant mortality is one of the best measures of how effective the health infrastructure of a country is. The two maps in Figure 2 compare global infant mortality rates and birth rates. Infant mortality rate is lowest in developed countries as is birth rate. The result is a relatively stable population. Where birth rates are high, infant mortality is also high .Now consider China – the birth rate is low due to the government's 'one child' policy. However, infant mortality remains high. A different situation occurs on the Indian sub-continent with high levels of both birth rate and infant mortality. As countries achieve higher levels of economic development, health provision tends to improve and infant mortality will decline. How will this affect population growth? What are the options for governments in terms of intervention?

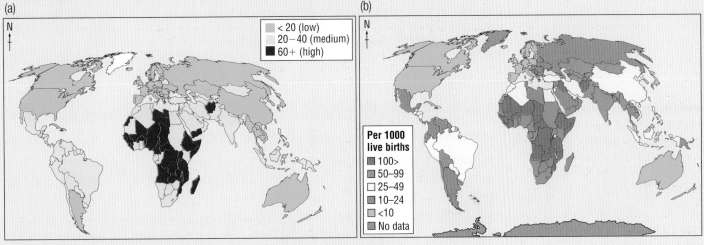

**Figure 2a** Infant birth rates per 1000 people; **b** mortality rates per 1000 live births

| Area of technological advance | Global effect on birth and death rate |
|---|---|
| Agriculture | Modern methods of food production, including the development of fertilisers, pesticides and selective breeding in plants and animals, have resulted in greater food availability than actual requirement. The result for some countries has been a surplus of food for export, and an ample quantity of nutritious food available to all. This has resulted in an increase in birth rate |
| | However, this food is unevenly distributed and the United Nations estimate that over one billion people live in poverty, are undernourished and do not have access to food. This inequality means that the advances in agriculture have yet to benefit all. In some countries the death rate is very high because of starvation and nutrient deficiency |
| Medicine | The last 30 years has seen huge advances in medical technology, such as improved vaccines and antibiotics, and methods of assisting and overcoming problems of infertility. As in agriculture, the benefits are very much country-specific with LEDCs, such as those in sub-Saharan Africa, being largely isolated from medical advances |
| | In other parts of the world, medical advances now mean that we are living longer, can receive donated organs and can repair damaged body parts with artificial equivalents |
| Infectious disease control | The treatment of infectious diseases and the ability to contain any spread has had a positive effect on survival – particularly in babies. Birth rates in some countries may have risen only slightly, but with perinatal and infant mortality decreasing, the result is an increase in population growth which could escalate as these surviving children reach reproductive age. However, this situation may not be maintained |
| | There is a rising threat from diseases such as HIV, SARS and even the re-emergence of malaria. The World Health Organisation estimate that over 40 million people in Africa are living with HIV and AIDS. These are still currently incurable and will lead to death. It seems that the advances in medical technology and disease control can only go so far in preventing a high death rate |

**Table 2** Factors affecting birth and death rates

## Questions

1 List five density-dependent and five density-independent factors that might affect the human population at any one time.
2 Look at what Thomas Malthus said about population growth and comment on his hypothesis.
3 Do you think it is possible for the human population to carry on growing exponentially?
4 Explain how technology in developed countries can alter the population growth curves.

Ecosystem is an interacting community of organisms (the biotic component plus the environment they interact with (the abiotic component.

## The role of nature and biodiversity

The human population is just one of the many populations of species which make up the global **ecosystem**. The natural world provides human beings with the resources needed for survival. For hundreds of thousands of years humans have lived by exploiting nature and using the resources available to them. The extra demands put upon natural resources by an ever-expanding human population threaten biological diversity. We are effectively in competition with other species for resources, and as we become ever more successful, we threaten some species with extinction. This is mainly due to the destruction of habitats for agriculture, mining or urban development, and rising levels of terrestrial and aquatic pollution. The concern over the loss of biodiversity is very real.

### Why care about loss of biodiversity and species extinction?

Extinction is forever! If we purposely or even accidentally drive a species to extinction we destroy a resource forever – perhaps without ever realising its value.

Humans depend on other species for food, fibres and medicines. It has been estimated that more than half of medical prescriptions contain a natural plant or animal product, or is a derivative of such a product. Just think of aspirin!

Humans also derive great aesthetic pleasure from interacting with other organisms. Town planners have appreciated for years the value of parks in cities. Also, consider how many people keep pets!

There are ethical issues associated with lifestyles that are detrimental to other species. The concept of humans having stewardship of the planet rather than ownership is common in all the world's major belief-systems.

Extinctions deprive us of the opportunity to study and understand ecological relationships, and this includes our own relationship with the rest of the ecosystem.

With such a large increase in the number of humans and predictions of an even greater population in years to come, there are inevitable conflicts between us and the rest of the ecosystem, particularly the biotic factors. These include:

- increased destruction of natural habitats as urban areas increase in size
- increased depletion of the Earth's fossil fuel resources to provide energy
- continued depletion of food stocks such as the ocean's fisheries
- overall decrease in biodiversity as wildlife cannot out-compete with most humans (limited space)
- increased problem of pollution and land degradation
- introduction of 'non-native' species which out-compete natural species and alter the habitat.

To help prevent this rise in human population having a negative effect on biodiversity, these areas can be managed. Sustainable development is essential so that future generations can enjoy the benefits we have now. Ecosystems can be managed in sustainable ways where biodiversity can be seen to have ecological, economic and scientific importance.

### Examples of sustainable practices

There are many examples of incentives or restrictions being imposed on agricultural practice that aim to promote and develop sustainable practice. This can be through fishing quotas, payment of subsidies for agricultural land that is 'set-aside', restricted development of 'green field sites' and the replanting of forests with native species to preserve large 'islands' of undamaged forest so surrounding areas can be recolonised following deforestation.

Costa Rica is a country in Central America that is one of the most biologically diverse in the world. Over 11% of its landmass, primarily rainforest, is legally protected and preserved for future generations. Most of the money for this protection has come from foreign pharmaceutical investment companies. Why? Well, these biologically rich areas might just hold the key ingredient in the production of a new drug. Scientists who search for such plants in these areas, working alongside local tribes are called ethnobotanists. It is important that in this search for possible cures for HIV or cancer that local tribes are not exploited, genetic resource rights are honoured and any profits obtained are shared.

## Fishing quotas

To ensure sustainable fishing, it is not only the quantity of fish taken from the sea that is important, but also their species, size, the techniques used in catching them and the areas where they are caught. Policies exist to regulate the amount of fishing, as well as the types of fishing techniques and gear used.

In the European Union, fishing, like agriculture, is a EU-wide responsibility in order to ensure security of the food supply.

The Common Fisheries Policy deals with the biological, economic and social dimension of fishing. It involves the conservation, management and exploitation of living aquatic resources, limiting the environmental impact of fishing and controlling and monitoring the markets for the catch.

Quotas or 'total allowable catches' (TACs) are a key element of the conservation of fish stocks. These quotas, based on scientific counts and estimates of fish populations, set an overall limit on the weight of fish which fishermen may land. This allows the resource to be managed in a sustainable way.

The setting of quotas is a contentious issue with many countries opposed to such regulation. For example, stocks of cod in the North Sea have been severely depleted for many years and environmentalists fear that without a complete ban on all fishing activities stock will be depleted beyond recovery. Politicians across Europe will not allow such a ban on fishing as this would compromise jobs and local economies. The aim of quotas is essentially to rebuild depleted stocks without economically crippling the fishing fleets.

### Farming and biodiversity

One of the biggest threats to biodiversity comes from intensive agriculture and the need to produce more food for the ever-increasing human population. It is surprising to consider that, with such a huge diversity of plant and animal life on the planet, we use only 30 species to provide 90% of all the food we eat worldwide. Furthermore, the agricultural practices we use to produce this food threaten many species and ecosystems, and the consequences could potentially prove disastrous for humans. These problems can be summarised in the following table:

| Agricultural practice | Threat to biodiversity |
| --- | --- |
| Use of pesticides | Will kill unintended species and can build up in food chains, killing organisms unrelated to the target pest. Overall reduction in biodiversity |
| Mechanisation and enlargement of farms | Farms, by increasing in size, often have large areas with only one type of crop grown – a monoculture. This eliminates other plants and the insects and birds which depend on them |
| Increased use of fertilisers | Over-fertilised soil leaches nutrients into surrounding water systems, leading to the problems of eutrophication and altering of these ecosystems |
| Draining of water systems | Larger farms and, more importantly, urban development and industry require ever-increasing quantities of water. The effect of this is to drain local rivers, streams and other water-courses and so negatively affect local ecosystems and biodiversity. Conversely, removal of forests in the catchment area of large rivers can lead to increased run-off and so-called 'flash' floods |

**Figure 1** An ethnobotanist

### Ecotourism

Many countries have begun to realise the value of their local environment and the fact that tourism can bring in more financial gain than if the land was used for agriculture or lost to logging companies. Ecotourism is big business, with benefits being brought to local communities both in terms of a regular income but also in the preservation of the local environment. Ecotourism can take many forms. Examples might include using decommissioned fishing boats for whale watching off the Scottish coast, and visiting the Galapagos Islands to study lizard populations.

## Questions

1  Explain some of the reasons why the setting of fish quotas might be inaccurate and so threaten the possibility of maintaining this resource sustainably.

2  Describe how the rise in human population might affect local and global biodiversity.

3  Suggest possible sustainable uses for rainforests other than logging and the development of monoculture.

## Global cycling of carbon

Carbon atoms are the building blocks of all carbohydrates, lipids and proteins and other organic molecules. They are estimated to account for nearly 20% of the mass of all living organisms. Living organisms take a carbon source from their environment and, for life to continue, this carbon must be recycled. Carbon is passed through organisms along food chains, returned to the atmosphere during respiration, stored in rocks as carbonate ions, locked away for millions of years as fossil fuels, and released as carbon dioxide when organic matter decays and decomposes. Figure 1 shows how carbon, in its many forms, is cycled on a global scale.

### Measuring carbon dioxide levels

Atmospheric carbon dioxide content has been measured at the Mauna Loa Observatory in Hawaii for over 50 years. The measurements are obtained by collecting gas using specialised weather balloons that rise up into the upper atmosphere and collect gases that can be analysed in the laboratory. The carbon dioxide values are measured in parts per million (ppm). The location was chosen because of its remoteness from any industrial influences that could contaminate and invalidate any results.

Scientists also measure samples of air trapped in the glacial ice of Greenland. These ice cores can be analysed using complex gas chromatography. The results can give clues as to what gases were present many thousands of years ago. These can be compared to what the global temperatures were and what the level of human activity was.

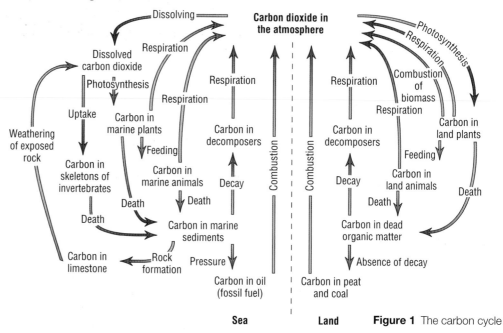

**Figure 1** The carbon cycle

### Human activity and the carbon cycle

The carbon cycle occurs in a closed system – planet Earth! Carbon is converted from one form to another over time. It is either trapped in fossil deposits, present in solution, fixed in organisms or released in gases such as methane and carbon dioxide. In the last 100 years there has been a change in the levels of carbon dioxide in the atmosphere, suggesting an imbalance in the system – more carbon dioxide is entering the atmosphere than is being removed. The two main reasons for this imbalance are primarily the burning of fossil fuels and deforestation.

Carbon dioxide levels fluctuate year on year but there has been an upward trend since measurements began. The period measured has seen a rise in carbon dioxide concentration of over 20% – an increase most likely to be **anthropogenic**, that is, caused by human activities. It does appear to coincide with a period of population growth and increased development and industrialisation. Figure 2 shows results from the Mauna Loa Observatory taken from 1958 to 2000 and shows how the concentration of carbon dioxide has risen during a time when the human population had risen exponentially.

### Burning of fossil fuels

Fossil fuels are the remains of prehistoric plants and animals. These have been buried under sediments between rocks or in peat bogs. They have fossilised rather than decomposed, and consist of hydrocarbons – stores of potential energy. When they are combusted, the energy is released and used to generate electricity or for heat energy directly. Fossil fuels are also used in making plastics and clothing, providing petrol for automobiles and even providing the plastic soles on our shoes!

## Deforestation

Mature forests form stable ecosystems and overall their carbon input equals their carbon output. Cutting down the forest causes a large release of carbon dioxide into the atmosphere. The levels of both photosynthesis and respiration drop. If trees are then removed, the remaining vegetation is often burned or left to rot, releasing further carbon dioxide. Global rates of deforestation are staggering. Estimates suggest that by 2030 over 80% of the Amazon basin will be lost forever. This could result in more than just an increase is carbon dioxide.

Deforestation also:
- destroys biodiversity
- destroys tribe lands and communities
- leads to soil erosion
- decreases soil fertility
- creates a drier local climate
- destroys local water systems
- has the potential to increase desertification.

**Figure 3** Deforestation

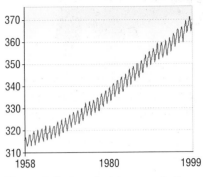

**Figure 2** Carbon dioxide concentration in parts per million of the air at the summit of Mauna Loa, Hawaii

The carbon trapped in fossil fuels represents a carbon 'sink' – carbon removed from the atmosphere. This is released when we combust these fuels, allow them to decompose or use them for some other purpose. Humans are currently burning up fossil fuels much faster than they are forming, and are using more now than at any other period in human history. The consensus is that human activity means that more carbon dioxide is entering the atmosphere than is leaving it, and that this is responsible for the rise in atmospheric concentrations.

### Climate change and carbon dioxide

There is much evidence to suggest that rising carbon dioxide levels are correlated to rising global temperatures. Data from Mauna Loa show clearly that the warmest years on record have occurred when the level of atmospheric carbon dioxide has been at its highest. Furthermore, data from ice-core analysis has shown that there appears to be a positive correlation between rising global temperatures and rising carbon dioxide levels.

Producing climate change models is difficult as there are other factors that affect global temperatures. These include the effect of other gases such as methane and water vapour, the degree of reflection from the Earth, the fraction of the Earth covered by snow and ice and the extent of cloud cover. As a result of these complications, some scientists question some of the claims, evidence and computer models that suggest that carbon dioxide is linked to global climate change, and that human activity is responsible for the rises in atmospheric carbon dioxide.

Scientists are clear on the fact that the Earth's climate is changing and that we are beginning to see the effects of global warming. The North Pole now has thinner ice sheets than ever before and by the year 2020 it is estimated that there will be no glaciers left in Europe. This is a worrying statistic when we consider that these glaciers supply the melt-water that feeds the major European rivers that are vital for farming, drinking water and transport.

The overall opinion based on the available scientific data is that a major contributor to climate change is the rise of carbon dioxide and that it is our duty to act now, reduce our impact on the planet, reduce gas emissions and help slow the rate of climate change. This will be examined again in spread 4.2.4.4.

### Key definition

**Global climate change** is the phenomenon whereby climates change and produce unexpected weather patterns and events. Such changes could cause global warming in certain geographical areas and global cooling in others. Climate is quite unpredictable but most scientists agree that our present stewardship of the planet has led to an overall change in global climate, with an average rise in surface temperatures over the planet.

## Questions

1 Describe how atoms of carbon can pass through food chains and return to the atmosphere as carbon dioxide.

2 Why do you think the graph for the atmospheric carbon dioxide levels taken at the Mauna Loa Observatory (Figure 2) shows a pattern of annual fluctuation?

3 Describe the influence of intensive farming on the rise in global carbon dioxide levels.

4 Explain the reasons why scientists continue to dispute the causes of and (to some) the existence of global climate change.

## What is a carbon footprint?

The carbon footprint is a measure of the impact human activities have upon the environment and the amount of greenhouse gases produced by these activities. The carbon footprint is measured in units of carbon dioxide – usually as tonnes per year.

Our awareness of the importance of reducing the human impact upon the environment and taking steps to reduce carbon dioxide emissions is a consequence of scientific evidence. There is evidence to suggest that climate change is primarily caused by anthropogenic or man-made pollution. The main pollutant that is linked to climate change is carbon dioxide, although methane is also a 'greenhouse' gas. Is there a 'debate' on human activity being responsible? The following extracts are taken from the Royal Society document 'Facts and Fiction about climate change'.

| Argument – on the one hand ... | Counter-argument – on the other hand ... |
|---|---|
| Many scientists do not think climate change is a problem | There are differences of opinion on details and the role of human contribution, but of 924 papers published between 1993 and 2003, 75% accepted the view that human activities have a major impact on climate change |
| There is little evidence that global warming is happening, some parts of the world are becoming cooler. Temperature rises could be due to urbanisation, if it is happening. It is not unique – Europe has been much warmer in the past | Based on worldwide measurements, average surface temperatures have risen by 0.6 degrees, the 1990s was the warmest decade since records began and the rise has been measured over land and sea, which rules out increased urbanisation as a cause |

**Table 1** Arguments on global warming

### Examiner tip

You should familiarise yourself with aspects of the 'climate change' debate. You should consider two key aspects – Is global warming happening? Is human activity responsible for global warming? Set out the arguments and be prepared to present comparative points. You need to consider the 'validity and reliability' of the evidence – particularly in the light of the vested interests of the groups who may be funding the research.

## Calculating a carbon footprint

In calculating your carbon footprint you need to consider the following:
- your annual energy usage (e.g. electricity, gas, oil, LPG)
- where your energy comes from (e.g. renewable/non-renewable sources)
- the number of people living in your house
- annual car/train/bus/coach mileage
- the number of airline flights you take in a year, and the destination of these.

Calculating personal footprints can be done using online, real-time calculators and businesses can buy services to carry out a thorough carbon audit. These services can produce an accurate picture of the current impact on the planet of your living or the services that your company offers. If you wish to calculate your own carbon footprint log on to www.direct.gov.uk and use the 'CO$_2$ calculator'.

### Reducing the impact of your carbon footprint

As an individual there are two components to the carbon footprint. One is termed the primary footprint and this is the direct impact that you have as an individual through everyday life. Then there is the so-called secondary footprint, which is the pollution caused by your buying habits. Both primary and secondary footprints can be addressed and reduced with a few small changes to buying habits and personal lifestyle choices.

| Reducing your primary footprint | Reducing your secondary footprint |
|---|---|
| Switch to a 'green' energy supplier which obtains energy from renewable sources | Reduce overall meat consumption |
| | Do not buy bottled water |
| Turn off lights and do not leave appliances on stand-by | Try to use local produce to limit food miles and the added impact of transportation of food |
| Turn down central heating and hot water | Buy organic as this food has a much lower energy impact and has far fewer negative environmental impacts |
| Replace old appliances to ensure they are as energy efficient as possible | |
| Replace all light bulbs with energy-saving equivalents | Recycle, reuse and reduce as much as possible and try to avoid unnecessary consumerism |
| Insulate the loft, walls and pipes and install double-glazing where appropriate | Think carefully about your leisure activities and consider their environmental impact |
| Car-sharing and using other modes of transport such as cycling | |

**Table 2** Reducing your carbon footprint

## Carbon offsetting

Carbon offsetting is a way of compensating for the emissions produced with an equivalent carbon dioxide saving. This involves calculating a carbon footprint and then buying 'carbon credits' from emission reduction projects. These projects will prevent or will have already prevented carbon dioxide emissions elsewhere, or removed an equivalent amount of carbon dioxide through planting trees.

The concept of carbon credits recognises the universal nature of carbon dioxide. That is, carbon dioxide is distributed all over the world and so it does not matter whether you make the carbon reduction in London or Lisbon – the positive effect on the environment will be the same.

Carbon offsetting involves two types of credits. There are certified and voluntary credits. The differences between these two schemes are outlines below.

## Certified credits

These are tradable and traceable credits that companies produce by adopting cleaner and greener methods of production. For example, a petrochemical company might purchase machinery that cleans its factory emissions and removes waste carbon dioxide. The benefit to the environment is paid for by a company's carbon credits, which will have gone on funding the machinery. The benefit to industry is that it now becomes economically difficult to pollute.

## Voluntary credits

Like certified schemes, these are also traceable but involve activities such as tree-planting programmes and so-called carbon sequestration schemes. Many low-cost airlines now adopt such schemes that allow you to pay for the carbon dioxide emitted during your flight by paying for a certain number of trees to be planted somewhere in the world.

Voluntary credit schemes are usually adopted by individuals wishing to reduce their carbon footprint and global impact, whereas certified schemes are primarily funded by companies. Government grants are available for companies to become carbon neutral and to have so-called zero emissions. This is obviously not possible without being part of a scheme that enables you to offset your emissions.

### Permits to pollute!

Any industry that extracts oil, coal and gas and produces carbon dioxide as pollution will soon be able to buy permits to pollute – like a carbon currency. Economists have the task of designing an auction, setting the price for such permits and ensuring that economic growth can take place without any future growth in carbon emissions. The idea is that permits would be based around the levels of carbon dioxide released into the atmosphere during 2006 and this date would be used as a benchmark. It is predicted that for every successive year the number of permits will decrease and this will drive forward energy efficiency measures and an associated cut in carbon dioxide pollution.

## Questions

1 Why is there a need to cut carbon dioxide emissions globally?

2 Suggest why one way to reduce your carbon footprint is to cut down on your intake of meat.

3 What do you think the carbon footprint of someone in a developing country would be, compared to that of someone from a developed country?

4 Why is the term carbon footprint scientifically misleading?

# Summary and practice questions

**1** Explain why it is important for gametes to be produced by meiosis.

**2** Describe the similarities and differences between male and female gametes.

**3** The following diagram shows a mature sperm cell.

Describe the part played by structure A in the processes leading to fertilisation.

**4** Explain the advantages and disadvantages of using Norplant as a contraceptive

**5** Compare the processes of spermatogenesis and oogenesis.

**6** Describe the actions of the hormones LH, FSH and oestrogen in the control of the menstrual cycle.

**7** Compare and contrast the actions of LH and FSH between males and females.

**8** Following fertilisation, the zygote divides by the process of ................... and forms a ball of cells called a ................... . This then migrates towards the uterus where it becomes ................... . It begins to produce a hormone called ................... which has the effect of inhibiting the action of ................... and ...................%. As the embryo develops into a fetus and towards the end of the pregnancy, other hormones are produced ................... will stimulate uterine contractions and ................... will stimulate the production of milk.

**9** Copy and complete the following:

| Problem | Possible treatment |
|---|---|
| Blocked fallopian tube | |
| Low sperm count | |
| Abnormal sperm | |
| Hostile cervical mucus | |

**10** Describe how intra-uterine insemination might increase the chance of pregnancy.

**11** Outline the processes involved in implantation.

**12** Explain how a multiple pregnancy might arise.

**13** Anti-oestrogen drugs such as Clomiphene can be used to encourage the hypothalamus to produce GnRH. What effect might this have?

**14** Using the following table:

| Age of woman / years | Chance of a live birth following IVF |
|---|---|
| Below 23 | Unknown |
| 23–35 | More than 20% |
| 36–38 | 15% |
| 39 | 10% |
| 40 and over | 6% |

Suggest why treatment for women over the age of 40 is not offered on the National Health Service.

**15** Explain how monoclonal antibodies might be used in pregnancy testing and why this method is so sensitive?

**16** List the key factors that affect the birth rate and death rate of human populations.

**17** Explain why it might not be accurate to state that advances in medical technology and disease control have resulted in a growing human population.

**18** Define the following terms: Ecosystem; Biodiversity; Deforestation; Habitat.

**19** Describe the effect of human activity on the carbon cycle.

**20** Complete the following:

| | Importance of biodiversity |
|---|---|
| Aesthetic reasons | |
| Medical reasons | |
| Agricultural reasons | |

**21** Explain the role of light in producing sugars and ATP in plants.

**22** Plants require nitrogen in the form of nitrates. These can be taken into the roots in solution from the surrounding ..................... or the plant can obtain ammonium ion which can be made directly in the roots by the action of ..................... bacteria. Nitrates are used to make ..................... in the plants. The action of nitrifying bacteria such as ..................... and ..................... is responsible for converting ammonium compounds from decaying matter in the soil to soluble nitrates.

**23** Explain what is meant by the following terms;
(a) Succession
(b) Deflected succession
(c) Ecosystem
(d) Carbon footprint

## Question 1

Aerobic training programmes result in changes to muscle composition. One such change can be an increase in the number of mitochondria present in muscle fibres.

**(a)** Figure 1.1 is a diagram of a mitochondrion.

    **(i)** On the diagram, identify the letters X and Y. [2]

    **(ii)** Using the information given in the diagram, calculate the actual length of the mitochondrion along the line A to B. [2]

**Figure 1.1**

**(b)** Molecules of pyruvate produced from the breakdown of glucose in glycolysis enter the mitochondria. Here pyruvate is completely broken down in the link reaction and the Krebs cycle.

    Name three products produced from the breakdown of pyruvate in mitochondria. [3]

**(c)** Mitochondria are also the site of ATP synthesis by oxidative phosphorylation. Explain how the structure of a mitochondrion is adapted to allow ATP synthesis by oxidative phosphorylation. [5]

**(d)** Suggest two changes that may occur within the mitochondria in response to aerobic training. [2]

## Question 2

The haemoglobin molecule consists of four polypeptide chains. Each polypeptide chain is attached to a haem group. In adults, two of the chains consist of alpha globin and two consist of beta globin. In the fetus, the two beta chains are not present and the molecule consists of two alpha chains and two gamma chains.

**(a)** With reference to the structure of haemoglobin, explain what is meant by the following terms:

    **(i)** quaternary protein structure.

    **(ii)** prosthetic group. [4]

**(b)** Outline how the sequence of bases in the gene for beta globin is used to construct a molecule of beta globin. [5]

**(c)** In the disease thalassaemia major, beta globin is not synthesised and the resulting haemoglobin molecules do not transport oxygen effectively. Researchers are attempting to treat this condition by 'switching on' the fetal gamma globin genes such that fetal haemoglobin is synthesised in adults.

    One problem with the proposed treatment is that fetal haemoglobin has a higher affinity of oxygen than adult haemoglobin.

    Figure 2.1 shows the oxygen dissociation curve for adult haemoglobin.

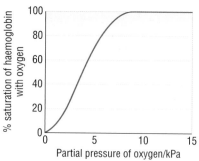

**Figure 2.1**

    **(i)** Sketch the shape and position of the dissociation curve for fetal haemoglobin on Figure 2.1. [2]

    **(ii)** Suggest why having fetal haemoglobin could cause problems during intense exercise. [2]

# Question 3

**(a)** Complete the table to compare the processes of spermatogenesis and oogenesis.

| Feature | Oogenesis | Spermatogenesis |
|---|---|---|
| Number of gametes produced from each germ cell | | |
| Number of polar bodies produced from each germ cell | | |
| Site of gamete production in ovary or testis | | |
| Name of accessory cell involved | | |
| Duration of process in life cycle | | |

[5]

**(b)** Explain how genetic variation is achieved during spermatogenesis. [4]

### Examiner tip

Yet again this is an example of a synoptic question. You covered meiosis in F222. Here it is again. There is a further synoptic link in this question in part (d). You covered base pairing in F221 and protein synthesis and RNA in this module. Part (d) combines aspects of both in a different context.

**(c)** Spermatogenesis halves the chromosome number. Fertilisation of the oocyte by the sperm restores this number. Before fertilisation can occur, the sperm must undergo a process called capacitation.
Describe and explain what happens to the sperm during capacitation. [4]

**(d)** Certain chemicals can inhibit capacitation in sperm. One chemical consists of a polynucucleotide. This binds to messenger RNA in the sperm cell mitochondria. The polynucleotide is known as an 'antisense' molecule.
  **(i)** AAUGCCAAGCCU is the base sequence of an antisense polynucleotide. Write out the base sequence of the mRNA that this would bind to. [1]
  **(ii)** State precisely how the two strands would be joined. [3]
  **(iii)** Suggest a possible use for such a chemical. [1]

# Question 4

Figure 4.1 is a diagram of the light-independent stages of photosynthesis.

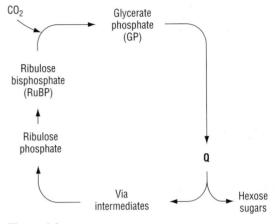

**Figure 4.1**

**(a)** Identify the compound labelled Q. [1]

**(b)** Identify the two products of the light-dependent reaction which are needed to convert GP to Q. [2]

**(c)** In addition to hexose sugars, plants can also synthesise amino acids. To do this, they need to take up nitrate ions from the soil. Outline the role of bacteria in maintaining levels of nitrate in soils. [4]

**(d)** In intensive agriculture, short-strawed varieties of cereals such as wheat are grown over large areas of land. These varieties respond to high levels of nitrate in the soil by producing more grain per plant. Outline the advantages and disadvantages of large-scale production of cereals using short strawed varieties. [4]

# ① Inheritance of human genetic disease

## What is genetics?

Genetics is the study of genes, their inheritance and their effects. Genes play an important part in our health. Some of the effects, such as the inheritance of hairy ears, are minor, but others such as cystic fibrosis and Huntington's disease have more serious consequences. In order to study these diseases and others more closely we need to remember some basic genetic definitions, which are given in the following table.

| Genetic term | Definition |
|---|---|
| Gene | A length of DNA that codes for the production of a particular polypeptide |
| Allele | One of the different forms of a gene which occupy the same locus on homologous chromosomes |
| Locus | The position on a chromosome at which a particular gene is found |
| Phenotype | A person's observable characteristics, resulting from an interaction between its genotype and its environment |
| Genotype | The genetic make-up of an organism. It describes all the alleles that the nucleus of a human cell contains. Genotypes can be homozygous or heterozygous |
| Dominant | A dominant allele always shows its effect on the phenotype |
| Recessive | A recessive allele shows its effect on the phenotype only when the dominant allele is absent |
| Autosomes | All the chromosomes except the sex chromosomes (X and Y chromosomes) |

### Examiner tip

Look at a picture of a normal karyotype and find chromosome 12. Refer to spread 1.1.3.2 to remind yourself about enzyme activity. Also look at DNA structure, spread 2.1.1.1.

## Gene mutation can cause genetic diseases

A **mutation** is an alteration of DNA sequence (**gene mutation**) or a change in the number or structure of chromosomes (**chromosome mutation**). A mutation will affect future generations only if it occurs in the gametes. A mutation is a very rare event. Most genes have less than one chance in a million of being incorrectly replicated.

**Phenylketonuria** (PKU) is one of the most common inherited disorders. This is caused by a mutation in a gene on chromosome 12. It is a genetic disease caused by a **recessive allele**. This means that an affected individual must have gained a mutant allele from each parent.

The gene codes for an enzyme called phenylalanine hydroxylase, which breaks down the amino acid phenylalanine into tyrosine. When this gene is mutated, the shape of phenylalanine hydroxylase changes and it is unable to properly break down phenylalanine. Phenylalanine builds up in the blood and tissue fluid and causes severe brain damage in young children. All babies born in the UK are tested for PKU at birth by analysing a tiny drop of blood from a heel prick. If the test is positive, the baby is put on a diet that does not contain phenylalanine, so avoiding brain damage.

**Cystic fibrosis** (CF) is one of the most common genetic diseases, affecting approximately 7500 people in the UK. 1 in 25 of the population carries the faulty cystic fibrosis allele. Cystic fibrosis is caused by a mutation in a gene called the cystic transmembrane conductance regulator (**CFTR**), which is found on chromosome 7. We have two copies of this gene, one from our mother and one from our father. The mutations are passed on from parent to offspring. The cystic fibrosis allele is recessive.

Cystic fibrosis is the result of a **deletion mutation** when nucleotides are lost from normal DNA. This alters the way the triplets of bases are 'read' and therefore changes the sequence of amino acids in the polypeptide formed. The most common mutation (70% of CF worldwide) is a deletion of three nucleotides that results in the loss of the amino acid phenylalanine. There are over 1400 other mutations that can produce cystic fibrosis. The mutations affect the CFTR protein in different ways.

The protein created by the CFTR gene spans the outer membrane of cells in the sweat glands, lungs, pancreas and other organs and acts as a channel connecting the cytoplasm to the surrounding fluid. This channel is primarily responsible for the movement of chloride ions.

When the CFTR protein does not work, the sodium chloride balance is upset. This imbalance creates a thick, sticky mucus layer that cannot be removed by cilia and traps bacteria, resulting in chronic infections. Recurrent infections scar the lungs. A person with cystic fibrosis will need daily therapy to help them cough up the mucus.

The mucus also blocks the secretion of digestive enzymes from the pancreas. The lack of digestive enzymes leads to difficulty absorbing nutrients, resulting in malnutrition and poor growth. Patients usually take pancreatic enzymes orally to help alleviate the problem. Thickened secretions can also lead to liver problems. At least 97% of men with CF are infertile, as the tubules that transport sperm from the testes may be blocked with mucus or may be missing.

The average survival for patients with cystic fibrosis is now over 30 years due to improved treatment of the disease.

**Huntington's disease** is a brain disorder that affects a person's ability to think, talk and move. Symptoms develop gradually over months or years. The age at which they first appear varies between about 30 and 50. Symptoms begin with memory loss, changes in personality and mood as well as uncontrolled muscle movements. Other symptoms of dementia appear as the condition progresses. Difficulties with speaking and swallowing, weight loss, depression and anxiety may also occur.

Huntington's is caused by a mutation in a gene on chromosome 4. It is a **triplet repeat mutation** (a 'gene stutter') when the CAG (cytosine, adenine and guanine) triplet is repeated again and again. In people who do not have Huntington's disease this triplet is repeated between 10 and 26 times. People with Huntington's disease have approximately 40 or more of these triplets. The more repeats they have, the earlier the symptoms seem to start.

The DNA sequence of CAG is the genetic code for the production of the protein huntington. Evidence suggests that the huntington protein in people with Huntington's disease tends to break into pieces, which then clump together. These protein clumps build up inside the nuclei of brain cells, which either impairs their functioning or kills them.

The mutation is inherited as a dominant allele – **autosomal dominance**. Most sufferers are heterozygotes, so they have a one in two chance of passing on the disease.

## Questions

1  Complete: An alternative form of a gene is known as an ....................., the position where it is found on a chromosome is known as the ..................... . A character is said to be ..................... if the allele controlling it only produces an effect when the two alleles in a pair of homologous chromosomes are the same.
2  What is a mutation?
3  In a patient with PKU, why are they unable to break down phenylalanine, and what is the consequence of this for a young child?
4  If a carrier of cystic fibrosis had a child with a person who was also a carrier, what are the chances of any offspring developing cystic fibrosis? Use a genetic diagram to explain your answer.

### Examiner tip

Have another look at membrane transport on spread 1.1.2.5 and why an ion needs a protein channel to cross the membrane.

### Case study: a poem

Oh here we go once again
They tip me upside down
And pat and poke me till I cough
And spit and almost drown
But hey it's not so often now
That we've got our new routine
It's the slickest operation
That this condition's ever seen
I have got my diet plan you see
My inhaler and my pills
Supplements and medicines
To sort out other ills
I do everything the Doctors
And Nurses say I ought
But best of all I like the bit
That involves me doing sport
So I'm outdoors most often
Throwing mud and climbing trees
Swimming and playing football
(after I've washed my knees)
But when things aren't quite so good
And it's nebuliser time
I like to write some poetry
And that just suits me fine!

*Chris Dawson*

(a)

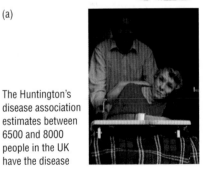

The Huntington's disease association estimates between 6500 and 8000 people in the UK have the disease

(b)  Inheritance of Huntington's disease – autosomal dominance

H – Huntington allele
h – normal allele

| Genotype | Phenotype |
| --- | --- |
| HH | Sufferer |
| Hh | Sufferer |
| hh | Normal |

**Figure 1** Huntington's disease and its inheritance

## Inheritance of sickle cell anaemia and ABO blood groups

In the previous spread we saw how one **dominant allele** could show its effect in the **phenotype** (Huntington's disease); if the allele responsible for the disease was **recessive**, then both recessive alleles had to be present for the illness to appear (cystic fibrosis). However, in some cases both alleles affect the phenotype. This is known as **codominance**. In this case the alleles are symbolised by an uppercase letter representing the gene and by suitable uppercase superscript letters, for example $W^A$ and $W^B$ for alleles A and B. You will see this when we discuss the inheritance of sickle cell anaemia and of the ABO blood groups.

### Inheritance of sickle cell anaemia

Sickle cell anaemia is the result of a mutation in the gene producing **haemoglobin** (see spread 5.1.1.1 for details of this mutation and its effect). The symbol **Hb** is used to represent the gene locus for the $\beta$-polypeptide chain of haemoglobin. There are two alleles of this gene; **$Hb^A$** is the allele for normal $\beta$-polypeptide and **$Hb^S$** the allele for sickle cell $\beta$-polypeptide. The results in three phenotypes:

| Genotype | Phenotype |
|----------|-----------|
| $Hb^AHb^A$ | All normal haemoglobin |
| $Hb^AHb^S$ | Half the haemoglobin is normal and half sickle cell haemoglobin – sickle cell trait. The alleles are codominant |
| $Hb^SHb^S$ | All sickle cell haemoglobin – sickle cell anaemia. These red blood cells also have a shorter life in circulation, only 15–30 days |

This means that when a couple both have sickle cell trait, each time they are expecting a child there is a 1 in 4 chance that their child will inherit the normal haemoglobin ($Hb^AHb^A$), a 2 in 4 chance she will inherit sickle cell trait ($Hb^AHb^S$), just like the parents, and a 1 in 4 chance she will inherit sickle cell anaemia ($Hb^SHb^S$).

The person who inherits the sickle cell trait in general will suffer no ill effects. But if they get abnormally short of oxygen for any reason, such as ascending to high altitudes in mountain-climbing or in ballooning, they may suffer episodes in which small blood vessels become blocked. Also, it is important that people with sickle cell trait should receive high levels of oxygen during general anaesthetics.

### Inheritance of ABO blood groups

Red blood cells contain a protein in their plasma membrane that determines the ABO blood group. There are two forms of this protein – antigen A and antigen B. The gene that codes for this protein has three alleles. The gene locus is represented by the letter I so the three alleles are; $I^A$ (allele for antigen A), $I^B$ (allele for antigen B) and $I^O$ (allele for no antigen) $I^A$ and $I^B$ are codominant. $I^O$ is recessive to both $I^A$ and $I^B$ so neither antigen is produced.

There are therefore six possible genotypes and four possible phenotypes.

| Genotype | Phenotype |
|----------|-----------|
| $I^AI^A$ | Group A |
| $I^AI^B$ | Group AB |
| $I^AI^O$ | Group A |
| $I^BI^B$ | Group B |
| $I^BI^O$ | Group B |
| $I^OI^O$ | Group O |

---

**Inheritance of sickle cell anaemia**

$Hb^A$ – the allele for normal $\beta$ polypeptide
$Hb^S$ – the allele for sickle cell $\beta$ polypeptide

If the parents are both heterozygous for this trait:

Parental genotypes: $Hb^A Hb^S$ $\quad$ $Hb^A Hb^S$
Parental gametes: $Hb^A$ $\quad$ $Hb^S$ $\quad$ $Hb^A$ $\quad$ $Hb^S$

Cross

| | $Hb^A$ | $Hb^S$ |
|--------|--------|--------|
| $Hb^A$ | $Hb^A Hb^S$ Normal | $Hb^A Hb^S$ Sickle cell trait |
| $Hb^S$ | $Hb^A Hb^S$ Sickle cell trait | $Hb^S Hb^S$ Sickle cell anaemia |

As equal numbers of each type of egg are produced, the chances of each of these four possibilities are equal. The probability of a child being $Hb^AHb^A$ is 0.25, the probability of being $Hb^AHb^S$ is 0.25 and the probability of being $Hb^SHb^S$ is 0.5.

**Figure 1** Inheritance of sickle cell anaemia

**Examiner tip**

This links with spread 4.1.2.5 and the role of antibodies. Remind yourself about the importance of matching blood groups.

**Unit F225**
Module 1  Genetics in the 21st Century
5.1.1 c, d, e

## Why doesn't the frequency of the allele for sickle cell decrease?

Children born with the genotype $Hb^SHb^S$ have sickle cell anaemia and are unlikely to survive long enough to be able to reproduce. So you would expect the $Hb^S$ allele to be removed from the population, or at least its frequency to be greatly reduced. But this is not the case, because the allele $Hb^S$ confers protection against malaria.

Malaria is caused by a small parasite (called *Plasmodium*) that lives inside red blood cells and causes recurrent fevers. The plasmodium parasite is transmitted from female *Anopheles* mosquitoes to us. The parasites spend part of their life cycle in the mosquito and part of it in the human host. Each year, more than 300 million people are infected with *Plasmodium* and 1.5 million of them die.

(a)

(b)

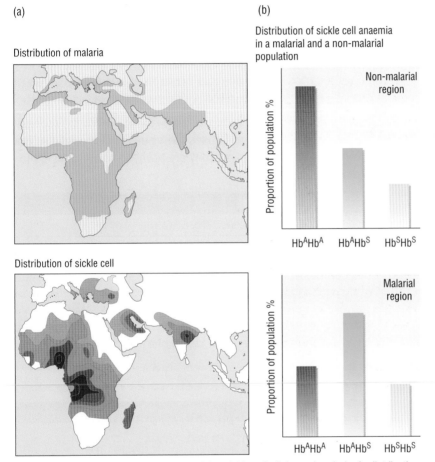

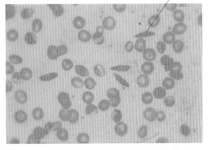

**Figure 3** SEM of red blood cells in sickle cell anaemia and blood smear

**Figure 2  a** Geographical distribution of the sickle cell allele and malaria; **b** distribution of sickle cell anaemia in a malarial and non-malarial population

There are several possible reasons why sickle cell has this effect. One explanation may be that $Hb^AHb^S$ causes the malaria infection to stay in the body for longer (the bulk of blood cells carrying malaria are destroyed quickly but a few may escape detection), so it allows the immune system to build up a proper defence against malaria. Recent research by the Wellcome Trust has shown that the protection against malaria was around 20% in the first two years of life compared with over 50% by the age of 10.

There is a causal link between the parts of the world where sickle cell anaemia is most common and the parts of the world where malaria is endemic; the mutant gene is advantageous and is selected for.

This means that having the genotype $Hb^AHb^A$ is a disadvantage in the areas where malaria is endemic, and a child with genotype $Hb^SHb^S$ is unlikely to survive to reproduce, so the best genotype to have in areas where malaria is endemic is $Hb^AHb^S$. In each generation children born with the genotype $Hb^AHb^S$ are most likely to grow up and reproduce, so both alleles are passed on to the next generation.

## Questions

1  What does it mean if both alleles are codominant?

2  If a child has a genotype of $Hb^AHb^S$, what are the likely genotypes of its parents?

3  Why do you think the percentage of individuals with sickle cell anaemia disease is much lower in the United Kingdom than in East Africa?

4  What are the chances of a child with blood group O being born to a heterozygous man with blood group A and a heterozygous woman with blood group B? Explain your answer by drawing a genetic diagram.

# Other causes of genetic diseases

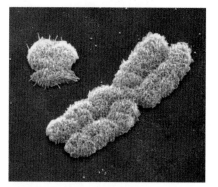

**Figure 1** Sex chromosomes

## Examiner tip

Review blood clotting – spread 1.1.3.1. There are also other disorders where blood clotting is affected. Thrombophilia occurs where the blood clots more readily. This can be the result of a mutation in the gene for prothrombin.

Unaffected male
Affected male
Unaffected female
Affected female
Person whose sex is not known
Marriage (mating)
Vertical line = offspring (in this case, son)
A family of four brothers and sisters. The last two are non-identical twins
Identical twins

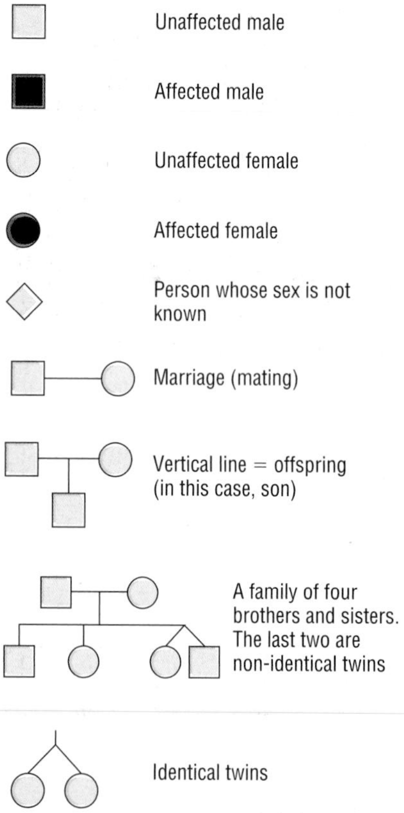

**Figure 2** Symbols used in pedigree charts

## Inheritance of sex-linked characters

In spread 5.1.1.1 we learned that we have autosomes and sex chromosomes. The autosomes occur in **homologous** pairs, but the **sex chromosomes** (X and Y) and are only homologous for part of their length (the part that allows them to pair in meiosis). Women have two X chromosomes, while men have one X and one Y. In humans the Y chromosome is much shorter than the X. The X chromosome carries a large number of genes, but the Y chromosome carries very few. Any gene that is carried on either the X or the Y chromosome is said to be **sex-linked**.

One example of a sex-linked disease is **haemophilia**. Haemophilia is a blood disease in which the blood fails to clot adequately. Sufferers may bleed to death even from a trivial wound. Bleeding also occurs into joints and other parts of the body, which is very painful. Haemophilia A is most common, and is due to a single recessive allele that results in the body producing less of a clotting protein called antihaemophiliac factor or **factor VIII**.

The **dominant allele**, H, codes for the production of factor VIII, while the **recessive allele**, h, results in a lack of the factor and causes the disease haemophilia A. Because these alleles are linked to the X chromosome they are always shown attached to the X chromosome: $X^H$ and $X^h$.

| Genotype | Phenotype |
|---|---|
| $X^HX^H$ | Normal blood clotting female |
| $X^HX^h$ | Normal blood clotting female but a carrier for haemophilia |
| $X^hX^h$ | Haemophiliac female, very uncommon |
| $X^HY$ | Normal blood clotting male |
| $X^hY$ | Haemophiliac male |

A normal man ($X^HY$) and a carrier woman ($X^HX^h$.) have children:

Parental genotypes $X^HY$ $X^HX^h$
Gametes $X^H$ Y $X^H$ $X^h$

| | $X^H$ | Y |
|---|---|---|
| $X^H$ | $X^HX^H$ Normal female | $X^HY$ Normal male |
| $X^h$ | $X^HX^h$ Carrier female | $X^hY$ Haemophiliac male |

## Predicting the inheritance of genetic disease

One useful way to trace the inheritance of a sex-linked disease such as haemophilia is to collect information about a person's ancestors. This information is then used to construct a **pedigree chart** – a sort of family tree, which shows the phenotypes of the various family members. Using this chart it is possible to predict the risks of a particular couple having a child who might suffer from a particular genetic disease.

For example, a woman may have a brother who has haemophilia, which would mean that her mother and her grandmother must have been carriers. If her mother is a carrier then there is a 50% chance that she is also a carrier and 25% chance that she will give birth to a haemophiliac son. A famous pedigree chart shows the inheritance of haemophilia from Queen Victoria in members of various European royal families.

Pedigree analysis can be used whenever the condition is caused by the allele of a gene that shows a simple inheritance pattern, for example, cystic fibrosis.

## Unit F225
### Module 1 Genetics in the 21st Century
#### 5.1.1 f, g, h

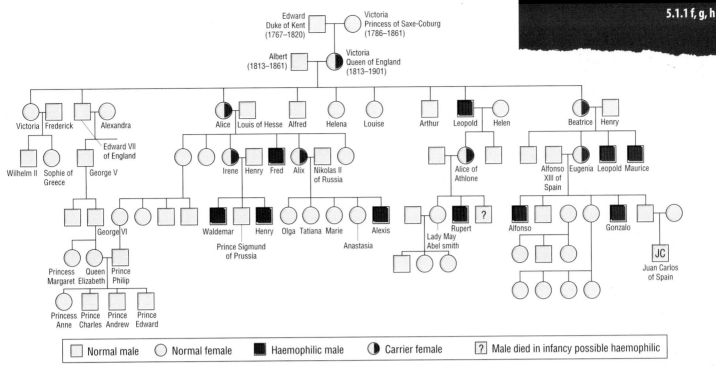

**Figure 3** A pedigree chart showing the distribution of haemophilia in the European royal families

Key:
- Normal male
- Normal female
- Haemophilic male
- Carrier female
- ? Male died in infancy possible haemophilic

## Autosomal linkage

At the start of this spread we talked about sex-linked chromosomes. Genes that have loci close together on autosomes can also be linked – they tend not to be separated by **crossing over** during **meiosis**.

An example of this is the inheritance of the ABO blood groups and **nail patella syndrome**. Nail patella syndrome is a rare genetic condition. People who are affected may have small or absent patellae (kneecaps), undeveloped nails (the nails, especially those on the thumbs, are typically absent or short and never reach the free edge of the finger), an inability to straighten the elbows fully and a greater risk of developing kidney disease.

Nail patella syndrome affects males and females equally, so it is not sex-linked. It is autosomal dominant and the locus of the gene that causes it is found on chromosome 9, very close to the ABO blood group gene. As they are very close they will tend to be inherited together, because the chances of a chiasma forming and crossing over to separate these is less likely.

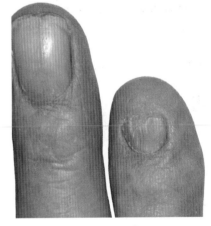

**Figure 4** Nail patella syndrome

## Questions

1 The 2 sex chromosomes are ................. and ................ .
   In the human female there are two ..................
   chromosomes. Haemophilia is an example of a
   ..................... . Haemophiliacs have less .....................
   than normal.

2 Why are males more likely to be haemophiliacs than females?

3 Figure 5 shows a pedigree chart of a family with albino members:
   (a) Is albinism sex linked? Explain your answer.
   (b) Is albinism caused by a recessive or a dominant allele? Explain your answer.
   (c) Do you think that the offspring in generation 2 are homozygous or heterozygous for albinism?
   (d) Explain why two albino children are born in generation 3.

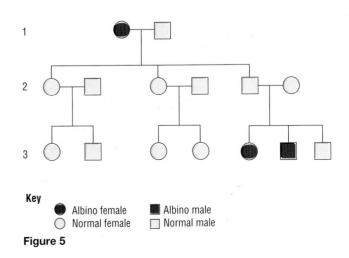

**Key**
- Albino female
- Normal female
- Albino male
- Normal male

**Figure 5**

## Mapping gene loci on chromosomes

In spread 2.2.12 on **meiosis** we saw that during **prophase** 1 the **chromatids** of a **bivalent** may break and reconnect to another **chromatid**, resulting in the exchange of **gene loci** between maternal and paternal chromatids. This is called **crossing over**. It results in different combinations of **alleles** in the gametes and therefore in the offspring. The new **genotypes** are called *recombinants*. A cross-over is more likely to occur between genes that are far apart – think of a piece of string: it is easier to tie a knot in a long piece rather than a very short piece.

The frequency of recombinants is the cross-over value:

$$\frac{\text{number of recombinant individuals produced}}{\text{total number of offspring}} \times 100$$

Using the data of cross-over values, we can map the relative positions of genes: two genes whose alleles give a cross-over value of 20% are said to be 20 units apart. In the last spread we saw that nail patella syndrome and the ABO blood group were closely linked. Very rarely, recombinants are seen, and when they are, the loci are found to be less than 0.1 of a unit apart on chromosome 9. By comparing one or other of these genes with another gene locus, the relative positions of more genes can be established.

## Diseases caused by translocation and non-disjunction

**Translocation** is another example of **chromosome mutation** where a piece of chromosome breaks off and is transferred to another chromosome. This can also cause Down's syndrome, which occurs when the end of the long arm of chromosome 21 joins another chromosome.

**Non-disjunction** (see spread 2.2.1.1) is an example of a **chromosome mutation** (see spread 2.2.1.5) when there is a change in the number of chromosomes. Non-disjunction of the **sex chromosomes** during meiosis (see spread 2.2.1.5) can result in either a gamete carrying two sex chromosomes or no sex chromosome instead of the normal one.

Turner's Syndrome is the most common sex chromosome abnormality of human females. It occurs in approximately 1 in 2500 newborns. It appears that only about 1% of zygotes with this abnormality survive to be born. Roughly half of the females with Turner's Syndrome have X chromosome **monosomy** – only one X chromosome (45,X). Most of the other individuals are **mosaics** (a term referring to a genetic situation in which an individual's cells do not all have exactly the same composition of chromosomes). This occurs because sometimes non-disjunction happens during an early cell division of the embryo, resulting in individuals with a mixture of 46,XX cells and 45,X cells. The exact mixture of the two cell types depends on when non-disjunction occurred. It is possible for a very early non-disjunction from a 46,XY zygote to be completely 45,X. If non-disjunction occurs later a 45,X/46,XY embryo will develop as a male (without Turner's Syndrome).

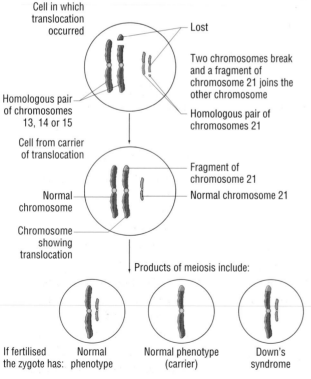

Cell in which translocation occurred

Lost

Two chromosomes break and a fragment of chromosome 21 joins the other chromosome

Homologous pair of chromosomes 13, 14 or 15

Homologous pair of chromosomes 21

Cell from carrier of translocation

Normal chromosome

Fragment of chromosome 21

Normal chromosome 21

Chromosome showing translocation

Products of meiosis include:

If fertilised the zygote has: Normal phenotype | Normal phenotype (carrier) | Down's syndrome

**Figure 1** Down's syndrome caused by translocation in chromosome 21

**Unit F225**
Module 1  Genetics in the 21st Century
5.1.1 i, j, k, l

# The use of karyotypes

To confirm the diagnosis of a genetic disease caused by a chromosome mutation, a karyotype is used (see spread 2.2.1.5). This is an organised profile of the physical appearance of the chromosomes of a single nucleus. Often the chromosomes are stained in some way, producing banding patterns that help to distinguish members of different pairs of homologous chromosomes.

In a karyotype, chromosomes are arranged and numbered by size, from largest to smallest. It is then possible to see if there are any chromosome abnormalities.

# Ethical issues

Ethics deals with choices that we can live by. Ethics is about moral judgements, and is an expression of attitudes, feelings and preferences about how we ought to live, and what decisions we ought to make. Modern bioethics often stresses four key principles: individuals should be able to make their own choices; do good and act in the person's best interests; avoid doing harm; and ensure fairness.

Ethical values have changed over time, and continue to change. Many of our daily actions and decisions are guided not only by personal feelings, but also by the demands of society and by what other people expect us to do.

Medical research sometimes seems to be concentrating its resources not on preventing babies from being born with a disability, but on preventing babies with a disability from being born, which is quite different.

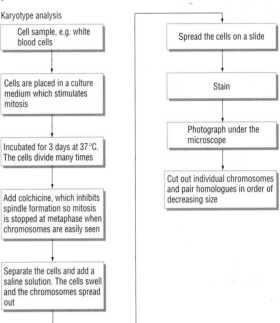

Karyotype analysis

Cell sample, e.g. white blood cells

Cells are placed in a culture medium which stimulates mitosis

Incubated for 3 days at 37 °C. The cells divide many times

Add colchicine, which inhibits spindle formation so mitosis is stopped at metaphase when chromosomes are easily seen

Separate the cells and add a saline solution. The cells swell and the chromosomes spread out

Spread the cells on a slide

Stain

Photograph under the microscope

Cut out individual chromosomes and pair homologues in order of decreasing size

**Figure 2** Karyotype analysis. In this karyotype, in the chromosome 21 homologous pair, part of one chromosome has broken off and become attached to another chromosome

**Down's syndrome** is the most common cause of mental retardation and malformation in a newborn. Down's syndrome occurs in about 1 in every 800–1000 babies. It affects an equal number of boys and girls. As a woman's age (maternal age) increases, the risk of having a Down's syndrome baby increases significantly. For example, at younger ages, the risk is about 1 in 4000. By the time the woman is aged 35, the risk increases to 1 in 400; by age 40 the risk increases to 1 in 110; and by age 45 the risk becomes 1 in 35. Maternal serum screening for Down's syndrome has been undertaken in the UK since the late 1980s. This is done in order to inform the mother – she may then decide to abort the fetus but it will be a decision based on information.

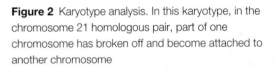

**Figure 3** Down's syndrome karyotypes

**Klinefelter's Syndrome** is one of the most common chromosomal variation found in humans. Klinefelter's Syndrome is found in about 1 out of every 500–1000 newborn males. The four most common symptoms are sterility, breast development, incomplete masculine body build and social or school learning problems. The most common of the four is sterility. Sufferers have normal sexual function, but they can't produce enough sperm for conceiving a child. While general comprehension is lacking, most Klinefelter's Syndrome patients excel in visual and mathematical (sequence) based tests. When it's discovered prenatally that a mother is carrying a child with KS should she abort the pregnancy, since all mothers want a 'perfect' child?

# Questions

1  The position of gene loci on a chromosome can be found by calculating ...................... .

2  Down's, Turner's and Klinefelter's Syndromes are all examples of .................... mutations.

3  A karyotype of a person with Klinefelter's Syndrome is recognised by ...................... .

4  Consider the situation where a karyotype indicates that a mother is carrying a child with trisomy 21. What options are available to the parents? List as many as you can and for each state the ethical issues involved.

Restriction enzyme forming sticky ends

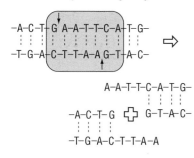

Bases T-T-A-A left exposed so these will readily join with their complementary bases of A-A-T-T

**Figure 1** Sticky ends and palindromes

## The role of restriction and ligase enzymes

**Genetic engineering** is an umbrella term which can cover a wide range of ways of changing the genetic material – the **DNA** code – in a living organism. Using genetic engineering, we can add, modify and replace DNA sequences.

In genetic engineering enzymes are used to cut up parts of the DNA of one organism, and to insert them into the DNA of another organism. In the resulting new organism the inserted genes will code for one or more new characteristics.

Although the details differ, genetic engineering involves three main stages:
- isolating and identifying the **gene** for the protein you want to make
- putting the isolated gene into another organism using a **vector**
- **cloning** the organism. This produces large numbers of the genetically engineered organism all containing the gene for the protein you want to make.

The first stage is to isolate the gene. **Restriction enzymes** (restriction endonucleases) will cut the DNA at a particular sequence of bases. Some restriction enzymes cut straight across both chains of DNA, forming **blunt ends**, but the most useful enzymes make a staggered cut in the two strands, forming **sticky ends**.

The cut ends are 'sticky' because they have short stretches of single stranded DNA that are **complementary** to each other. If two pieces of DNA have been cut with the same enzyme, the sticky ends will stick to each other by forming **hydrogen bonds** to complementary base pairs. We say the pieces have **annealed**. Restriction enzymes are highly specific – each restriction enzyme will cut DNA only at a specific base sequence, 4–8 base pairs long. Several of these 'recognition sites' may occur in a length of DNA, so fragments of DNA of varying length will be produced. These are called **restriction fragments**.

There are thousands of different restriction enzymes known. They are produced naturally by bacteria as a defence against viruses (they 'restrict' viral growth), and they are named after the species of bacteria from which they came. Two examples of restriction enzymes are EcoR1, which comes from *E. coli* strain R and was the first to be identified, and Hind III the third enzyme from *Haemophilus influenzae.*

Once the section of DNA has been cut out and annealed with the DNA of another organism, the broken DNA needs to be repaired. This is done by **DNA ligase enzymes**. DNA ligase repairs the DNA backbone by forming **covalent bonds**: it links up the sugar-phosphate backbones of the newly paired section. So restriction enzymes and DNA ligase are used together to join lengths of DNA from different sources.

### Recognition sites are palindromic

A palindrome is a word or phrase that reads the same in both directions, for example the word 'level'. Restriction enzymes cut DNA at **palindromic sequences** such as GG CCCCGG.

## Techniques of genetic engineering in microorganisms

First **restriction endonucleases** are used to cut out the required gene from the DNA of the donor organism – this may be a gene for human protein.

The second step is to prepare a vector molecule to carry this DNA into a host cell – often a bacterium. A bacterial plasmid (a small, circular strand of DNA often found in bacteria in addition to their main DNA) is often used as a **vector**. The vector is used to transport the DNA into the host cell. The same restriction enzymes are used to cut the bacterial plasmid. They leave sticky ends that correspond to those of the human gene.

**Unit F225**
Module 1 Genetics in the 21st Century
5.1.2 a, b, d

The third step is to join the new gene into the bacterial plasmid. The sticky ends are lined up and the gene is attached – *annealed by hydrogen bonding between base pairs.* Then DNA ligases join the pieces of DNA together to make a recombinant plasmid. The final step is to incorporate the engineered plasmid into the bacterium (**transformation**). Once the plasmid is inside the host bacterium, the genetically modified bacterial cells will need to multiply and quantities of the human protein can be produced. The protein will need to be separated from the host cells and then purified for clinical use. You can see this clearly in Figure 2.

### Marker genes

Only a small percentage of cells take up the recombinant DNA, so it is important to know which cells are genetically modified. Marker genes allow this. These are genes that express readily observed characteristics. They are inserted into the recombinant DNA that will be transferred to the vector. If this DNA is taken up and incorporated into the host cell's genome, the marker gene will express itself, so it is relatively easy to find later which organisms have been successfully engineered (transformed).

### Antibiotic resistance markers and reporter systems

The marker gene could code for resistance to a particular antibiotic. Growing them in a medium containing the antibiotic does this. Only the genetically engineered organisms will be resistant, and so only they will grow. The antibiotic will kill those that have failed to take up the plasmid. This works well, but has led to concerns about the spread of these 'resistance genes'.

The technology of using antibiotic-resistant genes has been superseded by the use of 'Reporter' systems. Widely used Reporter systems are those with the GUS genes. Examples of Reporter systems are GUS (β-glucuronidse, or GUS, is an enzyme that produces coloured or fluorescent products when provided with appropriate substrates). Bioluminescence resonance energy transfer (BRET) detection systems have recently been introduced. These systems are based on excitation energy being transferred from a bioluminescent donor molecule (luciferase) to a fluorescent acceptor molecule, like green fluorescent protein.

### Another route to isolating a gene

Although estimates of the number of genes vary, there are probably between 20 000 and 30 000 protein coding genes in the human **genome**. This makes isolating the correct gene challenging. A way round this problem is to use **complementary DNA**.

Complementary DNA (cDNA) is DNA copied from **mRNA**. The enzyme **reverse transcriptase** synthesises DNA from an RNA template. This enzyme is found in a group of viruses called **retroviruses**. The DNA made is single stranded but DNA **polymerase** can be used to produce a double stranded cDNA 'gene'.

For example, the beta cells of the pancreas make insulin, so make lots of mRNA molecules coding for insulin. This mRNA can be isolated from these cells and used to make cDNA of the insulin gene.

Microorganisms are the most commonly used organisms in genetic engineering because they are relatively easy, quick and cheap to culture and there are few ethical issues about their use. Unfortunately, bacteria cannot make all human proteins, and recently animals and even plants have also been used to make gene products. In the case of animals it is not appropriate to extract the product from their cells, so the product must be secreted in milk or urine. In plants, the product may be secreted from the roots. We shall look at this in the next spread.

Genetic engineering in microorganisms

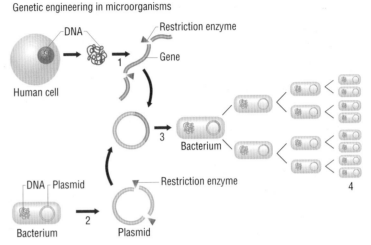

**Figure 2** Genetic engineering in microorganisms

## Questions

1 ..................... enzymes are used to cut DNA at specific sites. Some cut straight across both chains of DNA forming ..................... ends but most make a staggered cut in the two strands, forming .................. DNA ..................... completes the recombinant DNA backbone.

2 What is a bacterial plasmid?

3 When making recombinant DNA, why are marker genes often also inserted?

4 Explain why the plasmid is described as a vector.

## Techniques of genetic engineering using eukaryotic cells

In the last spread we looked at **genetic engineering** in microorganisms, but bacteria cannot always make human proteins. Recently animals and even plants have also been used to make gene products. In neither case is it appropriate to extract the product from their cells, so in animals the product must be secreted in milk or urine, while in plants the product must be secreted from the roots or produced by cultured protoplasts (plant cells without a cell wall).

Examples of human protein made in this way are anti-thrombin, an anti-blood clotting agent used in surgery, and AAT (alpha1-antitrypsin deficiency) an enzyme used to treat cystic fibrosis and emphysema. The host organism for anti-thrombin is a goat and for AAT a sheep.

The technique for using eukaryotic cell lines to produce a human protein can be seen by looking at how AAT is produced.

AAT is a human protein that inhibits **protease** enzymes like trypsin and **elastase**. There is a rare **mutation** of the AAT gene (a single base substitution) that causes AAT to be inactive. This means that the protease enzymes are uninhibited. This is most notable in the lungs, where elastase digests the elastic tissue of the alveoli, leading to the lung disease emphysema. It is possible to treat this disease by inhaling an aerosol spray containing AAT, which reaches the alveoli and inhibits the elastase. AAT for this treatment can be extracted from blood donations, but only in very small amounts.

The gene for AAT has been found and cloned, but AAT cannot be produced in bacteria because AAT is a glycoprotein. This means sugars have to be added to the polypeptide, and only animals can carry out this modification. AAT can now be produced by genetically modified sheep. It will be secreted in the sheep's milk. Figure 1 shows how this is done.

### Gel electrophoresis and polymerase chain reaction

**Gel electrophoresis** is a form of **chromatography** used to separate different pieces of DNA on the basis of their length. This can be seen clearly in Figure 2.

In order to see the DNA fragments you can stain them with a chemical such as azure A or ethidium bromide. The DNA shows up as a blue strand with azure A or fluoresces under ultraviolet light with ethidium bromide. A more sensitive way would be to label the DNA samples at the start with a radioactive isotope such as $^{32}$P. You place photographic film on top of the finished gel in the dark, and the DNA shows up as dark bands on the film. To locate a specific piece of DNA, you take a 'blot' of the fragments onto a nylon filter. Then you use a 'probe' with a base sequence that 'hybridises' to the DNA (complementary base pairing again!).

**Polymerase chain reaction** (PCR) is a technique developed in 1983 (Nobel Prize in 1993). It is a way of replicating DNA in a test tube. PCR can clone DNA samples as small as a single molecule.

The chemistry of PCR depends on the fact that the DNA bases are complementary. During PCR, high temperature is used to separate

| | |
|---|---|
| A female sheep is given a fertility drug to stimulate her egg production, and several mature eggs are collected from her ovaries | 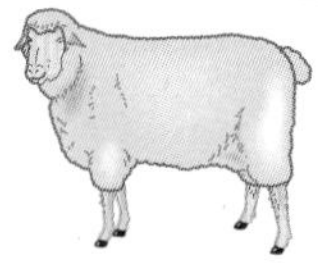 |
| The eggs are fertilised *in vitro* | 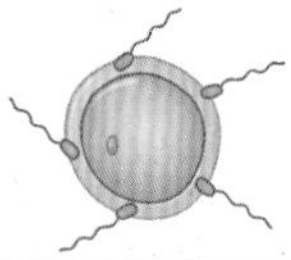 |
| A plasmid is prepared containing the gene for human AAT and the promoter sequence for b-lactoglobulin. Hundreds of copies of this plasmid are microinjected into the nucleus of the fertilised zygotes. Only a few of the zygotes will be transformed, but at this stage you cannot tell which | 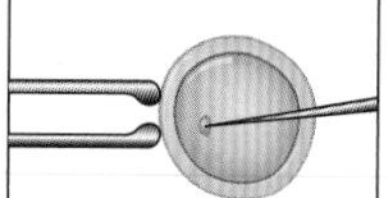 |
| The zygotes divide *in vitro* until the embryos are at the 16-cell stage | 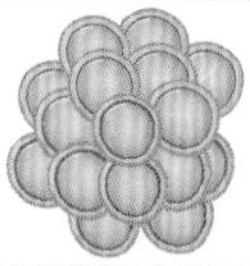 |
| The 16-cell embryos are implanted into the uterus of surrogate mother ewes. Only a few implantations result in a successful pregnancy | 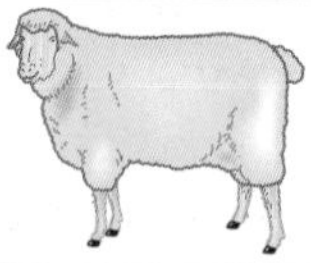 |
| Test all the offspring from the surrogate mothers for AAT production in their milk. This is the only way to find if the zygote took up the AAT gene so that it can be expressed. About 1 in 20 eggs are successful | 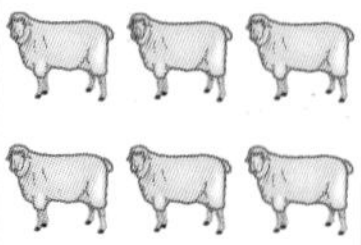 |
| Collect milk from the transgenic sheep for the rest of their lives. Their milk contains about 35 g of AAT per litre of milk. Also breed from them in order to build up a herd of transgenic sheep | 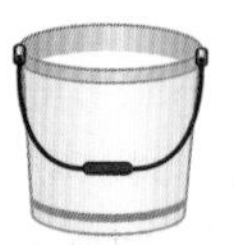 |
| Purify the AAT, which is worth about £50 000 per mg | 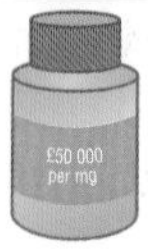 |

**Figure 1** How AAT is produced

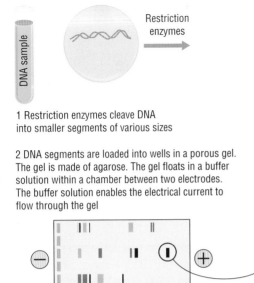
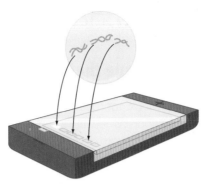

1 Restriction enzymes cleave DNA into smaller segments of various sizes

2 DNA segments are loaded into wells in a porous gel. The gel is made of agarose. The gel floats in a buffer solution within a chamber between two electrodes. The buffer solution enables the electrical current to flow through the gel

3 When an electric current is passed through the chamber, DNA fragments move toward the positively-charged anode. DNA is negatively charged because of the phosphates that form the backbone of the DNA molecule

4 Smaller DNA segments move faster and farther than larger DNA segments

**Figure 2** Gel electrophoresis

the DNA molecules into single strands – this disrupts the hydrogen bonds holding together the double helix, so the molecule unzips or 'denatures'. The DNA solution is allowed to cool, then the lengths of DNA are mixed with the four kinds of activated **nucleotides** (containing A, T, C or G) and the enzyme **DNA polymerase**. Complementary base pairs can reform and the double helix is restored. The DNA will be replicated as many times as the cycle is set for. This is a useful tool, for example, in forensics where only a few cells may appear in a sample.

**DNA profiling** is a form of genetic fingerprinting. It uses both the techniques described above. DNA profiling involves identifying the patterns of genetic material. Two humans will have the vast majority of their DNA sequences in common. Genetic profiling exploits the highly variable repeating sequences in DNA (**minisatellites**). Two unrelated humans will be unlikely to have the same number of minisatellites at a given **locus**.

In DNA profiling, a small sample of human tissue is taken and given a computerised numeric value in the form of a 'barcode'. The method was invented by Sir Alec Jeffreys at the University of Leicester and was announced in 1985. Professor Jeffreys was concerned that the initial genetic fingerprinting methods would not work in a criminal case because of the large amount of genetic material needed for the case to be successful. Fortunately, the advent of PCR increased sensitivity, allowing small amounts of DNA to be sampled, even from stored and partially degraded samples. Different repeating satellites called *microsatellites* have also been discovered, further increasing the accuracy of DNA profiling. Also called short tandem repeats, they are made up of a unit that can vary in length from one to seven bases.

DNA profiling is used in forensic science, paternity testing and matching organ donors.

# Questions

1 Why can't bacteria be used to make all human proteins?

2 How is the human protein extracted from genetically modified animals?

3 ....................... are involved in making human protein. Different lengths of DNA can be separated using ....................... . This works on the principle that as DNA is negatively charged if an electric current is applied, the DNA will be attracted to the ................... . The shorter lengths of DNA move the ..................... .

4 Which technique can clone DNA from a sample as small as a single molecule?

5 DNA profiling is an important technique and is used in ...................., .................... and .................... .

## The body in the carpet

Buying a house can be a stressful time, but surely not as stressful as the experience of a house-buyer in Cardiff who moved into his new residence, dug up the patio and found a buried carpet. Inside the carpet was a skeleton.

'It was a fairly safe bet this was a murder case,' says Professor Jeffreys. 'A facial reconstruction from the skeleton was put out in the press and on Crimewatch. Someone phoned Crimewatch and suggested that it was a girl called Karen Price who had gone missing about ten years previously.'

Working with Erika Hagelberg, an expert in bone DNA extraction and analysis, DNA was extracted from the skeleton and Professor Jeffreys compared it to DNA from Karen's mother and father; it was indeed their daughter.

This was enough to give the police a secure basis for identification. A murder enquiry was launched, which led to the identification of the murderer and accomplice. The evidence went to Cardiff Crown Court in 1991 – the first time PCR and ancient bone DNA evidence had been presented to a court in the UK. 'It was a lot of firsts for the court to handle, but the evidence was accepted and the perpetrators were convicted,' says Professor Jeffreys.

## The use of gene therapy to treat genetic disease

**Gene therapy** is an experimental technique that uses genes to treat or prevent disease. The most obvious approach involves inserting a normal **gene** to replace an abnormal gene. Other approaches include

- swapping an abnormal gene for a normal one
- repairing an abnormal gene
- altering the degree to which a gene is turned on or off.

The potential scope of gene therapy is enormous. More than 4200 diseases have been identified as resulting directly from abnormal genes, and countless others that may be partially influenced by a person's genetic makeup.

Although there is much hope for gene therapy, it is still experimental. Little progress has been made since the first gene therapy **clinical trial** began in 1990. A major blow came in January 2003, when the Food and Drug Administration (FDA) in the USA placed a temporary halt on all gene therapy trials using retroviruses as vectors in blood stem cells. FDA took this action after it learned that a second child treated in a French gene therapy trial had developed a leukaemia-like condition. The child had been successfully treated by gene therapy for X-linked severe combined immunodeficiency disease (X-SCID), also known as 'bubble baby syndrome'. This rare disease inhibits immune system functioning due to a mutation in the gene for a single enzyme adenosine deaminase (ADA). In April 2003 the FDA eased the ban on gene therapy trials using retroviruses as vectors in blood stem cells, but in July 2007 the FDA shut down a gene therapy trial for arthritis after a patient died.

### Problems with using gene therapy

- *Short-lived nature of gene therapy.* Before gene therapy can become a permanent cure for any condition, the cells containing the **recombinant DNA** must be long-lived and stable. Advances in stem cell research may make this more achievable.
- *Immune response.* The immune system is designed to attack foreign substances whenever they are introduced into human tissues. There is always a risk of stimulating the immune system in a way that reduces the effectiveness of gene therapy. Also, the immune system's enhanced response to invading substances it has seen before makes it difficult for gene therapy to be repeated in patients.
- *Problems with viral vectors.* Viruses are the carrier of choice for getting the new gene into the cell in most gene therapy studies. But they present a variety of potential problems to the patient. Also, there is always the fear that once it gets inside the patient the viral vector may cause disease.
- *Multigene disorders.* Unfortunately, some of the most commonly occurring disorders, such as heart disease, high blood pressure, Alzheimer's disease, arthritis and diabetes, are caused by the combined effects of variations in many genes. Multigene disorders such as these would be especially difficult to treat effectively using gene therapy.

### Recent developments in gene therapy research

Several cancers also have the potential to be treated with gene therapy. A gene therapy tested for melanoma (skin cancer) involves introducing a gene with an anticancer protein called tumour necrosis factor (TNF) into test-tube samples of the patient's own cancer cells. These cells are then reintroduced into the patient. In brain cancer, the approach is to insert a specific gene that increases the cancer cells' susceptibility to a common drug used to fight the disease.

## Ethical implications of genetic engineering in humans

**Somatic cell therapy** involves adding genes to cells other than egg or sperm cells. For example, if a person has a disease caused by a defective gene, a healthy gene can be

---

### Doctors test gene therapy to treat blindness

Tuesday May 1, 2007

London (Reuters) – A team of British doctors has carried out the world's first eye operations using gene therapy to try to cure a serious sight disorder, officials said on Tuesday.

The group from Moorfields Eye Hospital and University College London (UCL) has operated on a small number of young adults with Leber's congenital amaurosis, a type of inherited childhood blindness caused by a single abnormal gene.

The condition prevents the retina from detecting light properly, resulting in progressive deterioration and severely impaired eyesight. There is no effective treatment.

The new experimental procedure involves inserting normal copies of the faulty RPE65 gene into cells of the retina – the light-sensitive layer of cells at the back of the eye – using a harmless virus as a vector.

It will be several months before the success of the procedure can be properly assessed but medics said there had been no complications so far.

The move into human testing follows 15 years of laboratory and animal experimentation, including tests on dogs whose vision was restored to the extent they could navigate a maze with ease.

# Unit F225
## Module 1 Genetics in the 21st Century
### 5.1.2 f, g, h

added to the affected cells to treat the disorder. This is non-inheritable – the new gene will not be passed to the recipient's offspring.

**Germ cell therapy** would change genes in eggs, sperm, or very early embryos. The gene would be inherited, meaning that the modified genes would appear not only in any children that resulted from the procedure, but also in all succeeding generations. This application could open the door to the perpetual and irreversible alteration of the human species.

Some ethical issues for you to consider:

- We could choose to have changes made to us, but if the changes are carried out on the germ cells our children will inherit them, without them having any choice. Do we have that right, and how far should we take our ability?
- Conversely, is it responsible and ethically acceptable to leave the potential of our children to the chance effects of the 'genetic lottery', if we can develop the technological ability to make positive changes?
- If genetic engineering becomes the way of the future, will people whose parents could not afford to genetically 'modify' them while still an embryo, have a chance of being 'high-achievers' compared to the people who were modified to be perfect?
- Is it ethical to experiment on embryos? Do they have rights?
- Does genetic engineering constitute a misuse of our free will?

# Human Genome Project

The Human Genome Project was one of the most enterprising and challenging aspects of modern genetic research. This project was devised to map and sequence the entire **human genome** – that is, to locate every gene on every human chromosome – about 30 000 in all. When the project began in 1990 it was expected to take 15 years, but advances in technology allowed the work to finish ahead of schedule and under budget. Hundreds of scientists in the UK, China, France, Germany, Japan and the US were involved. The UK was the largest contributor to the project. One-third of the human genome was sequenced at the Wellcome Trust's Sanger Institute, which was the first centre to publish the complete sequence of an entire chromosome. In April 2003 researchers successfully completed the Human Genome Project.

### Today

Having the complete sequence of the human genome is similar to having all the pages of a manual needed to make the human body.

- The Human Genome Project has already fuelled the discovery of more than 1800 disease genes.
- As a result of the Human Genome Project, today's researchers can find a gene suspected of causing an inherited disease in a matter of days, rather than the years it previously took.
- There are now more than 1000 genetic tests for human conditions. These tests enable patients to learn their genetic risks for disease and also help healthcare professionals to diagnose disease.
- At least 350 biotechnology-based products resulting from the Human Genome Project are currently in clinical trials.

### Dilemmas raised by this information

- Should insurance companies be allowed to see the results of genetic tests carried out on individuals before they agree to give life insurance? Should insurance companies be able to insist that all those with a family history of a genetic disorder take a genetic test?
- Should companies be allowed to patent a gene after paying for research into that gene?
- How can we be sure that information on racial or ethnic background revealed by DNA sequencing is not used to discriminate against different groups?
- Should the information be used by individuals to select particular genes for their children?

## Human–animal embryo research approved

Wednesday September 5, 2007

London (Reuters) – Regulators decided on Wednesday to permit in principle the creation of hybrid human–animal embryos for research into illnesses such as Parkinson's, motor neurone disease and Alzheimer's. The Human Fertilisation and Embryology Authority (HFEA) agreed to allow a specific kind of inter-species hybrid, where human DNA is injected into a hollowed-out animal egg cell, a spokeswoman for the regulator said. The resulting 'cytoplastic hybrid' embryo, or 'cybrid' would be 99.9 per cent human and 0.1 per cent animal. Two teams of British scientists have applied to the HFEA for permission to create such hybrids to overcome a shortage of donated human eggs.

The researchers hope to use the hybrid embryos, which must be destroyed after 14 days, to create stem cells to help find new medical treatments for degenerative diseases. Opponents say mixing even a tiny amount of human genetic material with that of an animal is unnatural and wrong. Scientists in China, the United States and Canada have already carried out similar work.

## Questions

1 What does the term gene therapy mean?
2 What is the difference between somatic and gene cell therapy?
3 What was the aim of the Human Genome Project?

## Predicting the probability of genetic disease

Pedigree charts are often used to show the inheritance pattern of genetic conditions within a family. Such charts can help to predict the likelihood of a child being born with a particular genetic disease, as they can show whether a phenotype is controlled by a dominant, recessive or sex-linked allele. We looked at the details of pedigree charts in spread 5.1.1.3. You can revise your knowledge and use of pedigree charts by answering Question 1 at the end of this spread.

## The role of the genetic counsellor

If there is a possibility that a couple may have a child with a genetic disease, they are likely to be referred to a **genetic counsellor**. Genetic counsellors work with people concerned about the risk of an inherited disease. These could be new parents or couples planning a pregnancy, or family members concerned that they too may carry an allele for a disorder. Genetic counselling helps individuals understand their risk of having a genetic disorder. Counsellors also explain the risks of passing on a genetic disorder to the next generation and the choices available to people. Counselling enables a person to make an informed decision about the options available to them.

Counselling helps people understand the nature of a particular disease, and what having it will mean in practical terms. The counsellor can explain the options there might be for prevention/testing, the risks of recurrence and the implications for other family members.

Counsellors work as part of a wider healthcare team, involving clinical consultants, nursing and primary care teams. Most people are referred through a GP or hospital consultant after a diagnosis. Others seek advice after a genetic disease has been discovered in their family. They may want immediate help, or they may choose to let the diagnosis 'sink in' before seeking help.

The first meetings generally involve sorting out a family history and gaining extra diagnostic information, if necessary. Correct diagnosis is absolutely vital for genetic counselling to be effective, and people may be referred for further testing by a clinical geneticist. Once the diagnosis is clearly established, the counsellor can then tailor the sessions to meet the family's specific needs.

### Who is offered genetic counselling?

Some of the reasons for being offered genetic counselling include:

*   The couple already have a baby or child who has a physical problem or delay in development. The diagnosis may be uncertain and the doctors are wondering if there may be a genetic cause for the child's problems.
*   The couple have lost a baby during pregnancy or infancy.
*   One of the couple has, or carries alleles for, a particular condition, which might be passed on to the children.
*   There is a known genetic condition in one of the couple's family.
*   The couple are close blood relatives.
*   There is a strong history of cancer, for example, in the family.

## Ethical issues involved in genetic counselling

Ethical issues involved in genetic counselling are becoming more important now, because more genes are being identified and linked to diseases. One of the main ethical dilemmas arises from a conflict between the right of the individual to personal privacy and the interests of other family members. Should they be made fully aware of available information that could play a part in making important life decisions?

YORK
COLLEGE
Sim Balk Lane
York
YO23 2BB

### Examiner tip

You will be expected to outline ethical issues clearly. Avoid emotive phrases such as 'playing God' and concentrate on the goals, the rights and the duties of the individuals involved in making some very difficult decisions. Consider the possible consequences to individuals, to families and to society, as well as the rights of the unborn child.

**Figure 1** A genetic counsellor at work

## Unit F225
### Module 1  Genetics in the 21st Century
5.1.3 a, b, c

In the case of a pregnant woman, she and the counsellor are now faced with choices that were once left to fate. When genetic risks are high there is conflict between the desire to have a healthy child and the need to avoid problems for oneself, for the family and for society. Although 96% of counselling sessions end well with no or very few chances of the disease occurring, the remaining 4% of people in the high risk category are left with three options:

- prenatal diagnosis and abortion if required
- artificial insemination
- gene therapy.

Should the genetically defective be aborted? Do parents have a right to produce defective children?

Genetic counsellors learn many family 'secrets', such as previous abortions, previous abnormal births, and, occasionally, false paternity. These findings could be made after the birth of an affected infant when the family is genetically screened. The father may believe that he is a carrier, but then the test on him is negative. Should the family be protected from potential disruption caused by the disclosure that he is not the baby's biological father? Should the actual risk be revealed to them with no explanation? The result of a genetic test can create 'ripples' that can spread widely throughout a family. An equally difficult problem for the counsellor is presenting her knowledge in a non-judgemental manner.

When such ethical dilemmas arise, often the best that can be hoped for is a compromise so that the opposing views are respected and, personal conscience permitting, an individual's expressed wishes are respected. In practical terms, difficult ethical situations are not usually resolved quickly. Sometimes a course of action that causes least harm may be the best that can be achieved.

## Questions

1  Tamsin knows that her mother has Huntington's disease. She deduces that at least one of her maternal grandparents was a sufferer.

(a)  Is her reasoning correct?

(b)  She also deduces that at least one of her great-grandparents was a sufferer. Is she correct?

2  A 55-year-old, post-menopausal, retired school teacher with a family history of breast cancer in her mother and sister, noticed a painless hard lump in her right breast. A history and physical exam were followed by a mammogram that identified a 1 cm lump in the breast. Following surgery to remove the lump she was started on a course of Tamoxifen. The woman has two daughters, aged 35 and 37, who were strongly encouraged to have yearly physical exams and mammograms by their mother's doctor, and to perform monthly self-examinations. Why might genetic counselling be of value for this family?

3  A couple in which both partners may be carriers for cystic fibrosis consult a genetic counsellor to discuss the risks to their future. Each child of parents who both carry an abnormal CF gene has a 25% chance of inheriting CF. This means that, on average, in three out of four cases the prenatal test will show that the fetus does not have CF. What would the counsellor need to do in the first consultation and what do you think her advice would be?

### Case study

The parents of two boys aged six and 11 approach the genetic counsellor. The father has a dominantly inherited adult-onset genetic condition.

Even if the children have inherited this gene mutation from their father, they will not show any symptoms until they are in their 30s or 40s. Both parents are demanding that their children be tested and that they be informed of their children's genetic status. Should these children be tested? Does the parents' desire to know outweigh the rights of the children to wait until they are old enough to make their own decision? If the children are tested now, should they be informed of their genetic status and if so, when? Is it possible that the parents might treat the children differently if one were found to have the gene mutation while the other did not?

### Case study

A man was diagnosed with a mild form of adrenoleukodystrophy (ALD), an X-linked condition that can be carried by healthy females. He did not wish his diagnosis or the genetic implications to be discussed with his family. Seven years later, his niece gave birth to two successive boys who had a more severe form of ALD. Their illness only came to light when the older boy started to display symptoms. The mother's sister, the man's other niece, has also given birth to a son subsequently diagnosed with ALD. Both families are bitterly resentful that the medical services did not warn them of their genetic risk.

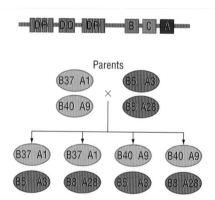

Parents

Possible combinations in offspring

**Figure 1** Inheritance of HLA haplotypes

## Why genetic compatibility is important in transplant surgery

When tissue from one person, the **donor**, is transplanted into a **recipient**, the recipient will produce an **immune response** against the transplanted tissue. This is known as **tissue rejection** because the tissue will eventually be destroyed. This is a reaction to the genetic differences between the two people, so it will not apply to identical twins. **Compatibility** (the ability to accept transplanted tissue) depends on the degree of genetic similarity at a particular **locus** between the individuals. Before considering a transplant, tests must be done to determine the likely compatibility between the intended donor and recipient.

First, a simple blood test is performed to find out the **blood group** of both the donor and recipient. Obviously, transplant donor and recipient must match in terms of blood group.

Next, **tissue typing** is carried out. All cells (except for red blood cells) carry **antigens** other than A and B in their cell surface membranes. When considering the possibility of a transplant, doctors are particularly interested in the **human leucocyte antigen system** (**HLA**). The genes that code for the HLA antigens are found on chromosome 6 and there are 6 **gene loci** involved. This area of chromosome 6 is called the **Major Histocompatibility Complex** (**MHC**). These loci are so close together that they tend to be inherited together (we say that they are linked). Each set of HLA antigens is referred to as a **haplotype** and every person inherits one haplotype from each parent. The loci are given letters to identify them. Each of the 6 loci has a large number of **alleles**, so you may have an HLA-A1, B8, DR17 while someone else has an HLA-A2, B7, DR15.

An important characteristic of the HLA system is the enormous number of HLA types that are possible. Each person has two each of the HLA-A, B, and DR antigens. There are roughly 25 different HLA-A antigens, 40 different HLA-B antigens and 20 different HLA-DR antigens. Most of these antigens can be further distinguished if you compare the DNA sequence that codes for each antigen (the molecular typings). When HLA antigens are typed using molecular typing methods, there are several hundred HLA antigens at each locus. So it is not an exaggeration to say that there are potentially millions of possible HLA types in the population.

For solid organ transplants, the important HLA antigen groups are the HLA-A, B, and DR antigens. A 6-antigen match is the best possible match for kidney transplants, but lesser matches have been used and have had successful outcomes.

For bone marrow or stem cell transplants, the important groups may include HLA-A, B, C, DR, DP and DQ.

Serological tissue typing has been phased out and replaced by DNA-based tissue typing (involving **PCR**) which has increased the accuracy and specificity of HLA typing. This has allowed for more precise HLA matching between donors and transplant patients.

## Potential sources of donated organs

Transplants are the best possible treatment for most people with organ failure. Spread 5.3.4.3 deals specifically with kidney transplants. The commonest organ transplanted is the kidney.

Although the first reported cornea transplant took place in Moravia as long ago as 1905, the history of transplantation as a clinical treatment really began only in the second half of the twentieth century. In 1954 there was the first ever kidney transplant operation on identical 23-year-old twins Richard and Ronald in Boston. The operation was a success, with Richard living another eight syears and Ronald, the donor, still alive today. Since then, transplants have become so successful in the UK that a year after surgery, 94%

**Unit F225**
Module 1 Genetics in the 21st Century
5.1.4 a, b, c

of kidneys in living donor transplants, 82% of heart transplants and 64% of heart/lung transplants are still functioning well. These figures are improving all the time.

### Sources of donated organs:

- *Cadavers (corpses)*: the number of available organs has fallen for several reasons. Improvements in road safety, medical advances in the treatment of patients and the prevention of strokes in younger people means that only a very small number of people die in circumstances where they are able to donate their organs. To make up for this, doctors have widened their criteria for accepting organ donors:
  - Older donors are considered. For example, livers from donors older than 70 years are now being transplanted.
  - Unsuitable hearts, which would not be used for transplant under normal circumstances, have been used as a 'bridge' in urgent situations until another heart becomes available.
  - Sometimes one liver can be split into two pieces for transplant into two patients. Kidneys may be donated by living relatives who can safely donate one of their two kidneys.
  - Based on this experience, and with a desperate need for organs, some transplant hospitals have removed part of the liver, lung or small bowel from living donors. 'Domino' transplantation allows a patient's healthy organ to be transplanted into another recipient. For example, when a patient needs a double lung transplant, he may receive a heart–lung transplant because the surgery of the combined transplant is technically easier to perform. The patient's own heart may still be healthy and may be transplanted into another patient instead of being discarded.
- Transplants involving kidneys from **living donors** have a better outcome than those from a cadaver. In the UK, for example, official statistics report that 95% of kidneys from living donors and 86% of kidneys from cadaver donors are still 'functioning well' after a year. Ideally, the donated organ is from a donor who is genetically identical to the recipient – an identical twin. The second-best alternative is an organ from a blood sibling. Failing that, a parent or another blood relative could be used. Currently, living donors can donate bone marrow, liver lobes, lung lobes and kidneys.
- Apart from fears about exploitation, some people say that the **sale of organs** is fundamentally wrong. Professor Delmonico, an adviser for human transplantation at the World Health Organization and professor of surgery at Harvard Medical School, said in an interview that 'it is physically improper for anyone to buy and sell organs as it violates the dignity of the human person'.

  Senior doctors have complained of growing strains on the NHS from botched transplant operations conducted abroad, while doctors in India see poor donors dying after selling one of their kidneys. In March 2007, China published new rules governing human organ transplants. It has forbidden the buying and selling of organs and requires that donors give written permission for their organs to be transplanted.
- **Xenotransplantation** (animal-to-human transplants) are being studied by some transplant centres, as pigs may be a possible source for transplant organs.

**Figure 2** Transgenic pigs

## Questions

1  How many haplotypes will you have in each of your liver cells?
2  What does the term compatability mean in the field of transplant surgery?
3  MHC stands for ..................... and is found on chromosome .................. . These genes code for ..................... .
4  Name four different sources of donated organs for transplant surgery. Which would be likely to have the best outcome? State one problem associated with each source.

## Producing a supply of embryonic stem cells

Stem cells have the remarkable potential to develop into many different cell types in the body – they are **pluripotent**. Serving as a sort of repair system for the body, they can divide without limit and **differentiate** to replace other cells. When a stem cell divides, each new cell has the potential to either remain a stem cell or become another type of cell with a more specialised function, such as a muscle cell, a red blood cell or a brain cell.

**Embryonic stem** cells are 'harvested' or collected from the very early stages of a fertilised **oocyte** called a **blastocyst**.

Blastocysts are like tennis balls – solid on the outside but hollow in the middle. Stem cells are found in a little lump inside that hollow ball. Once the stem cells are removed and placed in a sterile dish with nutrient liquid in the laboratory, they continue to make exact copies (clones) of themselves. They will only make exact copies of themselves and nothing else until a researcher adds different chemicals, or molecular signals, that trigger the stem cells to differentiate into muscle cells or neurones.

How can a researcher obtain embryonic stem cells?

- By using the blastocysts that are left over after *in vitro* fertilisation. For couples using *in vitro* fertilisation, the woman has eggs harvested, which are then fertilised by her partner's sperm in a dish in the clinic. Eggs that are successfully fertilised are allowed to divide for 3–6 days, forming blastocysts. After the woman has conceived, they must decide what to do with the leftover blastocysts. Some couples pay to keep their remaining blastocysts frozen while others are donated to provide stem cells.
- By cloning embryos and harvesting stem cells from them. The embryos are destroyed before they are 14 days old and are never allowed to develop beyond a cluster of cells the size of a pinhead.

The embryo is formed by transferring DNA from embryonic stem cells to an egg whose nucleus has been removed. The reconstructed egg cell containing the DNA from a donor cell is then treated with an electric current to stimulate cell division and blastocyst formation.

### Why are embryonic stem cells needed?

We were looking at organ transplants in the last spread and found that donated organs and tissues are often used to replace ailing or destroyed tissue, but the demand for transplantable tissues and organs far outweighs the available supply. Stem cells, directed to differentiate into specific cell types, offer the hope of a renewable source of replacement cells and tissues to treat diseases. Doctors hope that diseases such as Parkinson's and Alzheimer's, spinal cord injury, stroke, burns, heart disease, diabetes, osteoarthritis and rheumatoid arthritis will all be treatable. The ultimate aim is to create cloned stem cells from the DNA of a patient with a **degenerative disease**. These cells can then be turned into whatever types of cells are needed to repair or replace their damaged organs.

Human stem cells could also be used to test new drugs. For example, new medications could be tested for safety on differentiated cells generated from embryonic stem cells. They already use cancer cell lines to screen potential anti-tumour drugs.

Stem cells could be used to make what are often called 'designer babies'. The stem cells would produce an embryo whose hereditary make-up (genotype) would be purposefully selected ('designed'). The use of genetic techniques would ensure the best combination possible of their parents' **alleles**. Tests will allow the couples to select embryos free from a mutant allele that increases the risk of developing an illness.

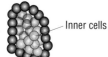

**Figure 1** Formation of a blastocyst

Morula (a ball of about 16 cells or more)  Blastocyst

Inner cells

Outer cells

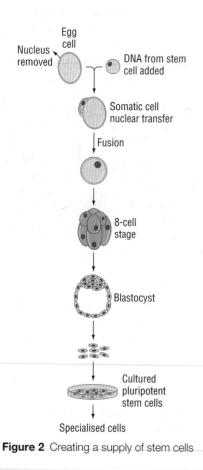

Egg cell

Nucleus removed

DNA from stem cell added

Somatic cell nuclear transfer

Fusion

8-cell stage

Blastocyst

Cultured pluripotent stem cells

Specialised cells

**Figure 2** Creating a supply of stem cells

# Ethical issues in cloning and in transplant surgery

The use of embryonic stem cells is controversial, with opponents arguing that all embryos, whether created in the laboratory or not, have the potential to go on to become a fully-fledged human. Others fear there are safety concerns.

Supporters of cloning say it could offer numerous benefits in the future, such as fighting disease, battling infertility or preserving endangered species. **Therapeutic cloning** is believed to have huge potential to treat disease and disability and is allowed in Britain. **Reproductive cloning** – the cloning of human embryos with the intention of creating a baby – was made illegal in 2001.

The UN recently voted in favour of a ban on all human cloning, but this was non-binding, which means the UK can continue to do therapeutic cloning.

Most opponents of this technology are particularly wary of its ability to lead to a new **eugenics**, where individuals are 'bred' or designed to suit social preferences such as above-average height, certain hair colour, increased intelligence or greater memory.

Is the use of genetic engineering in one's children, and the use of other techniques, defensible? Is it the moral obligation of parents to try to give their children the healthiest, happiest chance in life possible?

If designer babies are possible, it may only be the rich that have access to the technology. Might humans develop into two separate classes – one with a superior genetic make-up and the other being the underdogs, the 'ordinary people'?

## Ethical issues in transplant surgery

Renal transplantation (kidney transplants) is now a well established form of treatment for end-stage renal failure. The shortage of available kidneys, one of the major problems facing transplant surgeons, has led some to suggest using prisoners, street children and brain-damaged patients as organ donors. This suggestion is clearly unethical, but are there other ways of increasing the supply of kidneys that could be condoned?

Transplantation of an organ or tissue from a dead to a living person presents no ethical problem as such. Most religious groups have recognised the benefits of such transplants. Questions tend to arise from circumstances other than the transplant operation itself. These include brain death and elective ventilation of potential donors (potential donors need to be kept 'alive' after brain death until surgeons are ready to use organs). Some people fear that if a person is in a deep coma they may be classed dead and the organs removed. Should we opt in or opt out of organ donation? Should we use living related and unrelated donors? What about the sale of organs and the cost of transplantation in a world of finite resources? The use of fetal tissue in the treatment of conditions such as Parkinson's disease raises further issues.

Considering brain-stem death, the majority of operations using cadavers (corpses) as donors are carried out on brain-stem dead patients who are on life-support machines.

In order to determine that a patient is dead, two senior doctors have to perform a series of tests (irreversible loss of consciousness, inability to breathe plus absence of all other brain stem functions). Turning the ventilator off at this stage is *not* withdrawing treatment and allowing the patient to die, but it is ceasing to do something useless to someone who is already dead. Keeping the patient on a ventilator at this point raises hope where there is no hope.

## First designer babies to beat breast cancer

*The Times*
April 26, 2007

Tests will allow the couples to take the unprecedented step of selecting embryos free from a gene that carries a heightened risk of the cancer but does not always cause it. The move will reignite controversy over the ethics of embryo screening.

An application to test for the BRCA1 gene was submitted yesterday by Paul Serhal, of University College Hospital, London. It is expected to be approved within months as the Human Fertilisation and Embryology Authority (HFEA) has already agreed in principle.

Opponents say that the test is unethical because it involves destroying some embryos that would never contract these conditions if allowed to develop into children. Even those that did become ill could expect many years of healthy life first.

Some critics fear that the tests move society farther down a slope that will lead ultimately to the creation of 'designer babies' chosen for looks or intelligence.

## As news of the world's first face transplant is announced

*Consultant Facial Surgeon Iain Hutchison at Barts, chief executive of Saving Faces – The Facial Surgery Research Foundation, comments:*

'This is the first facial transplant of the modern era. All medical advances are to be celebrated, but this operation throws up many moral and ethical issues. This was a "quality of life" operation rather than a life-saving operation and has many implications for the recipient and donor's families.

'The recipient chose to take the risk of the operation failing if the blood vessels become blocked, there's a medium-term risk of the immuno-suppressant drugs failing to control rejection of the donor tissue, and a long-term risk of the drugs causing cancers. She could be back to square one without a face, needing further reconstruction operations.'

## Questions

1 What do you understand by the term 'pluripotent'?
2 Where are embryonic stem cells found?
3 What can embryonic stem cells be used for?
4 Which is legal in the UK – therapeutic or reproductive cloning?

1 A ..................... is a length of DNA which codes for the production of a particular polypeptide. The ..................... describes the genetic make-up of an organism. A homozygote has the two alleles of a gene ...... ..................... . A recessive allele only affects the phenotype when it is ...... ..................... .

2 A gene mutation is an alteration of ..................... whereas a chromosome mutation is a change in the ..................... or ..................... of chromosomes.

3 Huntington's disease is a ..................... mutation. It is inherited as a ..................... allele whereas cystic fibrosis and phenylketonuria are autosomal ..................... .

4 Sometimes neither allele is dominant to the other but both affect the phenotype. This is called ..................... .

5 If a child has a genotype of $I^AI^O$ what is its blood group?

6 If a couple with blood groups O and AB have a child what is the possible blood group of this child? Explain your answer by drawing a genetic diagram.

7 What do we mean if we say that a disease is sex-linked?

8 If we want to trace the inheritance of a genetic disease we can construct a ..................... .

9 Some genes have loci that are so close together that they are usually inherited together. They are said to be ..................... . This can be used to map the position of the loci on the chromosome by calculating the ..................... .

10 Translocation can cause Down's syndrome. It occurs when the long arm of chromosome ..................... joins another chromosome.

11 Klinefelter's and Turner's syndrome can be diagnosed by examining the patient's ..................... .

### The chi squared test ($\chi^2$)

This is a simple statistical test that can be used to tell if their results are significant or due to chance. The **observed** results are compared with those expected. The larger the difference between the observed and **expected** results, the greater the likelihood that the original prediction was wrong. We reject a hypothesis if the probability of our observations is a chance variation of what we expected to happen is less than 1 in 20 (5%). In other words a probability value of 0.05 is our level of significance

### Worked Example

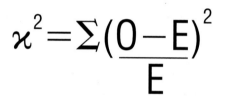

$$\chi^2 = \Sigma \left( \frac{(O-E)^2}{E} \right)$$

$\chi^2$ = the test statistic     $\Sigma$ = the sum of

O = Observed frequencies     E = Expected frequencies

Suppose that the ratio of male to female students in the Science Faculty is exactly 1:1, but in the Pharmacology Honours class over the past ten years there have been 80 females and 40 males. Is this a significant departure from expectation?

Work out the expected frequencies: We do this by adding up all the students, and dividing by the number of different categories we're looking at – which is two (male and female).

So the expected frequencies are (80 + 40)/2 = 60 in each area

| | Male | Female |
|---|---|---|
| Observed | 40 | 80 |
| Expected | 60 | 60 |
| O–E | –20 | 20 |
| (O–E)² | 400 | 400 |
| (O–E)²/E | 6.67 | 6.67 |

Chi-squared value = 6.67 + 6.67 = 13.34

## Unit F224
### Module 1  Genetics in the 21st Century
#### Summary and practice questions

Next work out the **degrees of freedom** (one less than the number of categories). This is a measure of the spread of the data. Degrees of freedom = 2 – 1 = 1.

We then look up the value of 13.34 with one degree of freedom in a table of chi-squared values. Using the $\chi^2$ table we find a critical value of 3.84 for $p = 0.05$. Our calculated value for $\chi^2$ exceeds the critical value for $\chi^2$

### Chi-squared table

| Degrees of Freedom | Probability, $p$ | | | | |
|---|---|---|---|---|---|
| | 0.99 | 0.95 | 0.05 | 0.01 | 0.001 |
| 1 | 0.000 | 0.004 | 3.84 | 6.64 | 10.83 |
| 2 | 0.020 | 0.103 | 5.99 | 9.21 | 13.82 |

In fact, in our example, the calculated value for $\chi^2$ (13.34) exceeds even the tabulated $\chi^2$ value (10.83) for $p = 0.001$. This means that there is a probability of less than 1 in 1000 that we could have got this result by chance. We can be 99.9% confident that some factor leads to a bias towards females studying pharmacology

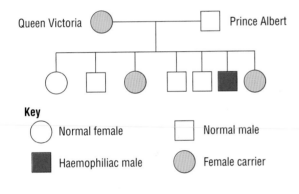

Queen Victoria — Prince Albert

**Key**
- ⚪ Normal female
- ⬜ Normal male
- ⬛ Haemophiliac male
- 🔘 Female carrier

**12** Using the above pedigree chart, state the genotypes of

Queen Victoria .....................

Prince Albert .....................

What percentage of their sons would you have expected to have inherited haemophilia. Explain your answer using a genetic diagram.

**13** How does the karyotype of an individual with Down's syndrome differ from that of a normal individual?

**14** A condition known as *Icthyosis gravior* appeared as a mutation in a boy in the early eighteenth century. His skin became much thickened and formed loose spines. When he grew up this 'porcupine man' married and had six sons, all with this condition, and five daughters who were all normal. For four generations this condition was passed from father to son. What can you suggest about the location of this gene?

**15** A patient is waiting for a kidney transplant. The blood group and tissue type must match. Explain what is meant by tissue typing. Consider the ethical issues involved in transplant surgery.

## The central and peripheral nervous systems

The actions of cells, tissues and organs within our body need to be coordinated. This means there needs to be a means of communication. This is the function of the nervous and endocrine systems. The central 'switchboard' is the **central nervous system** (CNS), which consists of the brain and spinal cord.

The **peripheral nervous system** includes all nerves not in the brain or spinal cord, and connects all parts of the body to the central nervous system. The peripheral (sensory) nervous system receives stimuli, the central nervous system interprets them, and then the peripheral (motor) nervous system initiates responses.

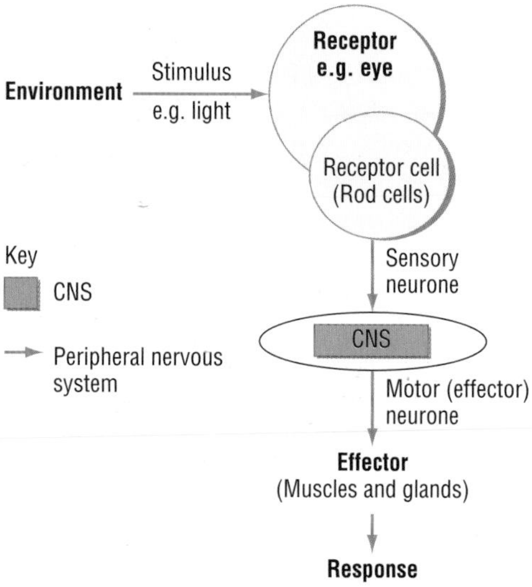

## The structure of the eye

### The structure of the retina

The retina covers about 65% of the eye's interior surface. In section the retina is no more than 0.5 mm thick. In the retina light energy is converted to nerve impulses (action potentials) that are carried along the optic nerve to the brain. There are two types of light-sensitive receptor cells – **rods** and **cones**. These lie in a layer that is outermost in the retina, against the pigment epithelium and choroid layer (Figure 3). A layer of **bipolar cells** and **ganglion** cells lie innermost in the retina, closest to the lens and front of the eye. This means that light has to travel through the thickness of the retina before it can strike and activate the rods and cones.

**Figure 1** Inter-relationships of components of the nervous system

| Name of structure | Structure | Function |
| --- | --- | --- |
| Sclera | A tough, white outer layer | Protects the structures within it and maintains the shape of the eye |
| Choroid layer | Richly supplied with blood vessels; inner part of this layer is made up of cells containing the dark pigment, melanin | Absorbs light and prevents it from being reflected inside the eye |
| Retina | Innermost layer of the eye. Contains the receptor cells – rods and cones | Receives the light stimulus |
| Fovea | Part of the retina where cones are concentrated | Provides maximum visual acuity (the resolution of the image that is perceived by the brain) |
| Conjunctiva | Thin layer at the front of the eye | Protects the surface of the eye. Is kept moist by a film of fluid secreted by the tear ducts |
| Cornea | Continuous with the sclera | Plays a major part in focusing light rays onto the retina – it refracts light rays as they enter the eye |
| Iris | A circle of tissue containing pigmented cells. Contains circular and radial muscles | Helps to control the amount of light passing into the eye |
| Pupil | A circular space (hole) in the centre of the iris | Its size is altered by the contraction and relaxation of the muscles of the iris, so it controls the amount of light passing into the eye |
| Ciliary body | Contains ciliary muscles | These muscles help to control the shape (diameter) of the lens |
| Lens | Made up of stacks of long, narrow, transparent cells. It is about 4 mm thick and is biconvex | Focuses rays of light onto the retina |
| Suspensory ligaments | Ligaments which run between the lens and the ciliary body | Hold the lens in place. Varying the tension on these ligaments, changes the shape of the lens (accommodation) |
| Vitreous humour | A gelatinous fluid found behind the lens | Maintains the shape of the eye by exerting outward pressure on surrounding tissues |
| Aqueous humour | A watery fluid which is less viscous than vitreous humour and found in front of the lens | Maintains the shape of the eye by exerting outward pressure on surrounding tissues – holds the front of the eye in shape |
| Optic nerve | Bundle of nerve fibres | Carries action potentials to the brain |

**Table 1** The structure of the eye

Rod and cone cells have very similar structures. A human retina contains about 125 million rod cells.

As you can see in Figure 3, rod cells have an elongated structure with the outer segment specialised for photoreception (receiving light). This outer segment contains many discs formed of membrane-enclosed sacks. The membrane contains the pigment, visual purple.

Cones are shorter, broader and more tapered than rods. The membranes or **lamellae** in cones are formed from one continuous folded surface, rather than separate disks as is found in the rods. The part of the rod or cone cell nearest to the outside of the eye is known as the *outer segment*, and the part closest to the inside of the eye is the *inner segment*. It is the end of the inner segment that forms **synapses** with other cells in the retina. The inner segment contains the **nucleus** and numerous **mitochondria**. Proteins are formed here before passing through the connecting region to the outer segment. This connecting region contains **microtubules**. In rod cells there is only one type of visual purple pigment, called **rhodopsin**. There are three types of cone pigments, each most sensitive to a certain wavelength of light: short, medium and long. They are all different forms of the pigment **iodopsin**. Once a cone pigment is bleached by light, it takes about 6 minutes to regenerate the pigment.

The highest concentration of cones is in the **macula**. The **fovea centralis**, at the centre of the macula, contains only cones and no rods. There is an area where there are no light-sensitive rods or cones. This is where all the neurones collect together and the optic nerve carries information from the eye to the brain. If a ray of light comes through the lens of your eye and is focused on this area, the brain receives no information. This is known as the **blind spot**.

**Bipolar cells** have a central body from which two sets of processes arise. The processes nearest the rods and cones are short, and branch into many endings that form synapses with either a number of rods or a single cone. The other process is longer, and forms synapses with a ganglion cell. **Ganglion cells** have numerous dendrites that form synapses with the bipolar cells and this is where action potentials are first generated in the retina. **Amacrine cells** have horizontally spreading processes rather than **axons**. They connect the axons of bipolar cells and the **dendrites** of ganglion cells.

Rod cells are able to work in low light intensity because the cell can respond to a single photon of light. Rods do not detect colour, which is the main reason why it is difficult to tell the colour of an object at night or in the dark.

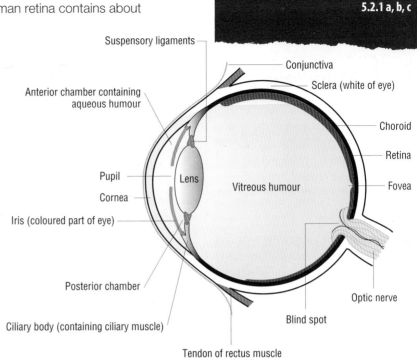

**Figure 2** The structure of the human eye

### Key definition

A **synapse** is a specialised junction where two neurones meet. See spread 5.2.2.4

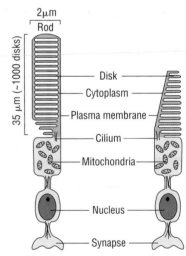

**Figure 3** The structure of the light receptor cells

## Questions

1. Name the surfaces within the eye at which light will be refracted.
2. State two differences between the structure of rod cells and cone cells.
3. If you look straight at an object when it is nearly dark you may find it difficult to see it but if you look to one side of the object it is clearer. Explain why this is.
4. Why do you think the retina is said to be inverted?

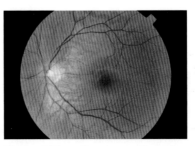

**Figure 4** The retina as seen through an opthalmoscope

## How a rod cell responds to a light stimulus

Rod and cone cells act as **transducers**, that is, they transfer energy in light to energy in **action potentials**. The action potentials are transmitted to the brain along nerve cells in the **optic nerve**.

A **resting potential** of $-40\,mV$ is maintained across a rod cell's surface membrane. This is achieved by the **sodium–potassium pump**, as it transports more sodium ions out than potassium ions in. Also, there is a constant flow of ions, which generates a circulating electric current. This is because open channels in the outer segment of the rod cell allow sodium ions to pass in through the plasma membrane, and open channels in the inner segment allow potassium ions to pass through. This means that sodium ions continually flow through these channels, down their concentration gradient into the outer segment, while potassium ions flow out of the inner segment. The rod cell forms a synapse with an adjacent **bipolar cell**. While the rod cell is in this resting state, the synaptic bulb secretes glutamate (a neurotransmitter), which diffuses across the **synaptic cleft** and prevents **generator potentials** being generated in the bipolar cell.

When light hits a **rhodopsin** molecule in a rod cell, the rhodopsin changes shape. The normal shape has a sharp kink at carbon 11 (this form is called 11-*cis*-retinal). Light causes this molecule to straighten (all-*trans*-retinal). This makes the whole rhodopsin molecule change shape, causing the sodium channels to close. This stops the circulating current, so the cell becomes even more negative inside (it is **hyperpolarised**). The result of this is that the rod cell stops secreting glutamate, so there is no inhibition of the bipolar cell and it becomes depolarised, creating a generator potential. This generator potential is transmitted across the synapse to a ganglion cell. From here an action potential is carried along the optic nerve to the brain.

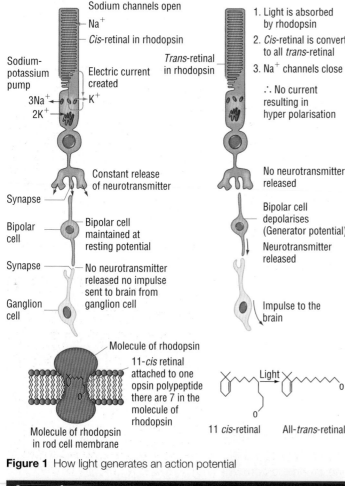

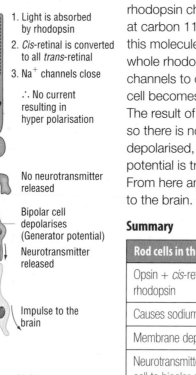

**Figure 1** How light generates an action potential

At rest
- Sodium channels open
- Na⁺
- Cis-retinal in rhodopsin
- Sodium-potassium pump
- Electric current created
- 3Na⁺  K⁺
- 2K⁺
- Constant release of neurotransmitter
- Synapse
- Bipolar cell
- Bipolar cell maintained at resting potential
- Synapse
- No neurotransmitter released no impulse sent to brain from ganglion cell
- Ganglion cell

When stimulated by light
1. Light is absorbed by rhodopsin
2. Cis-retinal is converted to all trans-retinal
3. Na⁺ channels close
∴ No current resulting in hyper polarisation
- Trans-retinal in rhodopsin
- No neurotransmitter released
- Bipolar cell depolarises (Generator potential)
- Neurotransmitter released
- Impulse to the brain

- Molecule of rhodopsin
- 11-cis retinal attached to one opsin polypeptide there are 7 in the molecule of rhodopsin
- Molecule of rhodopsin in rod cell membrane
- Light
- 11 cis-retinal   All-trans-retinal

## Summary

| Rod cells in the dark | Rod cells in the light |
|---|---|
| Opsin + *cis*-retinal ⟶ rhodopsin | Rhodopsin ⟶ opsin + *trans*-retinal |
| Causes sodium channels to open | Causes sodium channels to close |
| Membrane depolarised | Membrane hyperpolarised |
| Neurotransmitter released from rod cell to bipolar cell | No neurotransmitter released into synapse |
| Bipolar neurone hyperpolarised | Bipolar neurone depolarised |
| No neurotransmitter released into synapse between bipolar cell and ganglion | Neurotransmitter released into synapse between bipolar cell and ganglion |
| No generator potential | Generator potentials add together to form action potential |
| No action potential | Action potential travels along ganglion neurone |

### Case study

Pilots wear red goggles in strong light to be less influenced by changes in light intensity. The red light stimulates rhodopsin (in rods) so less rhodopsin is bleached – less time is needed for the eye to become dark-adapted.

The change in shape of the retinal molecule and the consequent change in shape of the **opsin** protein is called **bleaching**. The reverse reaction (*trans* to *cis* retinal) requires an enzyme reaction and is very slow. This explains why you are initially blind when you walk from sunlight to a dark room; in the light almost all your retinal was in the *trans* form, and it takes some time to form enough *cis* retinal to respond to the lower level of light indoors. The light intensity is also too low to stimulate the cones. It takes about 20 minutes for enough rhodopsin to reform for us to see properly.

## Using routine eye tests to assess receptor activity

Routine eye tests include tests for visual acuity, colour vision and pupil response.

**Visual acuity** means acuteness or clearness of vision. The word 'acuity' comes from the Latin *acuitas*, which means sharpness. The easiest way to measure visual acuity is by using a vision chart with different sized letters (a Snellen chart). The Snellen eye chart has a series of letters, or letters and numbers, with the largest at the top. You are asked to read down the chart. The smaller the letter you can read, the better your visual acuity. This measurement is probably the most important aspect of a complete eye examination. If you can resolve letters approximately one inch high at 20 feet you said to have 20/20 visual acuity. This is considered 'normal' acuity. If you have 20/40 acuity, you require an object to be at 20 feet to visualise it with the same resolution as an individual with 20/20 acuity would when the object was at 40 feet.

The results of this test will tell the optometrist what kind of lenses you need to help you see more clearly.

### Pupil response test

To do this test you sit in a dimly illuminated room and a bright light is shone in your eyes. This is usually done by swinging a torch so light shines first into one eye and then into the other. Light shone into one eye should make both pupils constrict equally. If there is a difference, the person is said to have a relative afferent papillary effect (RAPD). This could indicate damage to the optic nerve or the brain. It can also indicate that the person's nervous system is affected by alcohol or other drugs. This test can also be done on babies and even on an unconscious patient, as you do not have to say what you can see.

### How the blink response can be used to indicate levels of consciousness

The pupil response test relies on a reflex action and has the advantage that it does not require the patient to communicate. The blink reflex is the rapid closing of the eyelid when something threatening approaches the eye. The blink reflex is one of the last to be lost as unconsciousness deepens. If it is not present, this indicates that the person is in a coma. An eye opening indicates that the arousal mechanism in the brain is active. Eye opening may be spontaneous or triggered by speech (spoken name) or by a painful stimulus, or may not happen at all.

## Questions

1 The potential difference across the membrane of a rod cell is about .................... . This is maintained by the .................... pump which constantly transports .................... ions out and .................... ions in. A circulating electrical current is also set up as the membrane in the .................... segment is permeable to sodium ions and the membrane in the .................... segment is permeable to potassium ions.

2 When light falls on a rod cell it causes the .................... molecule to change shape from 11 *cis*-retinal to .................... . This means it no longer fits the binding site in .................... . This causes the sodium and potassium channels in the membrane to close so the .................... difference increases. The rod cell stops secreting glutamate (a neurotransmitter) so the .................... cell becomes depolarised which generates an .................... potential. This is transmitted to a .................... cell and then along the .................... nerve to the brain.

3 How does a routine eye test tell the optometrists something about your health?

4 What does it mean if someone says they are colour blind?

### Colour vision

About 8% of males and 0.5% of females have some version of 'colour blindness' from birth. Usually this is a genetically inherited trait, and is of the 'red–green confusion' variety. Reds, browns, olives and gold may be confused. Purple may be confused with blue, and pastel pinks, oranges, yellows and greens look similar. Total colour blindness (seeing in only shades of grey) is extremely rare. There is no treatment for colour blindness, nor is it usually the cause of any significant disability. However, it can be very frustrating for individuals affected by it. Being colour blind does keep one from performing certain jobs and makes others difficult. It is tested by looking at a collection of cards made up of different coloured spots which are arranged so that spots of a particular colour form images of numbers. Colour blind people cannot distinguish the numbers.

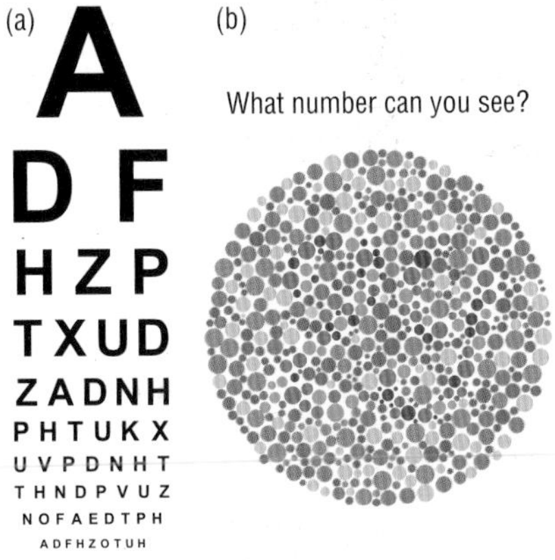
(a)
A
D F
H Z P
T X U D
Z A D N H
P H T U K X
U V P D N H T
T H N D P V U Z
N O F A E D T P H
A D F H Z O T U H
(b)
What number can you see?

**Figure 2a** A Snellen chart and **b** a test for colour blindness

### Case study

Thursday April 15, 2004 – *Guardian*

When a 39-year-old Belgian woman suffered a stroke and fell into a coma, doctors concluded that she was unlikely to regain consciousness and, after a time, diagnosed her condition as persistent vegetative state (PVS). One of the criteria on which they based their decision was her inability to blink or track a moving object with her eyes. It was only when they discovered that the stroke had damaged a cranial nerve, preventing her from opening her eyes, that they realised their error. If they opened her eyes for her, she followed their instructions. Having regained full consciousness soon after her stroke, she revealed she had overheard all the bedside discussions as to whether it was worth keeping her alive. At no point had she wanted to die.

Limits imposed by the size of the birth canal prevent the human brain from getting even bigger. At birth, an infant's skull is as large as it can be without endangering the life of the mother and child during childbirth. The difficulty experienced by humans in giving birth is unusual in the animal kingdom, as the head of the emerging infant has to be rotated as it passes through the birth canal. At birth, the human skull is rather soft, and it is moulded during its passage through the birth canal. It recovers its shape after the baby is born.

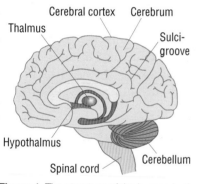

**Figure 1** The structure of the human brain

MRI scan

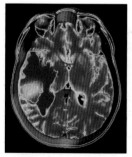

A coloured magnetic resonance imaging (MRI) scan of the axial section of the human brain showing a metastic tumour (yellow)

CT scan

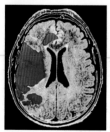

Cat scans are a specialized type of x-ray. The x-ray tube rotates around the patient and a computer collects the results. These results are translated into images that look like a "slice" of the person. Unlike CT the MRI scan uses magnets and radio waves to create the images. No x-rays are used in an MRI scanner.

**Figure 2** MRI and CT scans of the human brain

## The gross structure of the human brain

The **brain** is the most complex organ in the human body. It is the primary centre for the regulation and control of bodily activities, receiving and interpreting sensory impulses, and transmitting information to the muscles and body organs. It is also the seat of consciousness, thought, memory and emotion. The normal adult human brain typically weighs between 1–1.5 kg. The mature human brain consumes some 20–25% of the energy used by the body, while the developing brain of an infant consumes around 60%. Such heavy energy usage generates large quantities of heat, which must be continually removed to prevent brain damage.

The brain is protected by a bony covering called the **cranium**, which, along with the bones of the face, make up the skull.

Inside the cranium, the brain is surrounded by the **meninges** (connective tissue), a system of membranes that separate the skull from the brain. The meninges are made up of three layers of tissue:

- *pia mater* – the layer closest to the surface of the brain
- *arachnoid membrane* – the middle layer of tissue
- *dura mater* – the outermost layer.

Blood vessels enter the central nervous system above the pia mater. The cells in the blood vessel walls are joined tightly, forming the **blood–brain barrier**.

Histology of the brain: the brain is made up of two broad classes of cells, **neurones** and **glial cells**. Glial cells ('glia' is Greek for 'glue') form a support system for neurones. They create the insulating myelin. Glia are estimated to outnumber neurones by about 10 to 1. Astrocytes are characteristic star-shaped glial cells.

The brain can be divided into three parts: the **forebrain**, **midbrain** and **hindbrain**. The structure of the brain can be seen in Figure 1. Here it is clearer what it is – an expanded extension of the spinal cord. It contains **ventricles** filled with **cerebro-spinal fluid** (cerebral fluid), which is continuous with the fluid in the spinal cord. The shape of the ventricles is quite distinctive and can be seen clearly on a **MRI scan**. The meninges help to secrete **cerebral fluid** that also helps to protect and cushion the brain – it allows the brain to float.

### The cerebrum – the front of the brain

This is the largest part of the brain; in humans it covers most of the rest of the brain. It is made up of two cerebral hemispheres connected to each other by a bridge of tissue called the **corpus callosum**. The surface of each cerebral hemisphere is covered with a highly folded layer of tissue called the **cerebral cortex.** The cerebral cortex is made up of deep grooves (called *sulci*) and bumps or folds (called *gyri*). The folds/wrinkles make the brain more efficient, because they can increase the surface area of the brain and the number of neurones within it. Each hemisphere can be divided into four lobes, which are named according to the skull bone that they are most closely related to (i.e. frontal, parietal, temporal and occipital).

The outer part of the cerebrum is called grey matter and contains nerve cells. The inner part is called white matter and contains connections of nerves. Just below the cerebrum is the diencephalons, which contains the **thalamus** and the **hypothalamus.**

### The midbrain

The midbrain is located in front of the **cerebellum**. The cerebrum, the cerebellum, and the spinal cord are all connected to the midbrain. It lies behind and beneath the thalamus.

### The hindbrain or cerebellum

Behind the cerebrum at the back of the head is the cerebellum. In Latin, *cerebellum* means 'little brain'. But the cerebellum contains more nerve cells than both hemispheres combined.

Beneath the cerebellum lies the medulla oblongata. It forms the link between the brain and the spinal cord.

## Functions of parts of the brain

It is possible to allocate functions to different regions of the brain.

| Part of the brain | Function |
|---|---|
| Cerebrum | The cerebrum is concerned with higher brain functions, interpreting sensory impulses and initiating muscle movements. It stores information and uses it to process reasoning. It also functions in determining intelligence and personality |
| | The frontal lobe (front section) contains the centres for speech, memory, intelligence and emotion. It also initiates voluntary motor activity. There is a special region within the frontal lobe on the left side, called Broca's area, which helps control the patterns of muscle contraction necessary for speech |
| | The parietal lobe (behind the frontal lobe) is concerned with general sensation and the integration of senses |
| | The occipital lobe (at the centre back of the cerebrum) is concerned with vision |
| | The temporal lobe (at either side of the cerebrum) is involved in hearing. Wernicke's area, posterior end of temporal lobe, usually left hemisphere, is involved only in language comprehension. The amygdala, anterior part of temporal lobe, is an almond-shaped structure that plays an important role in motivation and emotional behaviour. The hippocampus is found deep in the temporal lobe. It is crescent-shaped and strongly involved in learning and memory formation |
| | The left half of the cerebrum controls the right side of the body; the right half controls the left side. This means that when one side of the brain is damaged, the opposite side of the body is affected. For example, a stroke in the right hemisphere of the brain can leave the left arm and leg paralysed |
| Cerebellum | It is involved in coordinating your muscles to allow precise movements and control of balance and posture |
| Medulla oblongata | It controls autonomic functions and relays nerve signals between the brain and spinal cord. It controls: respiration; blood pressure; heart rate; reflex arcs; vomiting; movements of the wall of the alimentary canal |

**Table 1** Brain functions

### Finding out which parts of the brain do what

Transcranial magnetic stimulation (TMS) is more than just a new way to treat mental disorders. It also allows scientists to probe tiny areas of the human brain to find out what they do.

Harvard neurologist Alvaro Pascual-Leone says images from brain scans can't do this. And it's essential for scientists trying to understand how the brain works. 'The only way we're able to really disentangle the functions of this very complicated organ is by tapping here or there,' Pascual-Leone says.

TMS gives researchers two different ways to tap on parts of the brain. One is by using an electromagnetic frequency that temporarily disables brain cells. Eric Wassermann runs a brain stimulation unit at the National Institutes of Health. He says researchers there suspected they had found an area of the brain that was essential to speech. But they didn't know for sure until they used stimulation on volunteers to disable that part of the brain. 'Suddenly speech would become garbled and stop,' he says. 'And when the stimulation was turned off, the speech would resume suddenly. It's very dramatic. I've had it done to me; it's a strange sensation.'

One interesting experiment with brain stimulation involved blind people reading Braille. Pascual-Leone says their brain scans show activity in places that usually process information from the eyes. 'The question was, is this activity in the visual areas really contributing to the ability of these subjects to read?' he says. Brain scans couldn't answer that question. So Pascual-Leone and other scientists tried brain stimulation. 'If you use magnetic stimulation to block the activity in the visual areas of the brain in these congenitally blind subjects,' he says, 'they make more errors when reading Braille by touch.' That meant the visual areas were critical.

To find out why, the team did another experiment. It involved waiting until just after a blind volunteer's finger touched a Braille symbol. Then, Pascual-Leone's team blocked activity in the brain's visual area. 'The subject knows that they have touched the finger, but can't come up with the Braille symbol that was presented,' he says, 'suggesting that the activity in the visual cortex was contributing to the decoding of the Braille symbol.'

## Questions

1 Which part of the brain controls the contraction and relaxation of skeletal muscles and so is responsible for movement and balance?

2 What is the function of the cerebrospinal fluid?

3 Which part of the brain controls breathing movements and the heart rate?

4 The ..................... is responsible for conscious thought, language and speech; it also integrates information from ..................... . The ..................... system is closely involved with emotions and the formation of ..................... .

## The autonomic nervous system

In the first spread of this module we saw that the nervous system is made up of the central and the peripheral nervous systems. The **peripheral nervous system** is subdivided into sensory and motor neurones. The motor neurones can be divided into two systems:

- somatic nervous system
- autonomic nervous system.

The autonomic nervous system carries nerve impulses from the central nervous system to the internal organs. It has two parts:

- **sympathetic**: generally acts to arouse the body, preparing it for 'fight or flight'
- **parasympathetic**: acts to relax and restore the body to normal levels of arousal.

The actions of the autonomic nervous system are largely **involuntary**. The word 'autonomic' means 'self-adjusting'.

## Neurones

These are specialised cells that are adapted to carry nerve impulses rapidly from one part of the body to another. Neurones consist of the following:

- A cell body which contains the nucleus and large amounts of rough **endoplasmic reticulum** and its associated **ribosomes** forming **Nissl substance**. There is also a well developed Golgi apparatus and large numbers of **mitochondria**. The presence of the Nissl substance can be explained by the fact that nerve cells are metabolically very active, and are involved in large amounts of protein synthesis.
- The **axon**, which is a long process that carries the impulse away from the cell body.
- The **dendron**, which is an extension of the cell body that branches into **dendrites**. There may be as many as 200 thread-like dendrites that allow many connections to be made with neighbouring neurones. They carry the impulse towards the cell body.

Many axons and dendrons are surrounded by **Schwann cells**, which wrap themselves many times around the axon, building up layers of their membranes. These membranes are rich in *myelin* (a lipid). They form a **myelin sheath** that covers the axon or dendron. The function of the myelin sheath is to act as an insulator and so to speed up the

> **Examiner tip**
>
> Remember that motor neurones occur in the autonomic and somatic nervous systems. The effectors for the somatic system will be skeletal muscles. What type of muscle will be the effectors in the autonomic system?

> **Examiner tip**
>
> Look back at your notes on the nucleus, rough endoplasmic reticulum, ribosomes, Golgi apparatus and mitochondria. As you learn how neurones function, you should be able to link the functions of these organelles with the specific needs of a neurone.

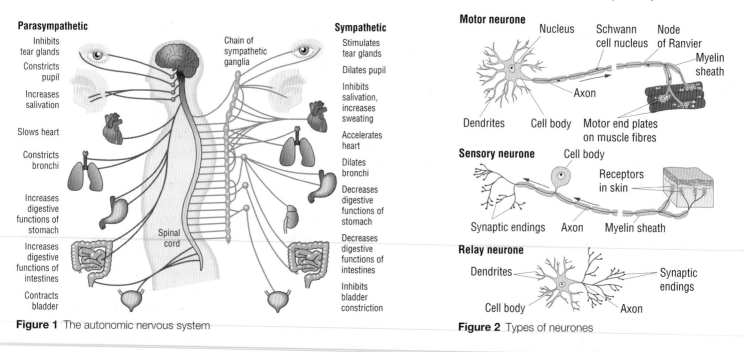

**Figure 1** The autonomic nervous system

**Figure 2** Types of neurones

transmission of a nerve impulse. The space between adjacent Schwann cells lacks myelin. These spaces are known as the **nodes of Ranvier**. In humans they occur about every 1–3 mm along the axon or dendron and are 2–3 μm long. These are the only regions where ions can enter or leave the axon, so they are very important in the transmission of the nerve impulse, as you will see later.

There are three types of neurones:
- motor neurones
- sensory neurones
- relay (intermediate) neurones.

**Key definition**

A **ganglion** is a nodule on the dorsal root of the spinal cord that contains cell bodies of sensory neurones.

### Motor neurones (efferent neurones)

These transmit impulses from the central nervous system to an **effector** (muscle or gland). The cell body of a motor neurone lies within the brain or the spinal cord. The neurone has a long axon and many short dendrites. The longest axon of a human motor neurone can be over 1 m long, reaching from the base of the spine to the toes.

### Sensory neurones (afferent neurones)

These neurones transmit impulses from a receptor to the brain or spinal cord. They have one dendron that brings the impulse towards the cell body and an axon that carries it away from the cell body. Sensory neurones have axons that run from the toes to the spinal cord, over 1.5 m in adults.

### Relay neurones

These neurones transmit impulses between neurones, for example, from a sensory neurone to a motor neurone in the spinal cord. They have numerous short processes.

### The reflex arc

A *simple reflex* is rapid and entirely automatic – it involves no learning, but is present from birth (*innate*). Examples of such reflexes include the sudden withdrawal of a hand in response to a painful stimulus, blinking as an object approaches the eye, or the jerking of a leg when the kneecap tendon is tapped. The simple reflex is a means of protecting the body from dangerous stimuli.

The reflex arc is the simple pathway by which nerve impulses are passed from a receptor to an effector via the central nervous system. It enables a rapid response to a stimulus, without the need to involve the decision-making processes of the brain. The brain is aware of the response only after it has taken place.

The response is rapid because the neurone pathway is short. Sensory cells (**receptors**) send impulses to the spinal cord along a sensory neurone. The impulse passes through the dorsal root ganglion into the spinal cord. From here it may pass directly to a motor neurone, or to a relay neurone and then a motor neurone. The impulse travels along the motor neurone to an effector such as a muscle or a gland.

Within the spinal cord, the impulse will also be passed on to other neurones, which will take the impulse up the spinal cord to the brain so that the brain is 'aware' of the reflex arc.

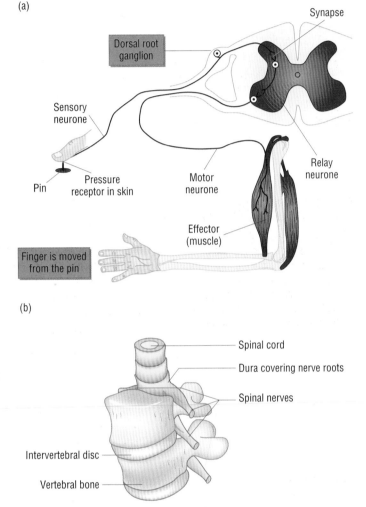

(a)

(b)

**Figure 3  a** A reflex arc, **b** a drawing of a vertebral column to show the relative position of the spinal cord and spinal nerves

## Questions

1. Where would you find the cell body of:
   (a) a sensory neurone?
   (b) a motor neurone?
2. What is an axon?
3. List the differences in structure and function between a sensory and a motor neurone.
4. What are Schwann cells and what is their function?
5. How does a reflex arc help to protect the body from damage?

### Examiner tip

Both Na⁺ and K⁺ are moved against their concentration gradients, so use energy from the **hydrolysis** of ATP. Remember also that membranes are partially permeable. Ions need protein channels to cross the phospholipid bilayer. And how will ATP be made?

## How action potentials are transmitted

Impulses travel along axons and dendrons and may pass on to the next neurone. The transmission or propagation of a nerve impulse along a neurone from one end to the other occurs as a result of chemical changes across the axon's plasma membrane. It is changes in the membrane that are responsible for the different events that occur in the neurone. In nerve and muscle cells the membranes are electrically excitable, this means that their membrane potentials can change, and this is the basis of a nerve impulse. The sodium and potassium channels in these plasma membranes are **voltage-gated**, which means that they can open and close depending on the voltage across the membrane.

When a neurone is not sending a signal, it is at 'rest' – the **resting potential**. The membrane of a resting neurone is **polarised** – there is a difference in electrical charge (**potential difference**) between the outside and inside of the membrane. The inside is negative with respect to the outside, at about –65 millivolts (mV). This is caused by an ionic imbalance as there are more sodium ions outside the axon, and more potassium ions inside. This ionic imbalance is maintained by the **sodium–potassium pump**, an example of active transport. The sodium–potassium pump removes three sodium ions from the cell for every two potassium ions it brings into the cell. Also, K⁺ diffuses back out again much faster than Na⁺ diffuses back in, giving an overall excess of positive ions outside the membrane compared with inside.

When a stimulus above a certain intensity or **threshold** arrives at a receptor or nerve ending, its energy causes a temporary and rapid reversal of the charges, or polarity, on the neurone membrane. The stimulus causes a Na⁺ channel to open. Since there is a higher concentration of Na⁺ outside than inside the axon, Na⁺ diffuses rapidly into the neurone. Sodium has a positive charge, so the inside of the neurone becomes more positive. The electrical potential changes to about +40 mV. This has the effect of closing the sodium channels and opening the potassium channels, allowing potassium ions to leave the cell. This causes the action potential to go back toward –70 mV (a **repolarisation**). The action potential actually goes past –70 mV (a **hyperpolarisation**) because the potassium channels are slower closing.

So first there is an influx of sodium ions (leading to massive **depolarisation**), then a rapid efflux of potassium ions from the neurone (leading to repolarisation). The resting membrane potential is restored by the Na⁺/K⁺ pump as excess ions are subsequently pumped in/out of the neurone. This transient switch in membrane potential is the **action potential**. The cycle of depolarisation and repolarisation is extremely rapid, taking only about 0.002 seconds, so it allows neurones to fire action potentials in rapid bursts.

### How the action potential passes along the axon

Once an action potential has started it is moved (*propagated*) along an axon automatically. The local reversal of the membrane potential signals the surrounding voltage-gated ion channels to open when the potential changes enough. A *local*

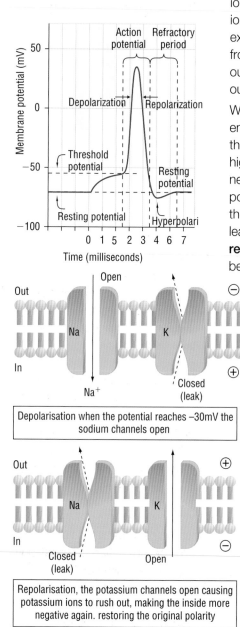

Depolarisation when the potential reaches –30mV the sodium channels open

Repolarisation, the potassium channels open causing potassium ions to rush out, making the inside more negative again. restoring the original polarity

**Figure 1** The action potential

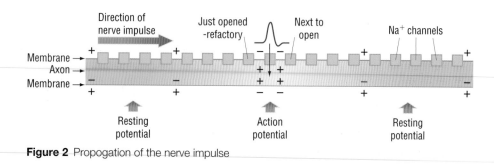

**Figure 2** Propagation of the nerve impulse

circuit is set up between the depolarised region and the resting regions on either side of it. Sodium ions flow sideways inside the axon, away from the positively charged region towards the negatively charged regions on either side. The axon membrane ahead will depolarise, generating an action potential.

### The refractory period

As the action potential passes a particular point on the axon, the axon membrane behind is still adjusting the position of the ions – it is in the refractory period – so no impulse occurs in the other direction. During this time (about 0.001 seconds), the $Na^+$ and $K^+$ channels cannot be opened by a stimulus. This means that *action potentials only move in one direction* along the axon.

### 'All or nothing' law

The action potential occurs only if the stimulus causes enough sodium ions to enter the cell to change the membrane potential to a certain **threshold** level. If the depolarisation is not great enough to reach the threshold, then no action potential (and hence an impulse) will be produced. This is called the **all or nothing law**. The ion channels are either open or closed: there is no half-way position. This means that the action potential always reaches $+40\,mV$ as it moves along an axon, and it is never reduced by long axons. Action potentials are always the same size, but the frequency of the impulse is determined by the intensity of the stimulus, i.e. strong stimulus = high frequency. A strong stimulus produces a rapid succession of action potentials; a weak stimulus results in fewer action potentials per second. Also, a strong stimulus is likely to stimulate more neurones than a weak stimulus.

The brain interprets the frequency of action potentials arriving along the axon of a sensory neurone, and the number of neurones carrying action potentials.

## The importance of saltatory conduction

### Saltatory conduction

In myelinated neurones the voltage-gated ion channels are found only at the nodes of Ranvier. Between the nodes, the **myelin sheath** acts as a good electrical insulator, preventing ions moving in or out. The action potential can therefore jump large distances (1 mm) from node to node. This is called **saltatory conduction**. The speed of propagation is increased dramatically due to this. Nerve impulses in unmyelinated neurones have a maximum speed of around $1\,ms^{-1}$ while in myelinated neurones they travel at $100\,ms^{-1}$.

The speed of an action potential is affected by:
- The myelin sheath, which acts as an electrical insulator, causing the impulse to jump from one node of Ranvier to the next (saltatory conduction).
- Axon diameter – the larger the axon, the faster it conducts. This is due to less leakage of ions from a large axon.
- Temperature, which affects the rate of diffusion of ions, so the higher the temperature, the faster the impulse.
- The number of synpses involved – the greater the number of synapses in a neurone pathway, the slower the conduction velocity.

## Questions

1 The resting potential is maintained by the ..................... pump in the plasma membrane. ...... sodiums are pumped out for every two ..................... pumped in.

2 When a neurone is stimulated an ........................ is generated. ..................... channels in the plasma membrane open which temporarily ................. the charge across the membrane. The membrane is ..................... .

3 Why is it important for neurones to have a threshold?

4 Why can an action potential only travel in one direction?

5 How does the presence of a myelin sheath speed up the transmission of an action potential?

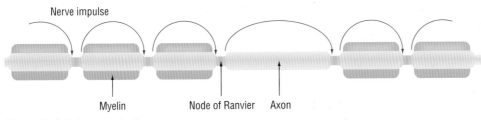

**Figure 3** Saltatory conduction

Nerve impulse · Myelin · Node of Ranvier · Axon

## The structure and function of a cholinergic synapse

A **synapse** is a junction between one neurone and another neurone or *effector*. The space between two neurone plasma membranes is known as the **synaptic cleft**. This space means that action potentials cannot travel across a synapse. To carry on the transmission across the synaptic cleft, a **neurotransmitter** is needed. Synapses in which the neurotransmitter is **acetylcholine (ACh)** are called **cholinergic synapses**.

The structure of synapses are shown in Figure 1. Synapses consist of:

* the presynaptic ending – the presynaptic neurone has a swollen portion of the axon at the end, known as the **synaptic bulb** (where neurotransmitters are made)
* the post synaptic ending (has **receptors** in the membrane for binding to the neurotransmitter)
* the synaptic cleft or space between the presynaptic and postsynaptic endings. It is about 20 nm wide.

The synaptic bulb contains numerous mitochondria, large amounts of endoplasmic reticulum and synaptic vesicles containing the neurotransmitter. When an action potential arrives at the membrane of the synaptic bulb, the voltage-gated **calcium channels** open, causing calcium ions ($Ca^{2+}$) to flow into the bulb. These calcium ions cause the synaptic vesicles containing ACh to fuse with the presynaptic membrane and release their contents by **exocytosis**. ACh is synthesised from choline and acetyl coenzyme A.

ACh diffuses across the synaptic cleft and binds to receptors on the postsynaptic membrane. The membrane protein receptors have a complementary shape to ACh. This changes the shape of the protein receptor, opening channels in the membrane through which sodium ions diffuse from the synaptic cleft into the postsynaptic neurone. This results in a depolarisation of the postsynaptic cell membrane, which may initiate an action potential if the **threshold** is reached.

The neurotransmitter is broken down by a specific enzyme in the synaptic cleft. The enzyme **acetylcholinesterase** hydrolyses the neurotransmitter acetylcholine. The breakdown products are taken into the presynaptic neurone by endocytosis and used to resynthesise more ACh, using energy and acetyl coenzyme A from the mitochondria. This prevents channels being permanently open on the postsynaptic membrane.

## The role of synapses

Synapses prevent impulses going in the wrong direction. An impulse can only cross the synapse in one direction. This is because synaptic vesicles are confined to the presynaptic side of the cleft, and protein receptors to the postsynaptic side.

Synapses provide the means through which the nervous system connects to and controls the other systems of the body such as muscles.

Synapses provide a means of interconnecting nerve pathways. They allow nerve impulses to diverge. This means that a single impulse along one neurone can be conveyed to a number of different neurones at a synapse, allowing a single stimulus to create a number of simultaneous integrated responses. The establishment of new synaptic links is the basis of learning.

Synapses allow the 'filtering out' of continual unnecessary or unimportant background stimuli. If a neurone is constantly stimulated (e.g. clothes touching the skin) the synapse will not be able to renew its supply of transmitter fast enough to continue passing the impulse across the cleft. This 'fatigue' places an upper limit on the frequency of depolarisation.

# Traumatic brain injury

A traumatic brain injury (TBI) is an injury to the brain caused by the head being hit by something or shaken violently. The damage may be caused by the head forcefully hitting an object such as the dashboard of a car (closed head injury) or by something passing through the skull and piercing the brain, as in a gunshot wound (penetrating head injury). The major causes of head trauma are motor vehicle accidents. Other causes include falls, sports injuries, violent crime and child abuse. The injury can change how a person acts, moves, thinks and learns. A traumatic brain injury can also change your behaviour and personality. The term TBI is used for head injuries that can cause changes in one or more areas, such as thinking and reasoning; understanding words; remembering things; paying attention; solving problems; talking; walking and other physical activities; seeing and/or hearing; and learning. The physical, behavioural or mental changes that may result from head trauma depend on the areas of the brain that are injured.

The term TBI is not used for a person who is born with a brain injury or whose brain injuries happened during birth.

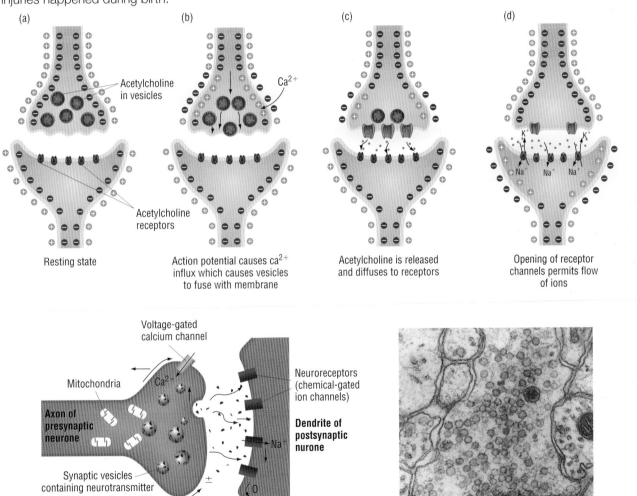

**Figure 1** Cholinergic synapse and electron micrograph of synapse

# Questions

1 Neurones do not directly connect with each other, but are separated by a tiny gap called a ................. .
2 List the sequence of events as a nerve impulse is transmitted across a cholinergic synapse.
3 Why do you think nerve impulses can only travel in one direction?
4 Suggest why, if action potentials arrive repeatedly at a synapse, the synapse eventually becomes unable to transmit the impulse to the next neurone.

## Imaging the brain

A **CT (computerised tomography) scan** is taken by an X-ray machine linked to a computer which takes a series of detailed pictures which build up into a three-dimensional picture of the inside of your body. Most people who have a CT scan are given a drink or injection to allow particular areas to be seen more clearly. The pictures can show tumours in the brain, spinal stenosis (narrowing of the spinal canal), a blood clot or intracranial bleeding (bleeding within the skull) in patients with stroke or brain damage from head injury.

The **magnetic resonance imaging (MRI) scan** is the diagnostic tool that currently offers the most sensitive, non-invasive way of imaging the brain, spinal cord, or other areas of the body.

(a)

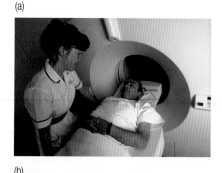

(b)

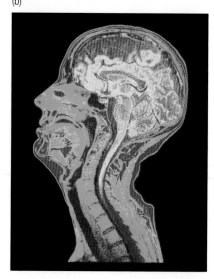

The MRI scanner is a tube surrounded by a giant circular magnet. Unlike a CAT scan or conventional X-ray, this type of scanning device does not use radiation. Instead, it uses a powerful magnetic field that makes the hydrogen protons in water molecules line up. Once lined up, they are then knocked out of line by radiowaves. When the radiowaves are stopped, the protons relax back into line, releasing resonance signals that are transmitted to a computer. Computer programs translate these data into cross-sectional pictures of the water in human tissue. Sometimes a special dye is injected to help show differences in the tissues of the brain. The resulting high-resolution image shows quite a lot of detail. Because there is no radiation with the magnet, the scans may be carried out on pregnant women. However, most scanners fit rather tightly around the patient, so some patients may feel uncomfortable. They may not tolerate lying in a tight tunnel for 45–60 minutes while the scan is being performed. The newer generation scanners are designed with more open space, although a more open tube does sacrifice the excellent detail provided by the tight tubes.

### What the MRI is used to measure

An MRI scan can be an extremely accurate method of detecting disease throughout the body. In the head, trauma to the brain can be seen as bleeding or swelling. Other abnormalities that often show up include brain aneurysms, strokes, tumours of the brain, and tumours or inflammation of the spinal cord.

Because MRI is particularly useful in detecting central nervous system demyelination (removal of myelin), it is a powerful tool in diagnosing MS. But with advancing age (probably over the age of 50), even in healthy people small areas may often be seen on MRI scans that resemble MS but are actually related to the ageing process, and are of no clinical significance.

(c)

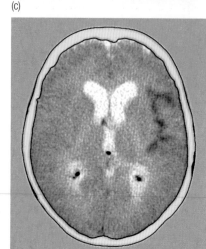

As MRI uses magnetic fields to detect subtle changes in brain tissue content, an increase of water content in the cells of brain tissue can be detected. This is one effect of a stroke. The benefit of MRI over CT imaging is that MRI is better able to detect small **infarctions** soon after stroke onset.

MRI scans are used to enable doctors to plan the treatment for a tumour in the brain or spinal cord. The doctors need to find out as much as possible about the type, position and size of the tumour. An MRI scan is used by neurosurgeons to evaluate the integrity of the spinal cord after a trauma to particular areas of the brain.

**Functional magnetic resonance imaging (fMRI)** is a relatively new procedure. fMRI of the brain uses the magnetic properties of iron in haemoglobin to produce real-time images of bloodflow to particular areas of the brain. MR imaging is used to measure the quick, tiny metabolic changes that take place in an active part of the brain. Physicians know the general areas of the brain where speech, sensation, memory and other

**Figure 1a** MRI scanner, **b** MRI scan and **c** a CT scan functions occur. However, the exact locations vary from individual to individual. Injuries and disease, such as stroke or brain tumour, can even cause functions to shift to other parts of the brain. fMRI not only helps radiologists look closely at the anatomy of the brain, but can also help them determine precisely which part of the brain is handling critical functions such as thought, speech, movement and sensation. This information can be crucial to planning surgery, radiation therapy, treatment for stroke, or other interventions to treat brain disorders.

A **nerve conduction velocity (NCV) test** is a diagnostic study during which both sensory and motor nerves are repeatedly stimulated in order to measure the speed at which nerve impulses are conducted. Electrodes are inserted into the muscle, or placed on the skin overlying a muscle or muscle group, and electrical activity and muscle response are recorded.

Unusually slow conduction velocities suggest damage to nerve fibres, for example, loss of the protective covering surrounding certain nerve fibres (demyelination), or other disease process. Patients referred for NCV tests suffer from nerve conditions that produce numbness, tingling, pain or loss of sensation, or from neurological diseases affecting primarily the peripheral nervous system.

## Why damaged neurones are unable to regenerate

While the peripheral and central nervous systems work together, they differ in one very important way: the peripheral nerves can regenerate, or grow back, if they are damaged, but damage to neurones in the brain or spinal cord is permanent. For example, if someone loses a finger or limb and it is reattached promptly, the severed nerves will usually grow back over time. Regeneration in the peripheral nervous system is very slow, about 3–5 mm a day.

In the central nervous system, damaged neurones are unable to regenerate at all. The paralysis that often accompanies spinal cord injuries is permanent. Brain damage is permanent too, although other parts of the brain can take over the function of the damaged areas. Doctors have observed this phenomenon in people who are recovering from a stroke, which is usually the result of a blockage in one of the blood vessels that feeds the brain.

### What prevents CNS neurones regenerating

Axon regeneration in the CNS fails for the following reasons:

- In the adult central nervous system, a standard response to injury is the formation of a glial scar, which quickly seals off the injured site from healthy tissue. This is beneficial because it prevents a cascading wave of uncontrolled tissue damage. But glial scarring also blocks the long-term regeneration of damaged axons. But *astrocytes*, a type of **glial cell**, enlarge and proliferate to form a scar and also produce myelin and inhibitory molecules that inhibit the regrowth of a damaged or severed axon. Glial scars represent a physical and molecular barrier to axon regeneration. To date no effective method has been described to remove an existing glial scar in a chronic injury.
- Most CNS axons make only a feeble regeneration response after they are cut.
- **Fibrinogen**, a blood-clotting protein found in circulating blood, has been found to inhibit the growth of CNS neuronal cells. Fibrinogen – contained in the blood that leaks at the site of injury – begins the process of inhibiting axon growth by binding to a **receptor**. This binding in turn induces the activation of another receptor on the neurone cells which, when activated, inhibits growth of the axon.

## Questions

1  Do CT or MRI scans use radiation to assess damage to the brain and spinal cord?
2  Which type of scan is better at showing soft tissues?
3  Which type of scan can be used to show brain activity?
4  A nerve conduction velocity test is used to assess injury to a
................. ..................
are placed onto the skin a certain distance apart on the nerve pathway. A mild
.................. shock is then applied through one of them. The time taken for the arrival of many action potentials to be picked up at the other electrode is measured. The speed of conduction is measured by dividing the ..................
between the electrodes by the ..................taken. If a nerve is damaged it conducts impulses
.................. than an undamaged nerve.

## Stem cells and neurone

**Stem cells** have the potential to regenerate body tissues. In no area of medicine is the potential of stem cell research greater than in treating diseases of the nervous system. The most obvious reason is that so many diseases result from the loss of nerve cells, but mature nerve cells cannot divide to replace those that are lost. In Parkinson's disease, nerve cells that make the chemical **dopamine** die. In Alzheimer's disease, cells that make acetylcholine in the brain die. In stroke, brain trauma and spinal cord injury, many types of cells are lost.

### Success of stem cell treatments

Stem cell treatments have been effective in curing spinal cord injuries in rats, although the treatment has yet to be successful in humans. Initial stem cell treatments will probably benefit only the recently injured.

But new experiments indicate that adding stem cells to spinal implants made of *hydrogels* can be used to treat older injuries as well. The hydrogels, a jelly-like lattice of amino acids, help fill the cavities that form over time in injured areas and create an environment in which neurones can grow. The gel supports more delicate cells and allows for the transmission of chemical signals that govern neural development. Scientists tested the process on 28 rats by removing spinal tissues and replacing them with hydrogel filled with stem cells from rat bone marrow. 4 weeks later, the rats displayed neural regrowth and recovered much of the limb functioning they lost. It is too soon to predict whether or not the treatment will work in humans, who have a much thicker spinal cord than rats.

Surgeons in London plan to use stem cells to treat patients with damage to their nervous systems. Stem cells from the noses of their patients will be used, which will mean there is no chance of rejection. If all goes well, the cells will patch a broken connection between nerves in the patient's arms and their spinal cord.

A stroke leaves a permanent gap in the brain that can destroy a person's ability to speak and move normally. Scientists believe it should be possible to use stem cells – the body's master cells with the potential to turn into any type of tissue – to fill the gap.

**Case study**

Neural stem cell treatment eases Parkinson's symptoms in primates

14 June 2007

Researchers from Yale University, Harvard University, University of Colorado and the Burnham Institute found human neural stem cells effective in treating primates suffering from severe Parkinson's disease. In the study, 5 of 8 monkeys with advanced Parkinson's disease were injected with human neural stem cells. The five monkeys had diminished tremors, were able to move and walk better, and showed better eating habits. 'Not only are stem cells a potential source of replacement cells, they also seem to have a whole variety of effects that normalize other abnormalities,' said D. Eugene Redmond, professor at Yale. 'The human neural stem cells implanted into the primates survived, migrated, and had a functional impact.' The results of the study are promising but it may take years to know whether similar treatment would have therapeutic value for humans with Parkinson's symptoms.

**Case study**

20 September 2005

Scientists have used injections of human stem cells to heal spinal injuries in paralysed mice, allowing them to walk normally again.

The research, which was funded by the Christopher Reeve Paralysis Foundation, suggests that stem cells could be used to repair spinal damage in people who have suffered damaging accidents or disease, although further studies, including safety tests, are needed before the treatment can go into human trials.

Stem cells were taken from the neural tissue of aborted fetuses. In tests, half of the 68 mice used were injected with around 75 000 stem cells above and below the injury site. The animals' behaviour was then assessed over the next few months.

'Animals that didn't get stem cells could only walk a little, and even though they improved slightly over the first two to three weeks, they were really struggling, but the mice that got stem cells go from stepping just occasionally to stepping all the time.'

Tests showed that the stem cells had formed new neurones and coatings that allow the nerves to send signals properly.

The 7 January 2007 issue of *Nature Biotechnology* reported the successful production of stem cells derived from the amniotic fluid removed during **amniocentesis**. With the proper culture conditions, they have been shown to be able to differentiate into a variety of cell types, including neural tissue.

## The effects of a stroke on short- and long-term memory

The physical damage caused to the brain by a stroke can have a wide range of effects. These will depend on the type of stroke and its severity, the part of the brain affected, the extent of brain damage and how quickly other brain cells take over the function of the damaged and dead ones. Around one-third of strokes are fatal. Strokes can have a detrimental effect on short-term memory (a working memory that stores information about things that have just happened and are currently happening). Long-term memory, the store of autobiographical information, language, acquired knowledge and much more, is usually well preserved in people who have suffered a stroke. After a stroke a person will show a noticeable loss of attention and concentration.

More short-term memory problems are seen in people with left-brain stroke. Long-term memory is usually preserved, but they may also have difficulty learning new information. They will probably need to have things repeated and be reminded over and over.

People with right-brain strokes have memory problems of another kind — they tend to get things out of sequence or misinterpret or confuse information. They may mix up the details surrounding an event. Usually, they can recall events but get confused about when they happened or who was involved. For example, they may think a family member visited this morning rather than last night.

### Techniques used to improve memory in stroke patients

- Rehabilitation therapy helps to re-train the brain to think and understand, and improve the patient's ability to speak, move and feel.
- Speech therapy re-educates the patient to speak, understand, read, write, solve problems and improve memory.
- Cognitive therapy.
- The use of new technology to motivate people to learn and to train their memory in a rehabilitation setting. This is achieved by projecting a Virtual Reality environment onto a 180° curved display screen that fills almost all of the user's view. The assistant or therapist selects and shows pictures, films, sounds or a mix of these different media for the patient, according to the patient's condition. The pictures, films and sounds are usually familiar to the patient. They show the family, home, familiar places and so forth.

People caring for a stroke patient can help them regain some of their short-term memory by:

- showing the patient photos and helping them identify people they used to know
- showing the patient photos in magazines or newspapers and helping them identify some of the well-known people shown
- playing a game with the patient in which a tray of objects is shown to them, then covered up, and they have to try to remember as many of the objects as possible
- having a large, clear calendar in every room so that they can keep track of the days
- leaving clear reminder notes around the house so that the patient knows what they should be doing and when.

## Questions

1 Strokes often affect ......................... memory.
2 Describe briefly two ways in which a carer could help to rehabilitate a stroke patient.
3 Research any recent advances in the use of stem cells for replacement of damaged neurones.

### Case study

Injuries and diseases that affect the nervous system not only demonstrate how important the system is to everyday life, but also show how the brain really determines who a person is as a human being. The actor Christopher Reeve, most famous for his movie role as Superman in the 1970s and 1980s, was thrown from his horse and his spinal cord just below his brain was crushed. Because of his spinal cord injury, he could not move or breathe on his own, and machines carried out some of his bodily functions. But because his brain was not injured, Reeve could talk, experience emotions, and plan and make decisions (for example, after the accident he directed a movie). Though being physically active was no longer a part of his identity, Reeve still enjoyed his roles as husband, father, friend and filmmaker. He still participated in acting and directing projects after his injury and became a spokesperson and advocate for spinal cord injury research. A book he wrote about his experiences since his injury is aptly titled *Still Me*. Christopher Reeve died aged 52 on 10 October 2004.

**Figure 1** Showing a stroke patient photos can help them identify people they used to know

### Brain chemicals may aid treatment of Parkinson's

Stanford Report, 14 February 2007

Marijuana-like chemicals in the brain may point to a treatment for the debilitating condition of Parkinson's disease. In a study published in the 8 February issue of *Nature*, researchers from the Stanford University School of Medicine report that endocannabinoids, naturally occurring chemicals found in the brain that are similar to the active compounds in marijuana and hashish, helped trigger a dramatic improvement in mice with a condition that mimics Parkinson's.

'This study points to a potentially new kind of therapy for Parkinson's disease,' said senior author Robert Malenka, MD, PhD, the Nancy Friend Pritzker Professor in Psychiatry and Behavioral Sciences. 'Of course, it is a long, long way to go before this will be tested in humans, but nonetheless, we have identified a new way of potentially manipulating the circuits that are malfunctioning in this disease.'

Malenka and postdoctoral scholar Anatol Kreitzer, PhD, the study's lead author, combined a drug already used to treat Parkinson's disease with an experimental compound that can boost the level of endocannabinoids in the brain. When they used the combination in mice with a condition like Parkinson's, the mice went from being frozen in place to moving around freely in 15 minutes. 'They were basically normal,' Kreitzer said.

But Kreitzer and Malenka cautioned that their findings don't mean smoking marijuana could be therapeutic for Parkinson's disease.

## What is a drug?

The word 'drug' has a range of meanings. It can mean medicines with beneficial therapeutic effects such as paracetamol, aspirin, penicillin and insulin. It can also mean everyday substances such as caffeine, nicotine and alcohol. Drugs also include illegal substances, such as cannabis, heroin and cocaine. A drug is any chemical you take that affects the way your body functions, how you think and feel.

Legal drugs are substances such as alcohol, tobacco and caffeine. The term includes prescription medicines that are intended to improve people's health. But legal drugs also carry risks of dependency and damage, and can be 'misused' in the same way that illegal drugs can.

Illegal drugs – in the UK, the Misuse of Drugs Act 1971, amended in 2005, lists the drugs that are subject to control and classifies them according to the level of harm associated with their misuse. Some drugs are not yet classified, but are still regulated through the Medicines Act (1968).

Drug abuse is the habitual misuse of a chemical substance including illegal drugs, prescription drugs and over-the-counter drugs. 'Abuse' refers to use that is problematic or harmful, either for the individual or for those around them.

## How drugs can be used to modify brain activity

Drugs flood into the synapses of a brain area via the bloodstream rather than through the carefully regulated axon terminals of interrelated neurones. Their high concentration and unregulated movements in and out of synapses can negatively affect us. Thus, the caffeine that keeps us awake (and perhaps alive) during the final stage of a long, late drive home will also probably delay our desired sleep because the effects of caffeine persist over three hours.

**Parkinson's disease** is caused by the progressive loss of neurones in a part of the brain that produces the chemical **dopamine**. Dopamine-neurones form networks of small clusters and are found in several regions of the brain. The main cluster of dopamine neurones responsible for controlling muscle movement is located in the midbrain, in an area called the *substantia nigra*. As the neurones die, less dopamine is produced and transported to the *striatum*, the area of the brain that coordinates movement. Symptoms such as tremors, stiffness, slow movements, and impaired balance and coordination develop as neurones die off and dopamine levels fall.

In addition, the enzyme in the synapse that breaks down the dopamine continues to deplete what little dopamine is left. The overall effect is a large loss of dopamine in the brain. This distorts the normal dopamine/acetylcholine balance – the level of acetylcholine remains normal, but there is not enough dopamine to balance it. The *basal ganglia* are a group of cells that regulate movement. When dopamine levels fall, they are prevented from modifying the nerve pathways that control muscle contraction. As a result, the muscles are too tense, causing tremor, joint rigidity and slow movement.

Most drug treatments for Parkinson's increase the level of dopamine in the brain or oppose the action of acetylcholine.

While many drugs are available to treat these symptoms during the early stages of the disease, the treatments become less effective with time. There is no cure for Parkinson's disease, although a lot can be done to relieve symptoms, especially in the early stages:

- **Dopamine replacement drugs** – the most effective treatment. These drugs are combinations of **levodopa**, which breaks down in the body to form dopamine, and a chemical that ensures there is the right amount of dopamine in the brain. They are effective at treating symptoms and can provide long-term improvement, although there are some side effects. People with Parkinson's disease can't simply take dopamine tablets or vegetable products containing dopamine (e.g. fava beans) to replace the missing dopamine, because dopamine taken by mouth does not get into the brain. Levodopa, on the other hand, does get into the brain and, once there, it converts to dopamine.

- **Drugs that mimic the action of dopamine** – commonly taken together with levodopa. They may be taken alone before using levodopa to reduce the long-term side effects of levodopa. Examples include bromocriptine and cabergoline (Cabaser).
- **Drugs that stop the breakdown of dopamine** – these can be used with levodopa and are usually given when dopamine replacement drugs start to lose their effectiveness. Examples include entacapone (Comtess), selegiline (e.g. Eldepryl).
- **Anticholinergic drugs** block the action of the acetylcholine.
- **Drugs that help to correct the balance between dopamine and acetylcholine** – doctors no longer prescribe these drugs often, because they are less effective than the drugs that replace dopamine. They are not usually used by people aged over 70 as they can cause memory loss, and urine retention in men. Examples include procyclidine.

### Use of drugs to treat severe pain

The ability to perceive pain is vital. However, faced with massive, intractable pain, it makes sense to use a drug that decreases the body's sensitivity to pain.

**Diamorphine** (heroin) is one of the longest established medications, with a 130-year history.

It can be used to relieve severe pain that can be caused by injury, surgery, heart attack or chronic diseases such as cancer. It is a chemical derivative of morphine, but it is more soluble and penetrates into the brain more rapidly than morphine.

Diamorphine hydrochloride belongs to a group of medicines called **opioids**. Opiates mimic the effects of naturally occurring pain-reducing chemicals (**endorphins**). They combine with the opioid **receptors** in the brain and block the transmission of pain signals. So even though the cause of the pain may remain, less pain is actually felt.

## How opioids block the transmission of pain

The word *endorphin* comes from the words 'endogenous morphine'. *Endogenous* means developing from within. Endorphins are similar to the **narcotic** morphine in their functions, and we produce them in our own bodies. Endorphins are produced at various sites in the body and function as the body's natural defence against pain. Endorphins act by attaching themselves to neurones and depressing their activity.

Opioids mimic the effects of naturally occurring pain reducing chemicals (endorphins) that are found in the brain and spinal cord. After they attach to a specific membrane protein called an *opioid receptor* they stimulate a chain of reactions that results in a depression of the normal activity of the neurone for a short time. They then leave the receptor and the normal function of the neurone returns. Opioid receptors are located on sensory neurone cell membranes in large concentrations in the brain and spinal cord, so the sensation of pain is temporarily halted.

All classes of opioid receptors share key similarities. First, the receptors have a common general structure. They are usually G protein-linked receptors imbedded in the plasma membrane of neurones. Once the receptors are bound, a portion of the G protein is activated, which allows it to diffuse within the plasma membrane. The G protein moves within the membrane until it reaches its target, which is either an enzyme or an ion channel. It may inhibit voltage-activated calcium channels so that the transmission of the pain impulse is prevented.

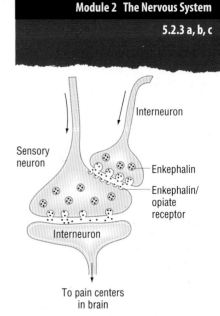

**Figure 1** How might opioids block the transmission of pain?

## Questions

1 A drug is a medicine with beneficial ...............effects such as paracetamol, antibiotics or insulin or it can mean everyday substances such as ............... and ............... . A drug could also be an .............. substance, such as cannabis, heroin and cocaine.

2 Drug abuse is the .............. of a chemical substance including illegal drugs, prescription drugs and over-the-counter drugs.

3 Which chemical is progressively in short supply in patients with Parkinson's disease?

4 Which drug can be prescribed to combine with the opioid receptors in the brain and block the transmission of pain signals?

## The use of cannabis

Cannabis is derived from the plant called *Cannabis sativa*, which is grown all around the world. It is a central nervous system (CNS) depressant. As a drug it is available in three main forms: as the dried leaves and buds, known as **grass** or **marijuana**, as a solid resin (**hashish** or **hash**) which is collected from the buds and flower heads, and also as a thick liquid prepared from the flowers or resin (**hash oil**).

The main mind-altering (*psychoactive*) ingredient in cannabis is THC (delta-9-tetrahydrocannabinol), but more than 400 other chemicals are present in the plant.

Cannabis is an illegal class C drug. The maximum penalty for possession is two years in prison. For supply you can get 14 years in prison. You can get an unlimited fine for both.

### Therapeutic use of cannabis

Cannabis has been used as a herbal medicine in many countries for a very long time. It has been used as a mild sedative or painkiller and for treatment of insomnia and gastric upsets. In the UK, it was legally prescribed up until 1928.

The beneficial effects of cannabis are as a mild **analgesic** and sedative which may relieve the symptoms of multiple sclerosis, hypoglycaemia and other disorders. In some cases it has been used as a medication for the terminally ill, where other treatments have failed to relieve distress.

But its possession or use in the UK is illegal at present and doctors are not able to prescribe cannabis in any form.

### Recreational use

The Indian vadas sang of cannabis as one of the divine nectars, able to give man anything from good health and long life to visions of the gods. The Zend-Avesta of 600 BC mentions an intoxicating resin, and the Assyrians used cannabis as incense as early as the ninth century BC. It was in ancient India that this 'gift of the gods' found excessive use in folk medicine. It was believed to quicken the mind, prolong life, improve judgement, lower fevers, induce sleep and cure dysentery. Because of its psychoactive properties it was more highly valued than medicines with only physical activity.

Cannabis is the most widely used illegal drug in the UK, with up to three million consumers per year. The effects of cannabis depend upon the amount used, its potency, the circumstances and the expectations or mood of the user.

The most common effects are talkativeness, cheerfulness, relaxation and greater appreciation of sound and colour. Colours and sounds play a big part in the state of users, as the perception of these is more pronounced and usually helps to relax the user. Perception of time, and occasionally of space, is altered. Cannabis users frequently report perceiving an enhanced performance for tasks involving creativity (art, music, etc.), although no scientific evidence indicates that the drug improves hearing, eyesight or skin sensitivity. Many users also experience a compulsion for binge eating (known as 'the munchies').

Like other drugs, cannabis has more dangerous effects, especially when higher doses are taken. Some side-effects include hallucinations and the user may become disoriented. This in turn can lead to the user becoming anxious or depressed and possibly suicidal. Some users will also become paranoid, especially if taking the drug at parties with a lot of other people around. Nausea and vomiting can occur when too much of the drug is taken at once. When smoked, cannabis usually acts quite quickly and sensations can last from 1–3 hours depending on the amount taken. Cannabis is also dangerous because of the damage caused to the lungs through smoking. While it has been reported to be non-addictive, this is certainly not the case for some people. Mixing cannabis with tobacco could also result in nicotine addiction and dependency on cigarettes.

The psychoactive effects of cannabis preparations vary widely, depending on dosage, the preparation, the type of plant used, the method of administration, the personality of the user, and the user's social and cultural background.

In relatively recent years, the use of cannabis as an intoxicant has spread widely in Western society – especially in the United States and Europe – and has caused concern in law-making and law-enforcing circles because of the social and health problems it causes. There is still little, if any, agreement on the scale of these problems or on their solution. Opinion appears to be divided. On one hand the use of cannabis is seen as such an extreme social, moral and health danger that it must be stamped out. On the other hand it is seen as an innocuous, pleasant pastime that should be legalised. Studies of the mental effects of cannabis show that the drug can impair or reduce short-term memory. It can alter sense of time and reduce the ability to do things which require concentration, quick reactions and/or effective coordination.

## Psychological and physical dependence on heroin and alcohol

*Drug and alcohol dependence* is a compulsion to continue drinking or taking a drug in order to feel good or to avoid feeling bad. When this is done to avoid physical discomfort or withdrawal, it is known as **physical dependence**; when it has a psychological aspect (the need for stimulation or pleasure, or to escape reality) then it is known as **psychological dependence**.

Physical dependence results from repeated, heavy use of drugs like heroin, tranquillisers and alcohol. Heavy and continual use of these drugs can change the body chemistry so that if someone does not get a repeat dose, they suffer physical withdrawal symptoms. Symptoms of withdrawal include restlessness, muscle and bone pain, insomnia, diarrhoea, vomiting, cold flushes with goose bumps ('cold turkey'), and involuntary leg movements. Major withdrawal symptoms peak between 24 and 48 hours after the last dose of heroin and subside after about a week. But some people show persistent withdrawal symptoms for many months. They have to keep taking the drug just to stop themselves from feeling ill, as their bodies have become adapted to functioning with the drug present.

Dependence is believed to result from adaptive changes in the nervous system, opposite in direction to the drug effects, which offset these effects when the drug is present and produce a 'drug-opposite' effect in its absence. Physical dependence is not the same as **addiction**, and can occur in non-addicted persons. People who are physically dependent on heroin usually develop a tolerance to the drug, making it necessary to take more and more to get the desired effects. Eventually, a dose plateau is reached after which no amount of the drug is sufficient. When this level is achieved, the person may continue to use heroin, but largely for the purpose of delaying withdrawal symptoms. Heroin withdrawal can cause death to the fetus of a pregnant addict. Sudden withdrawal by heavily dependent users who are in poor health can be fatal.

Alcohol abuse is a big problem. Alcohol is broken down into acetaldehyde in the liver by the enzyme **alcohol dehydrogenase**, and then from acetaldehyde to **acetic acid** by the enzyme **acetaldehyde dehydrogenase**. These two reactions result in diverting pyruvate from **gluconeogenesis**. This reduces the amount of glucose that the liver can supply to other tissues, especially the brain. This lack of glucose contributes to hangover symptoms such as fatigue, weakness, mood disturbances and decreased attention and concentration.

**Psychological dependence** is more common and can happen with any drug. In this case people use the drug experience as a way of coping with the world or as a way of feeling okay. They feel they could not cope without drugs even though they may not be physically dependent. People who are psychologically dependent on heroin or alcohol find that using it becomes far more important than other activities in their lives. They crave the drug and will find it very difficult to stop using it, or even to cut down on the amount they use.

Dependence will often include both physical and psychological factors. While the physical aspect will only be present with certain drugs, a psychological aspect will occur with any form of dependence.

### Are we turning into a nation of boozers?

The British, we've always been big boozers, right? Or have we? More people now die in Britain from alcohol-related illness than from drug abuse. Yet Britain lags behind France, Germany and many Scandinavian countries in terms of alcohol consumption. The increase in alcohol consumption in this country, and in particular the increasing incidence of binge drinking, has had a very significant impact on the NHS, not least on the acute services including Accident and Emergency departments. Up to 70% of attendees in A&E late on Friday and Saturday nights have been binge drinking and have developed a medical problem as a direct result, whether it is fight-related, due to a fall or due to the development of other symptoms, etc.

- Do you think that binge drinkers are dependent on alcohol, either physically or psychologically?

## Questions

1 Name one illness that cannabis could be used to treat.
2 Is cannabis a legal drug?
3 What is the difference between physical and psychological dependence?
4 Some people are able to drink large amounts of alcohol without developing dependency. If someone is dependent on alcohol and wishes to give up drinking, they will suffer from withdrawal symptoms. Once a person has been dependent on alcohol they can easily fall back into the same dependency unless they give up drinking alcohol. How can they be helped?

# Summary and practice questions

1 The nervous system contains several types of neurones. Of these ..................... neurones carry impulses to muscles and glands while ..................... neurones carry impulses from receptor cells to the central nervous system. The interior of a neurone has a lower concentration of ..................... ions than its surroundings, as a result of a ..................... in its membrane. This imbalance of ions creates a ..................... potential in the neurone, which is reversed during the passage of an impulse. When this happens, ..................... ions flood into the neurone, after which there is a compensating outward movement of ..................... ions.

2 Outline the sequence of events that take place when a nerve impulse arrives at a synapse.

3 Why do you think that it is important for neurones to have a threshold?

4 What effect does a myelin sheath have on conduction velocity?

5 Describe one function of each of the following regions of the brain.
   (i)    cerebellum
   (ii)   medulla
   (iii)  hypothalamus.

6 A head injury could damage the blood–brain barrier. What would be the affect of this?

7 Outline the nerve pathway for the pupil reflex.

8 The eyeball contains three outer layers; the ....................., choroid and ..................... . Incoming light rays are ..................... by the cornea and the lens of the eye. Rod and ..................... cells are stimulated by the effects of light on two pigments, ..................... and iodopsin. Light results in the *cis*-retinal being converted to ..................... which causes the ..................... channels to close, the ..................... neurone is depolarised, a neurotransmitter is released into the synapse between the ..................... and a ganglion. ..................... potentials add together to form an action potential.

9 A number of poisonous snakes produce venom which contains chemicals very similar in action to curare. South American Indians used to dip their arrow heads in curare. Curare caused death by respiratory failure. What is the advantage of this to a snake? Suggest how curare acts to cause respiratory failure.

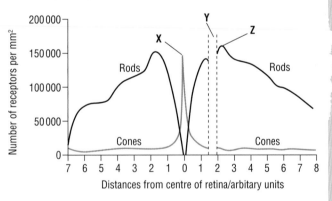

10 Use the graph above to explain why:
   (i)   No image is perceived when light is focussed on the retina at **Y**.
   (ii)  An image formed at **X** is perceived in more detail than an image formed at **Z**.

**11** The diagram below shows the changes in membrane potential at one point on an axon when an action potential is generated

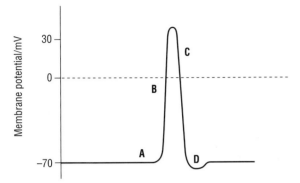

The changes shown in the diagram are due to the movement of ions across the axon membrane. Complete the table by giving the letter (**A** to **D**) that shows where each process is occurring most rapidly

| Process | Letter |
|---|---|
| Active transport of sodium and potassium ions | |
| Diffusion of sodium ions | |
| Diffusion of potassium ions | |

## The importance of homeostasis

To stay alive we must constantly adjust to changes within ourselves and to other changes in our environment. The ability to do this and maintain a relatively constant internal environment is called **homeostasis**.

Homeostasis is necessary because our cells are efficient but very demanding. If our cells are to function properly they need to be bathed in tissue fluid that can provide optimum conditions. They need regular supplies of oxygen and nutrients and the removal of wastes. Living cells can function only within a narrow range of such conditions of nutrient availability, temperature, pH and ion concentration, yet these and other conditions vary from hour to hour, from day to day and from season to season. There must therefore be a mechanism for maintaining internal stability in spite of environmental change. This is called homeostasis.

Claude Bernard (1813–1878), the French experimental physiologist, evolved the theory that carbohydrate is stored as glycogen in the liver and released, when necessary, as glucose into the blood. He also said that glucose levels vary in healthy individuals. Each of Bernard's findings convinced him that the body is constantly striving to maintain a stable, well-balanced internal environment, and one that is not affected too much by outside influences. Claude Bernard dismissed many previous misconceptions, took nothing for granted and insisted that facts were only any good if they were supported by evidence or raised further questions.

The term 'stable' is a little misleading as the conditions inside our bodies are not kept absolutely constant but are kept within a narrow range: that is, they fluctuate within a narrow range around an average called a **set point**.

(a)

(b)

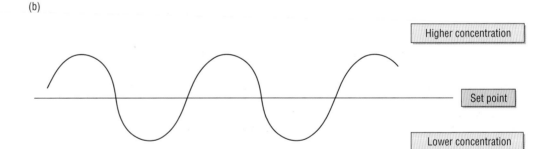

**Figure 1 a** Cartoon to illustrate homeostasis where average speed stays the same; **b** oscillation about a set point

## Do receptors, effectors and negative feedback form a part of homeostasis?

The usual way of maintaining homeostasis is by **negative feedback** – the body senses an internal change and then activates mechanisms that reverse that change. This involves a *receptor* (or sensor), an *effector* and an efficient means of communication between them. This is explained in Figure 2.

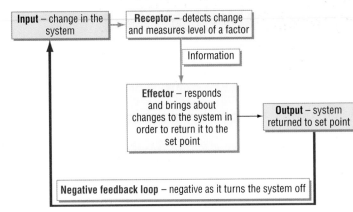

**Figure 2** Negative feedback

Continuous monitoring of the factor being controlled by the receptor produces continuous adjustments of the output. This results in oscillations around the set point. The set point for core body temperature, for example, is about 37.4 °C, but the temperature fluctuates within about 0.5 °C either above or below that value.

Claude Bernard concluded that the body must be under the control of one strong and central regulating force. We now know that the **hypothalamus** contains receptors that monitor the blood and also receives impulses from other body receptors. It coordinates this information and sends nerve impulses to the effectors.

### Role of the autonomic nervous system in homeostasis

The 'involuntary' or **autonomic nervous system** is responsible for homeostasis. In spread 5.2.2.2 you learnt about the two branches of the autonomic nervous system, the **sympathetic** and the **parasympathetic**. Nerve impulses from these two systems have opposite effects on the body. For example, an impulse travelling along a sympathetic nerve to the liver would result in glucose being released, whereas an impulse along the parasympathetic nerve would cause the liver to increase glycogen production slightly, thus working with the endocrine system to maintain blood glucose levels.

Many of the activities of the autonomic nervous system are regulated by the hypothalamus.

The hypothalamus contains many types of receptors that monitor factors such as water potential and blood temperature. It also receives information, in the form of nerve impulses, from receptors in other parts of the body.

Ultimately, the hypothalamus can control every endocrine gland in the body and alter blood pressure, body temperature, metabolism and adrenaline levels.

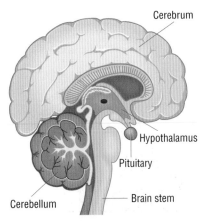

**Figure 3** The position of the hypothalamus

## Questions

1 Any self-regulating system such as homeostasis involves a series of stages:
   * ...................., the desired level at which the system operates.
   * receptor, this detects ....................
   * .................... responds and brings about changes to the system in order to return it to the .................... .
2 What is meant by homeostasis? Use examples in your answer.
3 Explain why negative feedback is frequently involved in homeostasis.
4 Why do you think homeostatic mechanisms are sometimes described as detection–correction systems?

### Why do we have to control our temperature?

We are able to maintain a constant core body temperature of around 37.4 °C, despite changes in our surroundings. This ability to regulate our body temperature is known as **thermoregulation**.

Thermoregulation is essential for our continued survival. It is one of the reasons we have been able to adapt to life in many different environments from the Antarctic to the tropics and the desert.

On a smaller scale we must be able to adjust to the difference between day and night temperatures, which could be as much as 18 °C.

As we are able to generate heat inside our bodies, we are said to be **endothermic**. Our body temperature is controlled by a balance between the production of heat inside our bodies and the loss of heat to the environment.

### Where the temperature receptors are

The *core body temperature*, which is the temperature of deep structures of the body such as the liver, is monitored in comparison to temperatures of peripheral tissues. In humans, optimum core body temperature is 37.4 °C but it does vary in different people, according to the time of day, the activity they are doing and, for women, the stage of their menstrual cycle. It also fluctuates about a **set point**.

Temperature-sensitive neurones (**thermoreceptors**) are situated in the hypothalamus. These are the sensors and they detect changes in the temperature of the blood flowing through the brain. The thermoregulation centre of the hypothalamus also receives information via sensory nerves from thermoreceptors located in the skin and internal organs. The hypothalamus connects with the rest of the body via the autonomic nervous system.

When the temperature sensors sense a change in either the external temperature or the core temperature, the hypothalamus sends impulses to the various effectors, such as the blood vessels in the skin, hair erector muscles and sweat glands, which will reverse the change.

### What happens when the body is too hot

- The arterioles supplying blood to the skin dilate, allowing more blood to flow close to the surface of the skin. This is called **vasodilation** and is the reason why pale people can look red when they are hot. Heat is lost to the air by radiation and conduction.
- Sweat glands secrete sweat. Sweat consists mainly of water, but also of many of the solutes dissolved in blood such as sodium ions, chloride ions and urea. As the sweat evaporates, it uses the heat in the skin to provide the energy to change the liquid into a gas (water has a high latent heat of vaporisation).

### What happens when the body is too cold

- The arterioles supplying blood to the skin constrict because the smooth muscle in the arteriole wall contracts, narrowing the **lumen** of the arteriole, so shutting off the blood supply to the capillaries that lie in the surface of the skin. This is called **vasoconstriction** and means that less heat is lost due to radiation.
- Erector muscles attached to the base of the hair follicles contract, pulling the hairs vertically up. This creates a pocket of air under each hair. Because air is a bad conductor of heat, it insulates the body. This is what also causes goose pimples, since humans don't have very much hair.

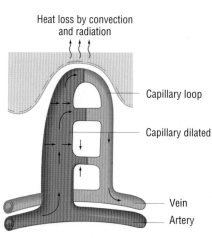

Heat loss by convection and radiation

Capillary loop

Capillary dilated

Vein

Artery

Warm environment

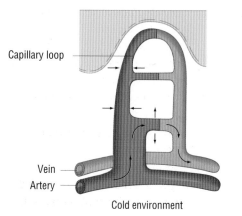

Capillary loop

Vein

Artery

Cold environment

**Figure 1** Vasodilation and vasoconstriction

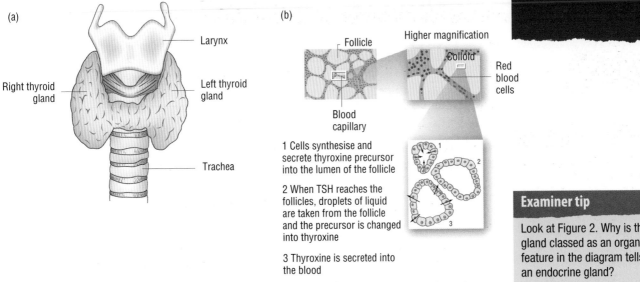

**Figure 2** Thyroid gland and a summary of the production of thyroxine

(a)

Larynx

Right thyroid gland

Left thyroid gland

Trachea

(b)

Follicle

Higher magnification

Colloid

Red blood cells

Blood capillary

1 Cells synthesise and secrete thyroxine precursor into the lumen of the follicle

2 When TSH reaches the follicles, droplets of liquid are taken from the follicle and the precursor is changed into thyroxine

3 Thyroxine is secreted into the blood

**Examiner tip**

Look at Figure 2. Why is the thyroid gland classed as an organ? Which feature in the diagram tells you this is an endocrine gland?

- Muscles can also receive impulses from the brain (the hypothalamus) to cause shivering. This produces heat because the cells in the muscle respire more as they do more work. Respiration is an *exothermic* reaction that gives out heat, so more heat is produced, which warms up the body.
- Sweat stops being produced.

### How temperature is controlled in the long term

The hormone **thyroxine** increases the **metabolic rate** (the metabolic rate is the rate at which all cells in the body carry out their biochemical reactions) and thus increases heat production.

Thyroxine is secreted by the thyroid gland, which lies in the neck on either side of the trachea.

When the hypothalamus detects a drop in temperature it secretes **thyrotrophin-releasing hormone** (**TRH**). TRH is secreted into the blood and carried to the anterior pituitary gland, where it stimulates the production of **thyroid-stimulating hormone, TSH** (thyrotropin). TSH stimulates the production of thyroxine by the thyroid gland. Thyroxine is secreted into the bloodstream.

If not enough iodine is available in the diet, then not enough thyroxine will be made to shut off the release of TSH. Prolonged stimulation of the thyroid by thyroid-stimulating hormone results in an abnormal enlargement of the gland, known as goitre, a condition that has been largely eradicated by the widespread use of iodised salt.

Thyroxine affects the metabolic rate by passing through the plasma membrane of its target cells to the nucleus, where it stimulates the **transcription** of a number of **genes**. One of the results of this is the production of more mitochondria. More respiratory enzymes are synthesised and glucose and fat metabolism is increased. The overall result is a greater rate of respiration and therefore more heat energy is released.

This is a slow response to the need for temperature control because it takes time for a gene to be transcribed.

We tend to secrete extra thyroxine when we are exposed to cold conditions for more than a few days. We are said to become *acclimatised*.

## Questions

1 What term is used to describe the dilation of blood vessels when we are hot?
2 Explain why food intake may increase in the winter compared with the summer months.
3 What changes happen in your skin when you go from a warm room out into a cold winter night?
4 Suggest one advantage of regulating blood temperature by the nervous system rather than by hormones.

Temperature control illustrates the principles of homeostasis.

## Techniques for measuring body temperature
As we saw in the last spread our core body temperature fluctuates about a set point.

### How temperature is measured
Temperature is measured using a **thermometer**. Thermometers generally used to be the mercury-in-glass type, which consists of a glass capillary that opens into a mercury-filled bulb at one end. If the temperature increases, the mercury expands and rises in the capillary. The height of the mercury is an indication of the temperature, and may be read on an adjacent scale. Today, most thermometers are rapidly responding electronic infrared sensors with a digital scale. Ear temperature is frequently used because its measurement accurately reflects **core body temperature**, since the eardrum shares its blood supply with the temperature control centre in the **hypothalamus**.

It is of great medical importance to measure body temperature because a number of diseases are accompanied by characteristic changes in body temperature. Also, we can monitor the course of certain diseases by measuring the body temperature, and in this way a doctor can evaluate the efficiency of a treatment. *Fever* (pyrexia) is defined as a body temperature that is higher than normal for each individual. It is a reaction to a disease, where the **set point** of the temperature control centre is varied to help the body's defences against the disease.

## Hypothermia and its treatment
**Hypothermia** occurs when the body's core temperature drops significantly below normal. This is usually considered to be below 35°C. As the core temperature falls, all metabolic reactions slow down. This is because the reactant molecules have less **kinetic energy** and therefore there will be fewer collisions.

The symptoms of hypothermia are as follows:

The first stage – *mild hypothermia* – is characterised by
- bouts of shivering
- grogginess and muddled thinking.

The second stage – *moderate hypothermia* – is indicated by
- violent shivering
- inability to think and pay attention
- slow, shallow breathing and a weak pulse
- slow and laboured movement.

The third stage – *severe hypothermia* – has set in when
- shivering stops
- the patient has difficulty in speaking
- there is little or no breathing and the pulse is weak, irregular or non-existent.

Those at risk of hypothermia can include:
- Anyone who spends a lot of time outdoors in cold weather is at risk. A strong wind will increase heat loss due to wind chill – wind blows away the layer of warm air next to the skin and so the rate of heat loss will increase. At particular risk are walkers and climbers trapped by injury or other difficulties on a windy, exposed mountainside. Being immersed in cold water or getting thoroughly wet increases the risk.

---

## Body temperature

Look at the graph in Figure 1a carefully and describe its pattern.
- What do you think the difference would have been if the temperature had been taken orally?
- Would this have been as accurate? What affects oral temperature?
- What do you think is the difference between accuracy and reliability?

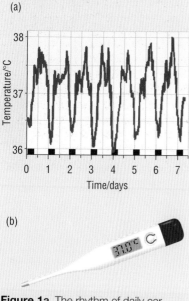

Figure 1a The rhythm of daily ear body temperature; **b** a digital thermometer

- The elderly are at risk because many older people are concerned about the costs of heating, so do not use it. They may not be very mobile, so spend much of the day sitting in one place. Also they may not eat a suitable diet.
- Babies under one year old are also at risk. Babies can become hypothermic by sleeping in a cold room.
- Anyone whose judgement may be impaired by mental illness or Alzheimer's disease.
- People who are intoxicated, homeless or caught in cold weather because their car has broken down.

Symptoms usually develop slowly. Someone with hypothermia typically experiences an inability to think and so may be unaware of the need for emergency medical treatment.

### Treatment

In order to care for someone with hypothermia, first call for emergency medical assistance and, while waiting for help to arrive, monitor the person's breathing. If breathing stops, start **expired air resuscitation** and if the pulse/heart stops, start CPR. Move the person out of the cold – if they cannot be moved indoors, cover their head and insulate their body from the cold ground.

Replace any wet clothing with warm, dry blankets. Do not apply direct heat – do not use hot water, a heating pad or a heating lamp. Instead apply warm compresses to the neck, chest wall and groin. Do not attempt to warm the legs and arms, because heat applied here forces cold blood back towards the heart, lungs and brain, causing the core temperature to drop. The person must not be given alcohol.

To prevent hypothermia, stay indoors as much as possible and limit your exposure to the cold. Eat regularly and include plenty of carbohydrates. Keep as active as possible (activity helps to generate warmth). Avoid alcohol – it causes dilation of peripheral blood vessels, increasing heat loss. Wear multiple thin layers of clothing that help to trap heat, rather than one thick jumper. If you go outside, always wear a hat (it can prevent as much as 20% of heat loss) and gloves.

## Hyperthermia

**Hyperthermia** occurs when the body's core temperature rises significantly **above** normal. The only way we have of reducing body temperature to below the temperature of the environment is by sweating. In humid conditions the air may already be saturated with water vapour, so sweat lies on the skin and cannot evaporate and cool the body.

### What are the symptoms?

Heat exhaustion is characterised by mental confusion, headache, muscle cramps and vomiting. Blood pressure may drop significantly from dehydration, leading to dizziness. As the blood pressure drops and the heart attempts to supply enough oxygen to the body, heart rate and respiration rate will increase. Some victims, especially young children, may suffer convulsions.

Hyperthermia in its advanced state is called **heat stroke**. This is a medical emergency requiring hospitalisation. Body temperatures above 40°C are life-threatening.

People at risk of hyperthermia can include:
- infants and children up to four years of age
- people of 65 years of age or older, because they may not compensate for heat stress efficiently
- people who are overweight, because they tend to retain more body heat
- people who over-exert during work or exercise
- people who are ill
- people who are on certain medication.

To prevent hyperthermia, do not leave infants, children or pets in a parked car. Dress infants and young children in cool, loose clothing and a hat. Limit exposure to sun during the midday hours. Make sure that infants, children and elderly people drink adequate amounts of liquid.

---

### Examiner tip

Make sure you do not confuse hypothermia with hyperthermia – think 'o' (hypOthermia, cOld).

---

### Case study

Beth is retired. She loves to work in her garden. Last summer during a very hot spell she continued to work outside. Every day the temperature was in the high 30s and the humidity was at least 90%. After a week of these conditions Beth sounded confused on the phone so her daughter came over. She found her mother passed out on the kitchen floor. An ambulance was called. Beth almost died as she had heat stroke, the most serious form of hyperthermia.

---

## Questions

1. Hill walkers can encounter extreme changes in environmental conditions. What precautions should they take to avoid hypothermia?
2. Why are the elderly more likely to suffer from hypothermia?
3. How can infants be protected from hyperthermia during a summer beach holiday?

# The pancreas as an endocrine gland

## Where is the pancreas and what does it do?

The pancreas is a large, pale-coloured gland that lies just behind the stomach. The pancreas is unusual as it is both an **exocrine** gland and an **endocrine** gland. As an exocrine gland it secretes digestive enzymes through the pancreatic duct into the duodenum. But we are interested at present only in its role as an endocrine gland. As an endocrine gland the pancreas produces the hormones insulin and glucagon, which pass directly into the blood capillaries that pass through the pancreas.

### Location of insulin and glucagon secretion

When you examine the pancreas under a microscope, you can see groups of endocrine cells scattered throughout it. These cells are known as the **islets of Langerhans** (after Paul Langerhans, who discovered the islets in 1869). The islets of Langerhans form about 15% of the pancreas. The islets have many blood capillaries associated with them. There are about one million islets in a healthy adult human pancreas.

The islets contain two types of endocrine cell:
- *alpha cells* (α cells), which secrete glucagon
- *beta cells* (β cells), which secrete insulin.

Both these hormones are involved in the control of blood glucose levels.

Alpha cells producing glucagon are usually distributed around the edge of the islet. Beta cells producing insulin are usually in the centre and account for about 80% of the islets. The islet cells secrete hormones directly into one of the many blood capillaries.

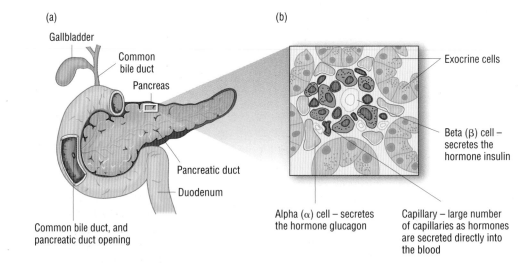

**Figure 1 a** The pancreas; **b** islets of Langerhans

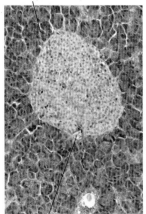

Pancreatic cells – secrete pancreatic juice

Islets of Langerhans

**Figure 2** Light micrograph of a section through the pancreas showing islets of Langerhans

### The structure of an islet of Langerhans

Compare the light micrograph in Figure 2 with Figure 1. You can see the islet of Langerhans with its many capillaries.

# Negative feedback controls blood glucose concentration

If blood glucose levels *rise* following a meal then:

- the α and β cells in the islets of Langerhans detect the rising concentration of glucose (**receptors**)
- α cells stop secreting glucagon, but β cells secrete insulin
- insulin binds to receptors in liver, fat and muscle cells
- this causes the uptake of glucose by these cells to increase (**effectors**)
- use of glucose in respiration is increased
- liver cells convert glucose to glycogen (**glycogenesis**), which is stored.

If blood glucose levels *falls* as a result of glucose being used rapidly by the cells or a diet lacking in carbohydrates then:

- the α and β cells in the islets of Langerhans detect the falling concentration of glucose (**receptors**)
- β cells stop secreting insulin
- α cells start to secrete glucagon
- less glucose is taken up by the target cells (**effectors**)
- the rate of use of glucose in respiration decreases
- **gluconeogenesis** occurs – amino acids, pyruvate and lactate are converted to glucose for respiration
- in the liver glycogen is converted to glucose and released into the blood.

The glucose levels in the blood fluctuate about a mean set point.

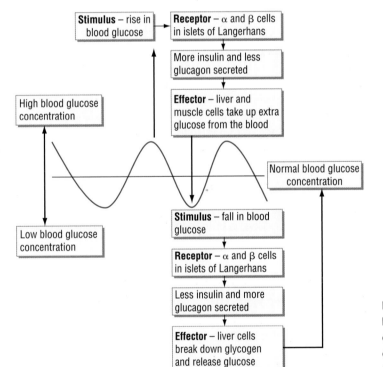

**Figure 3** Diagram to show how negative feedback causes the concentration of glucose in the blood to fluctuate about a mean

# Questions

1  What is the difference between an exocrine and an endocrine gland?
2  Where are the islets of Langerhans found?
3  What is the difference between glycogen and glucagon?
4  How is plasma glucose concentration lowered after a meal containing a lot of carbohydrate?
5  Why is it so important that glucose is changed into glycogen if the blood glucose level rises?

A form of diabetes that is characterised by early age of onset (usually less than 25 years of age) and autosomal-dominant inheritance (that is, it is inherited by 50% of a parent's children) with diabetes in at least two generations of the patient's family. MODY diabetes can often be controlled with meal planning, at least in the early stages of diabetes. It differs from type II diabetes in that patients have a defect in insulin secretion or glucose metabolism, and are not resistant to insulin. MODY accounts for about 2% of diabetes worldwide and six genes have so far been found that cause MODY, although not all MODY patients have one of these genes. Because MODY runs in families, it is useful for studying diabetes genes and has given researchers useful information about how insulin is produced and regulated by the pancreas.

## Role of the diabetes nurse

The job of the diabetes nurse is to provide education, advice and support to people with diabetes. This includes:

- crisis management advice (i.e. for newly diagnosed patients with hypoglycaemia (a low blood glucose level))
- support and advice during periods of change of treatment
- assist with day-to-day management of patients admitted to hospital.

## What causes the two types of diabetes?

Diabetes usually refers to **diabetes mellitus**. Diabetes is a chronic disease in which the blood glucose control mechanism has partly or completely broken down. The word 'mellitus' comes from Latin meaning 'honey', a reference to the sweet taste of the urine of diabetics. The ancient Indians tested for diabetes by observing whether ants were attracted to a person's urine, and called the ailment 'sweet urine disease'. In the UK around 1.6 million people have been diagnosed with diabetes.

There are two forms of diabetes mellitus ('sugar' diabetes) – type I and type II.

| Type I (insulin dependent) | Type II (insulin independent) |
|---|---|
| The body is unable to produce insulin | The body cannot produce sufficient insulin |
| This may be due to an autoimmune response, where the body's immune system attacks its own ß cells in the islets of Langerhans | May also be due to the target cells losing their responsiveness to insulin |
| Normally begins in childhood (also called juvenile-onset diabetes) | Usually arises in people over 40 years old (late-onset diabetes) |
| Develops quickly, normally over a few weeks | Develops slowly |
| Symptoms: a high blood glucose level; glucose present in the urine; increased thirst and hunger; weight loss; the need to urinate excessively; tiredness | Symptoms are normally less severe and may be put down to 'overwork' or 'old age' |

### Risk factors for developing diabetes

In most cases, Type I diabetes is thought to be an **autoimmune** disease. In this case the body makes antibodies that attach to the β cells in the islets of Langerhans and destroy them. It is thought that something triggers the immune system to make these antibodies. The 'trigger' is not known, but one theory is that a virus is the trigger. A recent discovery of a gene that can raise a person's risk of developing type I diabetes has raised hopes of developing more targeted treatments for young people.

Type II diabetes involves 'insulin resistance', or reduced sensitivity to insulin, and/or reduced insulin secretion. The loss of responsiveness to insulin probably involves insulin receptors in the plasma membranes.

A number of risk factors have been identified for type II diabetes:

- age – the risk increases with age, beginning after age 30
- obesity – fatter adults are more likely to have increased insulin resistance. Over 80% of people with type II diabetes are overweight at diagnosis: they have a BMI (body mass index) of over 27 (calculate your BMI by dividing your weight in kilograms by your height in metres squared)
- having an apple-shaped figure, where more fat is carried around the middle
- family history – the closer your relative with diabetes, the greater your risk of diabetes
- sedentary lifestyle
- high blood pressure
- genetic link – in 2000 a gene on chromosome 4 was found to be one of probably many genes that affect the risk of developing type II diabetes (see How Science Works box on the left)
- low birth weight – scientists have found that a lack of nutrients in the womb permanently damages key cells in the pancreas that secrete insulin
- gestational – occurs in late pregnancy and usually disappears, but is shown to increase the risk of developing type II later in life.

## Treatment of diabetes

The management of diabetes mellitus involves keeping blood glucose levels as constant as possible. The glucose levels have to be checked regularly.

| Type I diabetes | Type II diabetes |
|---|---|
| The patient needs to use insulin injections either twice or four times daily. Insulin cannot be taken by mouth because insulin is a protein and so would be digested. Insulin may be injected using a syringe | Often possible to manage by carefully controlling the diet, reducing intake of refined sugars. It is important to eat small meals at regular intervals so that the blood is never flooded with excess glucose |
| An insulin pen may be used. This is a small pen-sized device that holds an insulin cartridge. The amount of insulin to be injected is controlled by turning the dial on the bottom of the pen. The tip of the pen consists of a needle that can penetrate just under the skin and deliver the required amount of insulin | Increasing exercise and losing weight if obese or overweight may also help |
| | The first oral drug treatment tried is metformin. The reason for this is that, unlike the rest of the possible drugs, it does not cause weight gain. Metformin lowers the levels of glucose in blood in three different ways. First, it reduces the amount of glucose produced by the liver; second, it reduces the amount of glucose absorbed from food; and third, it improves the effectiveness of insulin in the body by reducing glucose already in the blood |
| An insulin pump is composed of a pump reservoir, a battery-operated pump, and a computer chip that allows the user to control the exact amount of insulin being delivered. The pump is used for continuous insulin delivery, 24 hours a day | If this does not work and the blood glucose levels are still too high, sulphonylureas with insulin are used to stimulate insulin production (sulphonylureas cause insulin to be released from the beta cells) |

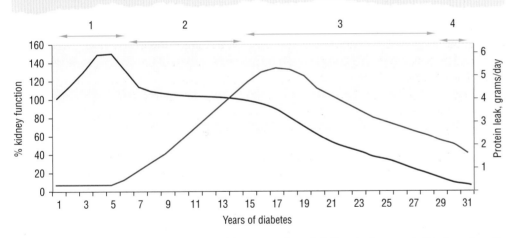

The diagram shows how kidney function reduces and the amount of protein in the urine increases in diabetic nephropathy

1 It begins with a tiny amount of protein appearing in the urine – this is called microalbuminuria. The kidney function may well be normal at this point

2 Over 10–15 years proteinuria increases

3 With the development of proteinuria, the kidneys' ability to remove poisons from the blood deteriorates such that 5–10 years later the kidneys are almost completely unable to remove these poisons from the blood

4 This is called 'end-stage renal disease', and, unless treated, the poisons can build up to fatal levels

**Figure 3** How kidney function declines and how the amount of protein in the urine increases in poorly controlled diabetes

## Why are regular checks important?

- Blood pressure: diabetics are twice as likely as the general population to suffer from coronary heart disease, and this is a major cause of death. The ideal blood pressure for a diabetic would be 130/80 or lower.

- Examination of the retina: retinopathy occurs when poorly controlled diabetes damages the tiny blood vessels in the retina. These blood vessels can become abnormally fragile and leak. This may cause blurring and occasionally loss of vision. An annual eye examination will detect any changes early enough for treatment to be successful.

- Kidney function tests: a diabetic may hold a lot of sugar in her urine, which will make her more prone to developing bladder infections such as cystitis. These infections can then spread up to the kidneys. In poorly controlled diabetes, high blood glucose and high blood pressure levels can affect the capillaries in the glomerulus (see the next spread). This is a leading cause of kidney failure in Europe and the USA.

## Questions

1 What is the other name for type I diabetes?
2 Name three risk factors for developing type II diabetes.
3 What is the treatment for type I diabetes?
4 Suggest why people suffering from diabetes are advised to eat their carbohydrates in the form of starch rather than as sugars.

## What is excretion?

Excretion is the removal of waste products of metabolism from the body. **Metabolism** is the term given to chemical processes in the body, such as respiration. There are many sequences of reactions in which the product of one reaction becomes the substrate for the next.

The main organs of excretion are:
- the kidneys, through which urine is excreted
- the skin, which excretes water and salts in sweat
- the lungs, which excrete water vapour and carbon dioxide.

**Elimination** (*egestion*) is the removal of waste products, such as fibre, that have never been involved in metabolism.

### The importance of excretion

Excretion is important because it removes toxic (poisonous) waste from the body. If these toxins build up, they will slow down and eventually stop important chemical reactions taking place.

One of the waste products of respiration is carbon dioxide.

$$\text{Sugar} + \text{oxygen} \longrightarrow \text{energy} + \text{water} + \text{carbon dioxide}$$

Carbon dioxide cannot be allowed to build up in the body, because it would increase the acidity of the blood (lower its pH). You would have a feeling of heaviness and would take more frequent and deeper breaths. Eventually you would become unconscious and may die.

We cannot store proteins or amino acids in our bodies, so any excess is broken down. The amino group has to be removed. It is removed in the liver by a process called **deamination**. Figure 1 describes what happens to excess amino acid.

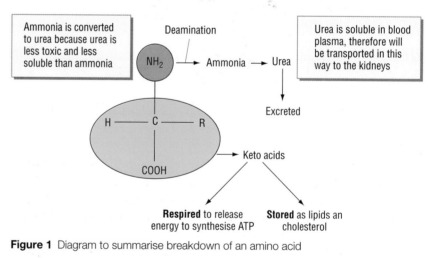

**Figure 1** Diagram to summarise breakdown of an amino acid

# Kidney structure

Most people are born with two kidneys, although you can lead a normal life with one healthy kidney. The kidneys are located near the middle of the back, one on each side of the spine and just below the rib cage. The kidneys are two bean-shaped organs about 10 cm long and about 6 cm wide. One quarter of your blood supply passes through your kidneys every minute. Each kidney receives blood through the renal artery, which branches off the aorta (main artery). When the blood has passed through the kidney and has been filtered, it returns to the bloodstream through the renal vein, which connects to the vena cava. A narrow tube, the ureter, carries urine from the kidney to the bladder where it is stored. The urethra takes the urine from the bladder to the outside.

**Examiner tip**

Learn the definition of excretion so you do not confuse it with elimination.

Can you label the longitudinal-section of a kidney? (These are easy marks, which you do not want to lose!)

(a)

(b)

**Figure 2  a** Position of the kidneys in the body; **b** longitudinal-section of a kidney

# Questions

1 Excretion is the removal of waste products of .................. . The three main excretory organs are the skin, the .................. and the .................. . Carbon dioxide is a waste product of .................... . Ammonia is formed when the .................. group is removed from an .................. . Ammonia is immediately changed into .................., which is excreted.
   The human kidney in longitudinal-section is seen to consist of an outer region called the ......... and a darker inner region called the .................. . Blood is brought to the kidney by the .................. .

2 Why would you die if excretion stopped? (Hint: consider the removal of both carbon dioxide and urea.)

## The gross structure and histology of a nephron

A microscopic examination of the cortex shows that the cortex and medulla of the kidney are made up of around one million tiny tubules or **nephrons**. The structure of a nephron is shown in Figures 1 and 3.

Wide afferent arteriole

Narrow efferent arteriole

Capillaries of glomerulus

Cavity of renal capsule

Nucleus of podocyte

Filtration slit

Podocyte in wall of renal capsule

Basement membrane

Endothelial cell

Capillary lumen

Circular pore

→ Filtrate is forced out of capillary by the hydrostatic pressure of the blood in the glomerulus. This pressure is sufficient to overcome the pressure of the fluid in the renal capsule

**Figure 2** Ultrafiltration in the renal capsule

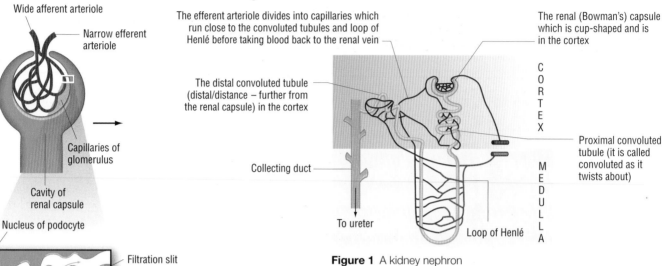

The efferent arteriole divides into capillaries which run close to the convoluted tubules and loop of Henlé before taking blood back to the renal vein

The distal convoluted tubule (distal/distance – further from the renal capsule) in the cortex

Collecting duct

To ureter

Loop of Henlé

The renal (Bowman's) capsule which is cup-shaped and is in the cortex

Proximal convoluted tubule (it is called convoluted as it twists about)

CORTEX

MEDULLA

**Figure 1** A kidney nephron

### Urine production by the nephrons

There are two stages in urine production:

- **Ultrafiltration** – involves filtering all small molecules out of the blood and into the renal capsule
- **Selective reabsorption** – involves taking back into the blood any useful molecules from the fluid in the nephron.

### Ultrafiltration

Inside the cup-shaped renal capsule is a **glomerulus**, which is a network of capillaries. Blood is brought in by the **afferent arteriole** (branches off the renal artery) and returned by the **efferent arteriole** to the renal vein. The efferent arteriole has a smaller diameter than the afferent so pressure is built up (think of a traffic jam). This **hydrostatic pressure** forces blood against a filter. This filter consists of three layers:

- the lining or **endothelium** of the blood capillaries; this has small holes in it through which **plasma** can escape
- the **basement membrane** acts as a molecular filter
- the cells of the lining of the renal capsule. These are called **podocytes**. These cells are lifted off the surface on little feet ('pod' means feet) so the filtrate passes beneath them and through gaps between their branches.

What can pass through this filter? No cells or proteins with a molecular mass greater than about 65 000 to 69 000 can get through, so they stay in the blood. The rate of filtration is high; about 125 cm³ per minute, but we produce only about 1 cm³ of urine per minute, which means that over 99% of the filtrate is reabsorbed.

### Selective reabsorption

Toxic compounds, excess solutes and water are removed and useful substances are reabsorbed.

All the glucose, amino acids, vitamins and hormones and 85% of the sodium ions, chloride ions, potassium ions and water are reabsorbed in the **proximal convoluted tubule** (PCT), either actively or passively. This greatly reduces the filtrate volume. The PCT cells are covered with **microvilli**, which greatly increase the surface area through which substances can be reabsorbed. These cells also have a large number of

Renal capsule

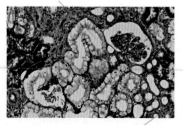

Proximal convoluted tubule

**Figure 3** Histology of the nephron

**Cross-section proximal convoluted tubule**

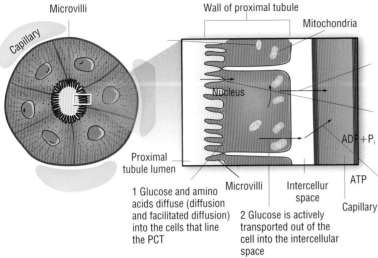

Na⁺ ions are actively transported out of the PCT cell and into the tissue fluid – this creates a lower water potential so water follows by osmosis

Sodium ions are actively pumped out of the tubule wall cell into the capillaries. This produces a gradient of sodium ions and so they move in from the filtrate in the lumen through a co-transporter membrane protein. This protein also brings in glucose and amino acids

3 Glucose diffuses through the wall of the capillary and is transported away

1 Glucose and amino acids diffuse (diffusion and facilitated diffusion) into the cells that line the PCT

2 Glucose is actively transported out of the cell into the intercellular space

**Figure 4** Selective reabsorption from the proximal convoluted tubule

mitochondria (indicating that they are involved in an active process). Na⁺ is actively transported into the cells of the PCT, and Cl⁻ ions follow passively. As the Na⁺ and Cl⁻ build up in these cells, water follows by **osmosis**, making the filtrate more concentrated.

In the descending limb of the **loop of Henlé**, more water is reabsorbed by osmosis. In the ascending limb more NaCl is reabsorbed by both active and passive processes, so the filtrate becomes more dilute again.

The **distal convoluted tubule** helps to regulate the NaCl and K⁺ concentration of the body fluids by secreting K⁺ into the filtrate and reabsorbing NaCl. It also helps in pH regulation by secreting H⁺ and reabsorbing hydrogen carbonate ions (HCO3⁻).

The loop of Henlé creates a region of high solute concentration in the medulla. The collecting ducts pass through this region and the **water potential** gradient between the inside of the collecting duct and the outside draws water out of the ducts by **osmosis**.

Reabsorption in the distal convoluted tubule and the collecting duct occurs as the cells of the distal convoluted tubule actively transport sodium ions out of the fluid in the tubule and into the tissue fluid. The sodium ions and the reabsorbed water then diffuse into the blood. Potassium ions are actively transported in the opposite direction. The rate at which these two ions are transported can be varied so their content in the blood can be 'fine-tuned' to maintain a homeostatic balance (see **homeostasis**, spread 5.3.1.1). As the fluid flows through the collecting duct it passes the loop of Henlé. In this region the water potential is low, so water flows out of the collecting duct due to osmosis and into the blood capillaries. The end result is a smaller volume of concentrated urine.

**Examiner tip**

Do not confuse urea with urine. Urea is formed as a result of deamination of amino acids in the liver. Urine is the fluid produced by the kidneys in which urea and other substances are dissolved.

## Questions

1 What are the four distinct regions of a nephron?

2 In which region of the nephron is all the glucose reabsorbed?

3 If the diameter of the efferent arteriole is decreased by muscle contraction, what effect will this have on the process of ultrafiltration?

4 Human kidneys process 1200 cm³ of blood every minute. Approximately 125 cm³ of fluid is filtered from this blood into the renal capsule, resulting in 1500 cm³ of urine being produced each day. Calculate the volume of filtrate, in cm³, produced by the kidneys in a day.

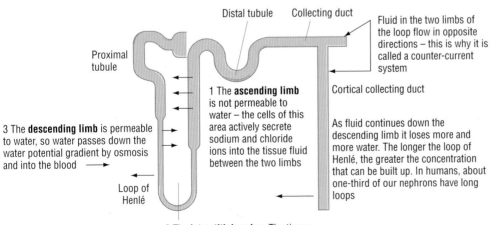

Fluid in the two limbs of the loop flow in opposite directions – this is why it is called a counter-current system

Cortical collecting duct

As fluid continues down the descending limb it loses more and more water. The longer the loop of Henlé, the greater the concentration that can be built up. In humans, about one-third of our nephrons have long loops

3 The **descending limb** is permeable to water, so water passes down the water potential gradient by osmosis and into the blood ⟶

1 The **ascending limb** is not permeable to water – the cells of this area actively secrete sodium and chloride ions into the tissue fluid between the two limbs

2 The **interstitial region**. The tissue fluid here contains a high concentration of sodium and chloride ions – a low water potential

**Figure 5** The loop of Henlé

**A Californian woman has died after taking part in a water-drinking contest, but why is too much water dangerous?**

The death of a woman in the US after taking part in a water-drinking contest shows you can have too much of a good thing. Jennifer Strange had taken part in the 'Hold Your Wee for a Wii' game, which promised the winner a Nintendo Wii. Afterwards she reportedly said her head was hurting and went home, where she was later found dead. Initial tests have shown her death is consistent with water intoxication.

If you drink too much water, eventually the kidneys will not be able to work fast enough to remove sufficient amounts from the body, so the blood becomes more dilute with low salt concentrations. The water then moves from the dilute blood to the cells and organs where there is less water.

Drinking too much water can eventually cause your brain to swell. The brain is inside a bony box (cranium) so has nowhere to go. The pressure increases in the skull, stopping it regulating vital functions such as breathing. Eventually you are likely to stop breathing and die. Those most at risk include people taking ecstasy, as the drug increases thirst and facilitates the release of anti-diuretic hormones so more water is taken in, but cannot be excreted.

But we lose water all the time, so we need to replace it. In normal circumstances we should aim to drink about one and a half litres every day. Dehydration can lead to bad breath, tiredness and a higher risk of bladder infections.

## Control of water balance by the kidney, hypothalamus and pituitary gland

In spread 5.3.1.1 we discovered the importance of **homeostasis** in maintaining a stable internal environment. The kidneys contribute to several homeostatic mechanisms, but one of the most important is the regulation of water and solute concentrations (**osmoregulation**). This involves a **negative feedback loop** consisting of a **receptor**, which monitors the amount of water (the receptor cells in the **hypothalamus**) and an **effector** that brings the water content back to normal when it is necessary (the *pituitary gland* and the *walls of the distal convoluted tubule*).

### Regulation of the water potential of the blood

Cells called **osmoreceptors**, which are in the hypothalamus of the brain, monitor the water potential of our blood. These cells detect a fall in the **water potential** of the blood flowing through the hypothalamus. It is still uncertain how this works, but we think that water is lost from these cells by **osmosis** when the water potential of the blood is low. This causes the osmoreceptor cells to shrink and it is this change which stimulates the **neurosensory cells** in the hypothalamus to secrete **antidiuretic hormone** (ADH). ADH passes along the axons of the neurosensory cells. These cells end in the **posterior pituitary gland**, which then releases ADH into the blood.

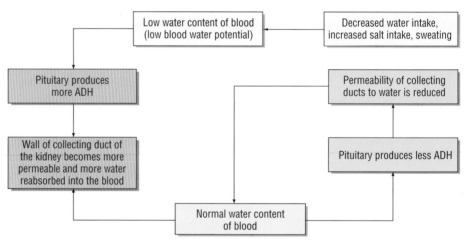

**Figure 1** The control of water balance

## How ADH affects the kidneys

- ADH travels in the blood to the kidneys. Here it increases the permeability to water of the plasma membranes of the endothelial cells in the walls of the collecting duct.
- Receptors on the plasma membrane pick up the ADH molecules, which cause an increase in **cyclic AMP** (cAMP) within the cell. This acts as a second messenger.
- cAMP leads to the insertion of **aquaporins** into the collecting duct plasma membrane.
- Aquaporins in the membrane increase the permeability to water.
- The fluid in the collecting duct has become more concentrated so a smaller quantity of more concentrated urine is produced.
- The osmoreceptors in the hypothalamus detect the result – the rise in water potential – so they stop sending impulses to the pituitary gland, which reduces the release of ADH.

### What are aquaporins?

Aquaporins are membrane pore proteins (intrinsic protein). They form pores through the **phospholipid bilayer**. They selectively conduct water molecules in and out, while preventing the passage of ions and other solutes. They are also known as water channels.

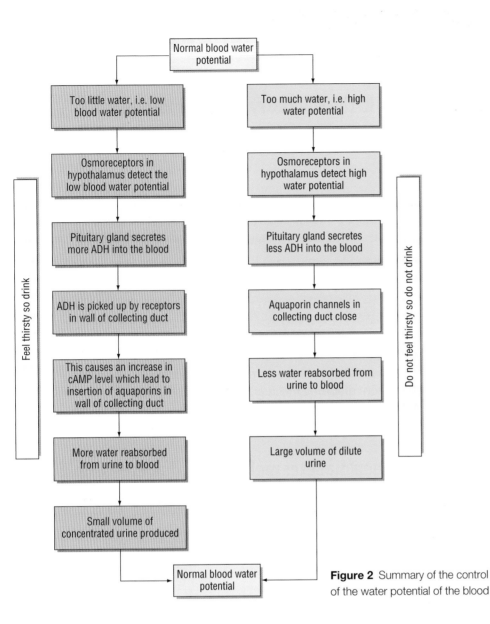

**Figure 2** Summary of the control of the water potential of the blood

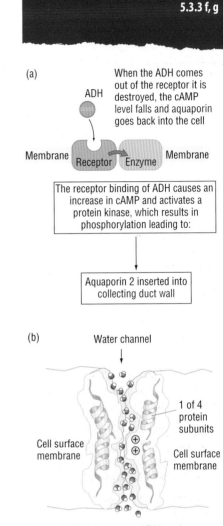

(a) When the ADH comes out of the receptor it is destroyed, the cAMP level falls and aquaporin goes back into the cell

ADH

Membrane  Receptor  Enzyme  Membrane

The receptor binding of ADH causes an increase in cAMP and activates a protein kinase, which results in phosphorylation leading to:

Aquaporin 2 inserted into collecting duct wall

(b)  Water channel

1 of 4 protein subunits

Cell surface membrane

Cell surface membrane

**Figure 3** Link between ADH and aquaporins

## Questions

1  Which gland contains cells that act as osmoreceptors?
2  ADH makes the walls of the collecting duct more permeable to water: true or false?
3  If the water concentration in the blood decreases:
   **(a)** will the reabsorption of water increase or decrease?
   **(b)** will the volume of urine produced increase or decrease?
4  A student drank 1000 cm³ of distilled water. Her urine was collected immediately before she drank the water and every 30 minutes for the next 3 hours.

| Time/mins | Vol of urine/cm³ |
|-----------|------------------|
| 0 | 100 |
| 30 | 198 |
| 60 | 320 |
| 90 | 210 |
| 120 | 150 |
| 150 | 100 |
| 180 | 60 |

**(a)** Describe the changes in urine output during the 3 hours.
**(b)** What process is responsible for what you have described.
**(c)** Explain the difference in volume of urine collected at 60 minutes and at 120 minutes.

### Can a hangover be explained?

If you drink too much alcohol you are likely to wake up with a 'hangover', which is due to dehydration rather than to the toxic effects of the alcohol. Alcohol inhibits the production of ADH, so you produce lots of urine and your blood becomes very concentrated. The 'hangover' feeling is caused by the effect of the very concentrated blood on your brain cells. If you know that you have drunk too much alcohol, have a long drink of water before going to bed.

One Clinistix which is dipped into the urine sample

The stick is coated with a mixture of the enzymes glucose oxidase and peroxidase and a blue chromogen dye

The glucose oxidase catalyses the reaction between glucose and oxygen, producing gluconic acid and hydrogen peroxide. The test is based on the specificity of this enzyme. The peroxidase then catalyses the reaction between hydrogen peroxide and the blue chromogen dye. The dye changes from:

Dark pink ← Light purple ← Dark purple

Increasing glucose content →

**Figure 1** Clinistix test for glucose

## What do changes in the chemistry of urine indicate?

Doctors can learn a lot about us from our urine. For example, they can tell whether we have kidney disease or diabetes, whether we are pregnant or if we have been abusing drugs. A urine sample is taken midstream to avoid collecting any contaminants present in the urethra.

Urine analysis is thought to be the oldest clinical chemistry test: in medieval times this was one of the very few tools that a physician could use to diagnose an illness. Sometimes physicians even tasted the urine, because sweetness would indicate diabetes!

These days, however, test strips are used to test for glucose in the urine. The strip is dipped into the urine sample and the colour change indicates the concentration of glucose present.

Normally there is no glucose in the urine because it is reabsorbed by active uptake in the kidney tubule. Glucose enters the urine only when the level in the blood reaches above 180 mg/dL. At this point the concentration of glucose in the blood is so high that no more glucose can be reabsorbed in the proximal convoluted tubule. This indicates that the patient is suffering from **diabetes mellitus**. Urine can also be tested to see if a patient is suffering from **diabetes insipidus**. In this case, the patient will excrete large volumes of dilute urine, even when the blood water level is low.

## Differences between diabetes mellitus and diabetes insipidus

| | Diabetes mellitus ('sugar' diabetes) | Diabetes insipidus ('water' diabetes) |
|---|---|---|
| What it is | Inability to control blood glucose levels | Cannot control the balance of water in the body – may pass over 3 litres of urine in 24 hours |
| Test | Use a test strip, e.g. Clinistix, to test glucose levels in urine | Dilute urine with a low specific gravity. Specific gravity reflects the concentration of substances in the urine |
| Causes | Type I diabetes: the pancreas does not produce sufficient insulin<br><br>Type II diabetes: the pancreas does not secrete enough insulin and the body cannot use it properly | Damage to hypothalamus so malfunction in ADH secretions<br><br>Genetic disease; caused by a mutation in one of the genes coding for the production of **aquaporins** or for the **receptors** on the collecting duct membrane |
| Treatment | Type I diabetics use insulin injections<br><br>Type II diabetics careful control of diet and possibly insulin injections | Give the patient synthetic ADH<br><br>Give the patient a low sodium diet and medication to reduce the amount of urine produced |

**Table 1** Comparison of diabetes mellitus and diabetes insipidus

### Other homeostatic mechanisms in the kidney

- **Renin (angiotensinogenase)** is a circulating enzyme which is released by the kidney in response to low blood volume. It raises blood pressure by constricting blood vessels, increasing the secretion of ADH and **aldosterone**, and stimulates the **hypothalamus** to activate the thirst reflex.
- **Erythropoietin** is a hormone produced by the kidneys, which stimulates the body to produce more red blood cells (**erythrocytes**) in the bone marrow when oxygen levels in the blood are low.

### Examiner tip

**Aldosterone** is a hormone released by the adrenal glands. Its function is to regulate blood pressure. It increases the reabsorption of sodium and water along with the excretion of potassium in the distal tubules of the kidneys. This action raises blood pressure.

### What do changing levels of renin and erythropoietin indicate?

The production of renin increases in diseased kidneys because it is produced in response to a low blood pressure. A low blood pressure will result in less hydrostatic pressure in the glomerulus. This will decrease ultrafiltration and so results in poor kidney function.

Cancers and cysts in a kidney may result in raised levels of erythropoietin production. A damaged kidney exhibits low oxygenation levels and this would trigger the production of erythropoietin. But the usual response in a diseased kidney is for erythropoietin levels to fall, resulting in anaemia.

## Kidney failure

When the kidneys start to fail, toxins, including urea and other nitrogenous wastes, are not filtered out of the blood, and so start to accumulate in the tissues.

**Acute renal failure** may occur as a result of:
- a serious illness or operation, particularly those complicated with severe infection
- a sudden loss of large amounts of fluid (blood or tissue fluid)
- a sudden blockage to the drainage of the urine from the kidney by, for example, a kidney stone
- a side effect of some medications
- bacterial infection in the kidneys.

Most cases of acute renal failure can be treated and the kidney function will then, in time, return to normal.

**Chronic renal failure** may be caused:
- as a complication of diabetes mellitus
- by uncontrolled high blood pressure
- by an inflammation affecting the kidney tissue
- as a result of certain inherited diseases, such as polycystic kidney disease, etc.

Often, the cause has occurred many years earlier and can no longer be clearly identified. If it is not noticed at an early stage, chronic kidney damage is usually not reversible.

### Diagnosing kidney failure

The most characteristic symptoms of kidney failure are a reduction in the volume of urine and **oedema** (excessive accumulation of fluid in the tissues). As a result, the feet, hands and the area around the eyes may swell and become puffy. The urine may be cloudy or bloody in appearance. However, you will remember that in diabetes insipidus the urine cannot be concentrated, so a large amount of dilute urine is produced.

The most common diagnostic tool is analysis of blood and urine samples.

*Urine samples* may show erythrocytes (red blood cells), leucocytes (white blood cells) and proteins are present. A patient with diabetes or high blood pressure should have the amount of protein present in his urine checked at least once a year.

*Blood samples* may show high levels of **creatinine** (a waste substance that is found in the bloodstream) Creatinine levels increase as kidney disease progresses. A blood sample can also be analysed for estimated *glomerular filtration rate* (GFR), which is a measure of how well the kidneys are removing wastes and excess fluid from the blood. It may be calculated from the **serum** creatinine level, allowing for age, weight, gender and body size. The lower the GFR, the less effective the kidney function. There may also be an increase in the levels of potassium ions.

**Ultrasound scans** and **CT scans** of the kidneys will identify any structural abnormalities or blockage (such as kidney stones) of the kidneys.

**Key definition**

The **haematocrit** measures how much space in the blood is occupied by red blood cells. It measures the ratio of the volume occupied by packed red blood cells to the volume of the whole blood. Check 1.1.1.2 and the use of the haemocytometer. Think about the advantages and disadvantages of each technique.

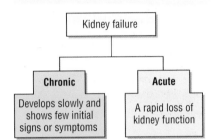

**Figure 2** Two main types of kidney failure

Studies have found that cardiovascular disease is independently linked with the development of kidney disease and vice versa: kidney disease is an independent risk factor for cardiovascular disease. The increase in blood pressure caused by the increased production of renin is a risk factor in cardiovascular disease. Renin acts by causing vasoconstriction and in this way leads to *hypertension*. Increase in erythropoietin production leads to an increase in *haematocrit* values, which in turn leads to a reduction in heart rate.

## Questions

1 State two causes of kidney failure.
2 When a person has kidney failure why must the amount of sodium chloride (salt) in their diet be carefully controlled?
3 What is the role of erythropoietin?
4 Explain why a blood sample is taken when a doctor suspects a diseased kidney.

Many years ago kidney failure was inevitably fatal, but modern medicine offers three major treatment options: haemodialysis, peritoneal dialysis and kidney transplant. In **haemodialysis** an artificial kidney machine carries out the functions the kidneys can no longer perform. The patient is connected to the machine by plastic tubing that attaches to special blood vessels in the arm or leg. The treatment can be carried out at home or at a dialysis unit in a hospital. In **peritoneal dialysis**, waste products from the blood are flushed from the body with fluid instilled and drained through a catheter that has been surgically placed in the abdomen. Once the catheter is in place, this technique is usually done at home.

## Haemodialysis

Haemodialysis uses a special filter called a *dialyser* that functions as an artificial kidney.

Since its introduction in the 1960s the composition of the dialyser and dialysate has improved, but the main design has remained the same.

Arterial blood from the patient is usually taken via a *shunt* from a vein in the forearm and is pumped through the dialyser. The dialyser contains partially permeable plastic filters which are known as artificial capillaries. While blood flows inside these capillaries, the dialysate, a special fluid, flows around the outside in the opposite direction. In dialysis molecules are exchanged between the dialysate and the blood. The composition of the dialysate is carefully controlled so that there is net movement of urea, water, glucose and salts such as potassium, sodium and chloride out of the blood. Blood cells and protein molecules are too large to pass through the membrane so they remain in the blood. The blood is returned to a vein in the arm.

Only a small amount of blood (about 200 cm³) is outside the body at any one time.

If the proper diet is followed, most of the side effects such as muscle cramps and hypotension (a sudden drop in blood pressure) can be avoided. The diet should contain balanced amounts of food high in protein such as meat and chicken. Salty foods should be avoided. The amount of potassium (found in salt substitutes, some fruits, chocolate and nuts) should be controlled. The patient also needs to limit the amount he drinks.

Haemodialysis can be carried out in a special unit in a hospital or at home.

(a)

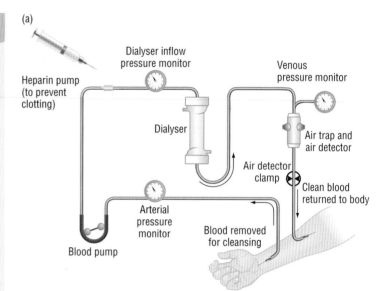

(b) Blood inlet via a pump, which keeps the blood moving through the dialyser and at a higher pressure than the dialysate

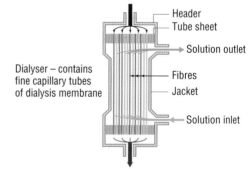

Blood flows through the middle of the artifical capillaries, while the dialysing fluid flows along the outside in the opposite direction

**Figure 1** Haemodialysis

|  | Hospital treatment | Home treatment |
|---|---|---|
| Pros | Trained professionals present, can get to know other patients | Do not have to travel (so less tiring), can do it at the hours you choose, gain a sense of independence |
| Cons | Must travel to the hospital for treatment | Helping with the treatments may be stressful to the family, need training, need space for storing machine and supplies |

**Table 1** Hospital dialysis versus home dialysis

## Peritoneal dialysis

This type of dialysis uses the lining of the abdomen (the peritoneal membrane) to filter the blood. Peritoneal dialysis works on the principle that the peritoneal membrane that surrounds the intestine can act as a partially permeable membrane. If a specially formulated dialysis fluid is instilled around the membrane then dialysis can occur by diffusion. Excess water can be removed by **osmosis** by altering the concentration of glucose in the fluid. Dialysis fluid is instilled via a catheter, which is placed in the patient's peritoneal cavity through a stoma (hole) near their navel. Peritoneal dialysis can be done in the patient's home, workplace or almost anywhere. All that is needed is a clean area, a way to elevate the dialysis bag and a method of warming the fluid.

There are two main types of peritoneal dialysis:

**Continuous ambulatory peritoneal dialysis (CAPD),** the most common type. Exchanges of fluid are done throughout the day, usually four exchanges a day. The dialysis solution passes from a plastic bag through the catheter and into the peritoneal cavity, where it stays for 4–6 hours with the catheter sealed. The time period that the dialysis solution is in the abdomen is called the *dwell time*. Next, the dialysis solution is drained into an empty bag for disposal. The abdomen is then refilled with fresh dialysis solution.

**Continuous cyclic peritoneal dialysis (CCPD)** uses a machine called a *cycler* to fill and empty the peritoneal cavity 3–5 times during the night. In the morning an exchange is started with a dwell time that usually lasts the entire day.

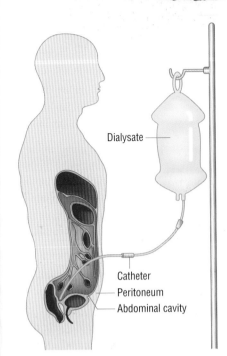

**Figure 2** Peritoneal dialysis

| Advantages of peritoneal dialysis | Disadvantages of peritoneal dialysis |
| --- | --- |
| Can be done at home | Needs attention to cleanliness while performing exchanges |
| Relatively easy to learn | |
| Easy to travel with, bags of solution are easy to take on holiday | Possible complications such as infection where the catheter enters or even infection of the peritoneum (peritonitis) |
| The diet can contain more fluids and salts, although must still be limited | Must cut back on number of calories in the diet as there are calories in the dialysis fluid |

**Table 2** Advantages and disadvantages of peritoneal dialysis

**Examiner tip**

Remember this is a synoptic module, so review the role of EPO from 4.1.2.4.

**Examiner tip**

Remember this is a synoptic module so refresh your memory about blood clotting (1.1.3.1).

### Use of recombinant erythropoietin in kidney failure and dialysis

In kidney failure the production of **erythropoietin** is affected and new **erythrocytes** (red blood cells) are not made at the right rate. This leads to anaemia, a complication of kidney failure. Another problem is blood loss. Blood can be lost as residual blood in the dialyser or blood lines, through frequent blood testing or through a dialyser catheter when blood is discarded routinely at the start of dialysis. These problems can be overcome by giving the patient recombinant (genetically engineered) erythropoietin.

## Questions

1 In haemodialysis, why do blood and dialysate flow in opposite directions?
2 In haemodialysis, why is heparin added to the blood?
3 In haemodialysis, why is heparin not given in the last hour of treatment?
4 In continuous ambulatory peritoneal dialysis, why would it be no use to leave the dialysate inside the peritoneal cavity for longer than the recommended dwell time?

## Dialysis is not a cure

**Haemodialysis** and **peritoneal dialysis** are treatments that help to replace the work that the kidneys did. These treatments help the patient to feel better and live longer, but they do not cure kidney failure.

## Kidney transplants

Kidney transplantation surgically places a healthy kidney from another person into the patient's body. The transplanted kidney does the work that the two failed kidneys used to do.

The surgeon places the kidney in the lower abdomen and connects the artery and vein of the new kidney to the patient's artery and vein. Unless they are causing infection or high blood pressure, the failed kidneys are left in place.

If a medical evaluation shows that you are a good candidate for a transplant but you do not have a family member who can donate a kidney, you will be placed on the transplant waiting list to receive a kidney from someone who has just died, a *deceased donor*. When a deceased donor kidney becomes available it is given to a suitable candidate. Suitability is judged on two factors:

- Blood type (A, B, AB or O) – must be compatible with the donor
- HLA factors – from a previous spread (5.1.4.1). In this module you will remember that you inherit a **haplotype** from your mother and one from your father. A higher number of matching **antigens** increases the chance that your kidney will last a long time.

Generally speaking, there is a 95% chance that a kidney transplant from a living donor will be functioning two years after the transplant. A deceased donor kidney transplant will usually continue to function in 85% of patients after the one-year mark.

### Examiner tip

Look back over immunity (2.3.2.1) and the role of the Human Genome project (5.1.2.3).

### A case study

Susan is being treated for a serious kidney disease. She is currently on a dialysis machine, but treatment is steadily less effective. Before her condition declines any further, the doctor suggests family members undergo tests to determine tissue compatibility, in order to transplant a kidney. Only the brother shows a degree of compatibility high enough to be considered a candidate. The doctor meets the brother alone to discuss the risks and benefits of the operation. Although agreeing to be tested, the brother decides not to donate a kidney after weighing up the various alternatives because of the risks, and because, as he puts it, he doesn't 'feel he and his sister have ever been close enough that they would ever take that kind of risk for each other'. The doctor repeats a full explanation of the risks involved, and urges him to rethink his decision because of the serious nature of his sister's illness with increasingly little time to spare. The brother remains adamant in his refusal. What should the doctor tell his kidney patient?

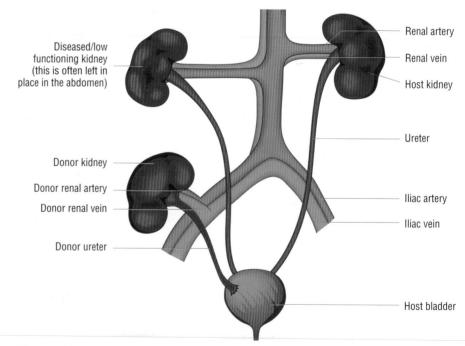

**Figure 1** Kidney transplantation

## What are the advantages and disadvantages of kidney transplantation?

| Advantages | Disadvantages |
|---|---|
| Freedom from dialysis and freedom from the time commitment it takes | Need to take medicines (immunosuppressive drugs) for the life of the kidney |
| Increased energy | Need to undergo a major surgical procedure under a general anaesthetic |
| Diet is less limited | Risks associated with the surgery include infection, bleeding and damage to surrounding organs |
| Feeling better physically | Continual monitoring for signs of organ rejection |
| A better quality of life | Side-effects of the medicines – the anti-rejection medicines cause fluid retention and high blood pressure; immunosuppressants can increase your susceptibility to infections |
| No longer seeing themselves as chronically ill | |

**Table 1** Advantages and disadvantages of kidney transplantation

### Some points for you to consider:
- Only 1 in 4 people sign the donor register. England's Chief Medical Officer has suggested 'opting out, not opting in', but this would take away an individual's fundamental right to decide. Most European countries use the system of 'presumed consent'.
- Doctors have to approach family members who have just watched someone die and then ask them if they can have their body parts. Perhaps having an opt-in system is emotionally kinder on the relatives – it would save them from the shock of a doctor approaching them.
- It's not the highest priority in everybody's life to think about when they're going to die and what's going to happen to them afterwards – it's not something you go around thinking about day in, day out.
- Some people argue that the supply problem could be solved if supplemented by a legal, regulated market for human transplant organs purchased from live donors.
- Research is continuing into growing new organs to take the place of damaged or diseased ones. Scientists have previously shown that embryonic tissue transplants can be used to grow new kidneys inside rats. Is this a very important first step for finding out how to grow new kidneys in humans?
- A new technique has been developed in the United States that allows the transplant of a kidney from a living donor who does not have a matching blood group. This involves washing the blood of the recipient three times a week for 3 months to remove all the antibodies that would damage the transplanted kidney. Do you think that this will prove to be a successful alternative?

## Questions
1 Outline some ideas about what could be done to encourage more people to become organ donors.
2 What are the ethical arguments against 'opting out, not opting in'?
3 What are the arguments against selling organs from living donors?

### Ethical issues involved in kidney transplants

On average, patients who are listed for a deceased donor kidney transplant wait approximately 3 years. Kidneys are allocated based on, among other considerations, the match between the donor and the recipient blood groups and the genetic or tissue type (HLA type). The length of time that a patient waits for a transplant will depend on how often donors with kidneys that are well matched with the patient become available. This means that someone with a rare tissue type is likely to have to wait longer than someone with a more common tissue type. Every person waiting for a deceased donor kidney is registered with the Organ Donor Register.

The number of people waiting for organ transplant has reached a record high, statistics reveal.

There has been an increase in demand for kidney transplantations because the number of people with severe kidney disease is increasing in all western societies. But the number of kidneys available in the UK has not really changed in the last 10 years, so the waiting list has grown.

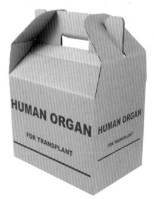

**Figure 2** A human kidney being transported for transplant

# Summary and practice questions

1  What do you understand by the term 'homeostasis'?

2  Change from a set point is detected by a ..................... and an ..................... responds to bring the system back to the set point. An example of homeostasis is control of temperature. When the body is too cold blood vessels in the skin ..................... . This is called ..................... . When we are too hot the blood vessels ..................... . Heat is now lost to the air by ..................... and ..................... . We also sweat. This cools the skin when the sweat ..................... . Longer term temperature control is brought about by the hormone ..................... .

3  What is the difference between hypothermia and hyperthermia?

4  The islets of Langerhans are an ..................... gland found in the ..................... . The alpha cells secrete ..................... . After a meal containing carbohydrate ..................... detect the rise in glucose. The ..................... secretes insulin which stimulates the conversion of glucose into ...... ..................... and ..................... the use of glucose in respiration. There are ............... forms of sugar diabetes, Diabetes mellitus. Type ...... is insulin dependent and normally begins in childhood. Type 2 usually starts in people over ..................... .

5  List three risk factors for developing Type 2 diabetes.

6  What is the difference between excretion and elimination?

7  Ammonia is formed when the ..................... group is removed from an amino acid. Ammonia is immediately changed into urea and excreted. The human kidney consists of around one million tiny tubules or ..................... . The two stages of urine production by the nephrons are ultrafiltration and ..................... . The four distinct regions of a nephron are the renal capsule, the ....................., the ..................... and the ..................... . The podocytes are found in the ..................... . The afferent arteriole in the glomerulus has a ...... ..................... diameter than the efferent arteriole therefore ..................... pressure.

8  Where does most of the selective reabsorption occur?

9  Where there is a difference in water potential, water passes out of the tubule by ..................... .

10  ADH stands for ..................... . ADH is secreted by ..................... ..................... in response to a fall in the water potential of the blood which is detected by the ..................... in the ..................... . ADH travels in the ...... ..................... to the kidneys where it increases the permeability to water of the cells in the ..................... ..................... so the urine becomes ..................... concentrated. In Diabetes insipidus the patient cannot control the balance of ..................... in the body.

11  Name two hormones produced by the kidneys.

**12** Treatment for kidney failure:- devise a plan to show the pros and cons of each of the present day treatment for kidney failure.

For example:

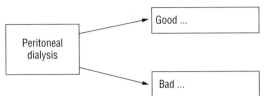

**13** The flowchart below summarises the homeostatic regulation of water potential of human blood.

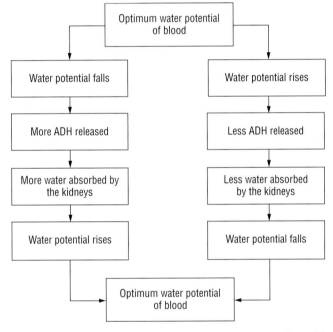

**(a)** Where is the change in water potential detected? [1]

**(b)** How is ADH carried to the kidneys? [1]

**(c)** Explain why the above is an example of negative feedback. [2]

**(d)** Alcohol inhibits the release of ADH. Suggest how the volume of urine produced would change if a person drank a lot of alcohol. [2]

**14** State three differences between nervous and hormonal control in a human. [3]

**15** Explain how a cold environmental factor such as that in the Antarctic increases thyroxine production.

**16** The table below shows the composition of fluid in the kidney.

| Component | Concentration/g 100cm3 | | |
|---|---|---|---|
| | Blood plasma entering glomerulus | Filtrate in renal capsule | Urine in collecting duct |
| Water | 90–93 | 97–99 | 96 |
| Proteins | 8 | 0.0 | 0.0 |
| Glucose | 0.1 | 0.1 | 0.0 |
| Urea | 0.03 | 0.03 | 2.0 |

**(a)** Why is there no protein in the filtrate in the renal capsule? [1]

**(b)** The concentration of glucose is the same in the blood plasma entering the glucose as it is in the filtrate in the renal capsule but there is no glucose present in urine. Why? [2]

**(c)** Explain why the concentration of urea is greater in the urine than it is in the filtrate. [1]

## The ageing reproductive system

**Why do we age?** This question has baffled scientists for nearly 150 years and remains a mystery. Many people are surprised to find that in the twenty-first century there is still major scientific disagreement over such a fundamental issue. Most medical advances have been the result of experimentation. But ageing, because it is a relatively long-term process, is a difficult subject for experimentation. An experiment to determine if a particular drug increases longevity in humans could take years or even many decades to perform.

**What is ageing?** Ageing refers to deterioration over time, including increased weakness, increased susceptibility to disease and adverse environmental conditions, loss of mobility and agility, and age-related physiological changes. Reduced reproductive capacity is also usually considered to be part of ageing. This includes many subtle but important changes in the *reproductive process for women*. The impact of age on fertility is becoming increasingly more significant as more women choose to delay childbearing until later in life. Today as many as 50% of the patients requesting infertility treatment face the problem of reproductive ageing.

It is important for all women and their partners to understand that a woman's reproductive potential declines with age (Figure 1). Most women are not even aware that this decline begins around the age of 30. Even though a woman may continue to have regular **menstrual cycles** until she reaches **menopause**, the potential to have children may be lost several years before menopause. Nearly one-third of couples that include a woman aged 35 or older will have problems conceiving. Fewer than 30% of women aged 40 and over are able to conceive naturally. In addition, a woman's chance of having a miscarriage also increases with age. The main factor associated with infertility in women is the quality of the **oocyte**. This declines as women age. Furthermore, **FSH** levels increase as a woman ages. Women with abnormal FSH levels often have difficulty conceiving and if a conception occurs there is an increased chance of a miscarriage.

Around menopause changes in the reproductive organs and genitalia occur rapidly. Menstrual cycles stop, and the ovaries stop producing oestrogen. After menopause, there is a thinning or *atrophy* of tissues in the labia minora, clitoris, vagina, and urethra. This can result in chronic irritation, dryness, and a discharge from the vagina. Vaginal infections are more likely to develop. Also, after menopause the uterus, Fallopian tubes and ovaries become smaller.

Age also has an impact on *male fertility*, though the result is not as severe as it is in women. As men age, the number of sperm, the motility of sperm and the percentage of normal sperm all decrease slightly. When the male partner is 50 or above, pregnancy and birth rates decline and miscarriage rates increase. Despite the fact that some men can become fathers at the age of 80 or older, couples should consider the effects of ageing in both partners when planning to become parents.

It is not clear whether it is ageing itself or the diseases associated with ageing that cause the gradual changes that occur in male sexual function. The frequency, length, and rigidity of erections gradually decline with increasing age. Levels of the male sex hormone, **testosterone**, decrease also, reducing sex drive (**libido**). Bloodflow to the penis decreases. Other changes include decreases in the sensitivity of the penis and in the volume of the **ejaculate**.

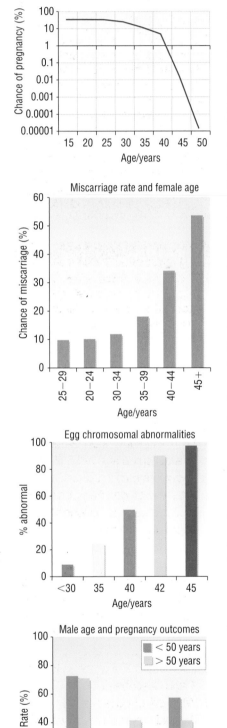

**Figure 1** The effect of ageing on the reproductive process

# Changes in physiology associated with the menopause

Menopause is a normal event in a woman's life cycle. Her periods stop and she is no longer able to conceive. It is often called the 'change of life'. During menopause, a woman's body slowly makes less oestrogen and **progesterone**. This often happens between the ages of 45 and 55 but for a few women it can start as early as 35 or as late as 60. It is interesting to note that as life expectancy for women is now 80 years, one-third of a woman's life occurs after menopause. Women go through different phases of menopause, including the peri-menopausal, menopausal, and post-menopausal periods. During the *peri-menopausal period*, the regular cycle of menstruation is disrupted and menstruation becomes irregular. This phase may last anywhere from 6 months to a year. During the peri-menopausal period, oestrogen production decreases and eventually stops. Menopause is defined as the cessation (ending) of the menstrual period. Women are described as *post-menopausal* when they have gone 1 year without a menstrual period.

The lack of oestrogen and progesterone causes many changes in a woman's physiology that affect her health and wellbeing. These changes include:

- Elevated levels of total **cholesterol** and **LDL cholesterol**. This increases the risk of **coronary heart disease** (CHD) in women. During the reproductive years, oestrogen prevents increased levels of blood cholesterol and maintains the activity of oestrogen receptors in women, thus preventing the risk of CHD.
- *Calcium* loss from the bones increases in the first 5 years after the onset of menopause. This results in a loss of bone density. Bone loss then tapers off until about the age of 75, when calcium loss accelerates again. This makes older women more likely to suffer from **osteoporosis** and bone fractures.
- The body composition of menopausal women also changes, with the percentage of body fat increasing and muscle mass decreasing. The increase in body-fat percentage is believed to be partly due to decreased physical activity.

Some other transient but unpleasant symptoms of menopause include hot flushes, fatigue, anxiety, sleep disturbance and memory loss.

## The use of HRT (hormone replacement therapy)

HRT is often used to treat the symptoms of the menopause. The therapy is designed to increase the declining levels of the female hormone, oestrogen.

### Benefits
HRT is the most effective treatment for the relief of symptoms such as hot flushes and sleep disturbance that can both affect physical and mental health. As well as treating menopausal symptoms, HRT was believed to be beneficial in reducing the risk of heart disease and bone fractures caused by osteoporosis (thinning of the bones).

### Which HRT?
- Women with a uterus need oestrogen and progestin (synthetic progesterone). Progestin counteracts the effect of oestrogen and against cancer of the uterus.
- An oestrogen-only pill has been launched for women under 50 who face a premature menopause because of womb surgery. This can trigger an early menopause even if the ovaries are retained. The new low-dose pill, Premarin, contains 0.3 mg of oestrogen, less than half the standard HRT preparation of 0.625 mg. The hormone combats symptoms such as hot flushes and night sweats. Women will still get benefit from reduced risk of side-effects such as thrombosis and breast tenderness. It also combats bone thinning, which affects women after the menopause and can lead to osteoporosis and broken bones.

### Summary
Safety concerns rose in 2002, when the US Women's Health Initiative study claimed patients using HRT were at higher risk of heart disease and strokes – contrary to previous research that suggested oestrogen could protect them from heart problems. But earlier this year re-analysis of the data found the extra risks may apply only to patients in their 60s and 70s who do not normally use HRT. The key to deciding to take HRT is to weigh the risks associated with taking it against a particular woman's risk of heart disease or osteoporosis without taking HRT. HRT remains an effective treatment for the short-term relief of menopausal symptoms such as hot flushes, night sweats and mood swings.

## Case study

According to a 2006 study in the *Proceedings of the National Academy of Sciences*, the genetic quality of sperm declines as a man ages. Previous studies have indicated that ageing negatively affects sperm counts and sperm motility (the sperm's ability to move). This new research suggests that poor sperm motility is linked to DNA fragmentation, which increases the chances for male infertility and the likelihood of fathering children with genetic problems. Researchers found that the rare genetic mutations associated with dwarfism increased by 2% with every year of a man's age.

## Examiner tip

Do not confuse osteoporosis with **osteomalacia** or 'soft bones', which is associated with problems with vitamin D metabolism

## Questions

1 Women usually stop ovulating at around the age of 50. This is known as the ................. . After this stage of a woman's life ................. is no longer secreted from the ................. . No ................. are produced, making the woman infertile. The lack of ................. may cause hot flushes, vaginal dryness and a loss of bone density that can lead to ................. .

2 Look at the first graph in Figure 1 and describe it.

3 What effect does ageing have on the male reproductive system?

4 HRT involves providing the body with oestrogen, usually combined with artificial progesterone. What is the main benefit of a low dose of HRT given for a limited time?

## Cyclic and continual HRT therapy

### Types of HRT

- Combined HRT: tablets and patches are forms of combined (**oestrogen** and **progesterone** together) HRT. Using combined HRT usually involves taking a form of oestrogen continuously and progesterone for 10–14 days of every month. When the progesterone is finished a period or withdrawal bleed will occur, but these are often lighter than the natural menstrual periods previously experienced.
- Continuous Combined HRT: continuous combined HRT involves taking both hormones every day without breaks. In theory, therefore, no monthly bleeds should occur. However, it may be 2–3 months before they disappear. In younger women spotting or bleeding may be a persistent problem, and they may prefer the non-continuous regime with its regular withdrawal.
- Cyclic hormone therapy is often recommended. With this therapy, oestrogen is taken in pill form for 25 days, with progestin added somewhere between days 10–14. The oestrogen and progestin are taken together for the remainder of the 25 days. Then no pills are taken for 3–5 days. There will be monthly bleeding with cyclic therapy.

**Figure 1** Tablets are the most commonly prescribed form of HRT

HRT is also available as a *patch* applied to the abdomen or the thigh. This patch allows the oestrogen to be absorbed through the skin into the bloodstream and it does not have to be broken down by the liver.

Some women prefer this method because they do not have to take pills. These preparations are often favoured for women with diabetes, epilepsy, gallstones or kidney disease. The patches are changed once or twice a week on regular days. Patches allow oestrogen to be slowly absorbed through the skin, providing a more gradual and, therefore, natural release than a tablet.

Vaginal cream containing oestrogen may be given to women for vaginal dryness. The cream is usually given along with one of the other forms of HRT because the cream may not relieve many of the other symptoms and does not appear to protect against bone disease.

**Mirena** is an alternative method of delivering the progesterone part of HRT. It is an **intrauterine system** (a type of IUD) that contains a small amount of progestin. Mirena works just inside the womb rather than affecting the whole body, and may be a safer way of getting progesterone.

### Examiner tip

Remember, oestrogen is a **steroid** hormone. It would be useful to remind yourselves of the properties of steroids, especially their lipid solubility. This explains why they can cross the cell membranes of skin cells and be absorbed.

## Side effects of taking HRT

HRT can cause **side effects** such as feeling or being sick, cramps in the abdomen, bloating, weight changes, tenderness and enlargement of the breasts, and changes in liver function, rashes and pigment changes in the skin. PMS-type symptoms can also occur, along with fluid retention.

There are other more serious but rare side effects:
- There is a slightly increased risk of **breast cancer** in women taking HRT. For example, about 45 in every 1000 women aged 50 years not using HRT will develop breast cancer during the following 20 years. This is the 'baseline' risk. In women using HRT for 5 years, this figure rises to 47 in every 1000 women. In those using HRT for 10 years there would be an extra 6 cases – or 51 in every 1000. And for women using HRT for 15 years, an extra 12 cases or 57 per 1000 women could be expected to develop breast cancer in the following 20 years. This increased risk seems to disappear within about 5 years of stopping HRT.
- There is also an increased risk of **endometrial cancer**. However, the inclusion of progestin in combined HRT lessens this risk.
- In April 2007 scientists at Cancer Research UK reported that a link between HRT and **ovarian cancer** may have led to the death of 1000 women in 14 years.

- HRT also increases the chances of having a **blood clot** in the legs (a *deep vein thrombosis*) or lungs (a *pulmonary embolism*). This means women on HRT are also at increased risk of having a blood clot on a long-haul plane flight, or in other situations where they are in prolonged cramped conditions.
- Slightly increased risk of **heart attack** or **stroke**, both in women with **cardiovascular disease** and in healthy women. This effect may be associated primarily with HRT using continuous combined oral oestrogen and progestin (Prempro) or the daily use of an oral progestin such as Provera with oestrogen.
- **Osteoporosis** HRT reduces hip fractures by reducing the normal rate of loss of bone density. For example, in women aged 50–59 there are 1–2 cases of hip fracture for every 1000 non-HRT users, compared to 0–1 cases for every 1000 HRT users.
- HRT may reduce a woman's risk of **Alzheimer's disease** if begun before the age of 65. New research has found that younger HRT users were 50% less likely to develop Alzheimer's or other forms of **dementia** than women who did not have the treatment. But it was a very different story for women taking HRT after 65 – their risk of dementia was dramatically increased, in some cases doubled.

## Alternative methods of treating the menopause

A large clinical trial found that combined hormone replacement therapy (HRT) increased the risk of invasive breast cancer, strokes, heart attacks and blood clots. Many menopausal women decided to investigate alternative options. In the majority of cases, scientific studies supporting the effectiveness and safety of alternative treatments for menopause are limited.

### Phyto-estrogens

(POs) are naturally occurring compounds found in plants. They share some of the same biological activities with oestrogens produced in the body. There are three types of phytoestrogens – isoflavones (found in beans, particularly soy beans), lignans (found in cereal grains and fruits) and coumestans (found in germinating seeds such as alfalfa). Levels tend to be greater when plants are germinating or producing seeds.

*Evidence* – the benefits of dietary **phytoestrogens** in menopause have been largely based on studies comparing the experiences of Asian women (who have a diet high in isoflavines) with those of Western women. One study showed that women in Western countries have an 80% incidence of hot flushes, while Asian women living in China have an incidence of only 20%. However, other factors could be involved, such as cultural differences in how women describe menopausal symptoms.

When a woman is pre-menopausal, some POs compete with her own oestrogens for the same **oestrogen receptor** sites. POs can therefore compete with endogenous oestrogens and thus have an inhibiting effect on their action. This is known as **competitive inhibition**. It may help to reduce the incidence of oestrogen-dependent cancers, such as many types of breast cancer. POs can also slow down the production of oestrogen by fat cells. When a woman is undergoing the menopause the POs bind to receptor sites and cause weak oestrogen-like responses that can reduce some of the symptoms of the menopause, such as hot flushes and vaginal dryness.

### Antioxidants

Antioxidants are substances that may protect your cells against the effects of *free radicals*. Free radicals are molecules produced as by-products of metabolism, or by environmental exposure to tobacco smoke and radiation. Free radicals can damage cells, and may play a role in heart disease, cancer and other diseases. Antioxidant substances include *beta*-carotene, selenium, Vitamin A, Vitamin C and Vitamin E. Antioxidants are found in many foods such as fruits and vegetables, nuts, grains, and some meats, poultry and fish. Sufficient epidemiological studies have shown that eating foods high in antioxidants can decrease the risk of many types of cancer. Based on this evidence, it is suggested that everyone eats at least five servings of fruits and vegetables each day. A natural approach to menopause should ideally involve a complete lifestyle approach, incorporating dietary changes, exercise and stress management strategies.

## Questions

1 Use the figures showing the risk of breast cancer in women taking HRT to plot a graph (number of cases against length of time HRT taken).

2 Some HRT programmes involve continuous administration of the hormones, whereas others are ................., in which there is a change in the hormones in a 4-week cycle.

3 Describe two alternative methods of treating menopausal symptoms.

4 List three risks of taking HRT.

## Effects of ageing on the brain

Like our bodies, our brains also age physically as we get older. From data collected there is no doubt that as we grow older so do our brains, but recent research indicates that most mental functions are untouched. Many of our abilities, including problem solving, improve as we age. Given the dramatic rate in which the older population is increasing (the number of those 85 and over in the UK is expected to double to more than 1.8 million by 2028) it is important that we learn more about the effects of age on the brain. Other health problems that people either have already or develop as they grow older – like high blood pressure, diabetes, or deafness – affect how their brains change with time. Short-term memory is not affected by ageing until most persons are in their 80s and 90s. Forgetfulness in the 50s and 60s is generally related to other factors such as stress.

Research has shown the following effects as the brain ages or a person gets older:
- There is a **decrease in brain weight and brain volume.** Certain brain areas shrink more than others, including the frontal lobe (important for mental abilities) and the hippocampus (where new memories are formed). But although brain weight can decrease by 10% at age 90, the size or number of cells does not correlate to mental function.
- **The outer surface thins:** the cortex – the heavily ridged outer surface of the brain – thins slightly with age. This thinning is not, as scientists once believed, the result of widespread loss of brain cells. Instead, the thinning of the brain's outer surface is probably due to a decrease in synaptic connections (the number of synapses between neurones), a process that starts when we're about 20 years old.
- **The white matter decreases:** many studies have linked aging with decreases in the brain's white matter (the myelinated neurones). Myelin helps to improve communication between brain cells. Research shows that changes in myelin are linked with reduction in speed of **cognitive processing.** Cognitive processing includes memory, attention, action, problem-solving and decision-making abilities.
- The **ventricular system** enlarges.
- The **brain generates fewer neurotransmitters** (chemicals like serotonin, which diffuse across synapses) and plasma membranes have fewer receptors that lock onto these messengers. This change may have an effect on memory.

Loss of mental function and ageing do not go hand-in-hand. Mental capacity does show a decrease in middle age (45) and a steeper drop after 65, but individual variation is great. There is a certain percentage of people who continue to function normally even when they age. This suggest that the problems accredited to an ageing brain are due to a combination of other factors such as insufficient mental stimulation, unhealthy diet, lack of novel experience or lack of sufficient social interactions.

## Effects of ageing on the peripheral nervous system

As people age, **peripheral nerves** may conduct signals more slowly. Usually, this effect is so minimal that no change in function is noticeable. The peripheral nervous system's ability to respond to injury also declines. When the axon of a peripheral nerve is damaged in younger people, the nerve is able to repair itself as long as the cell body is undamaged. This self-repair is slower and can be incomplete in older people, making them more vulnerable to injury and disease.

The most noticeable changes with age in the peripheral nervous system are in our **senses**.

**Hearing problems:** an ongoing loss of hearing linked to changes in the inner ear. Sensory hair cells in the cochlea die, and so do neurones in the auditory nerve. These are is the most common causes of hearing problems in older people. Speech becomes difficult to understand, especially when there is background noise. Certain sounds become annoying or loud. Most people lose ability to hear high-pitched sounds – this can start in their 20s.

**Sight problems:** poor vision in older people can result from specific eye conditions, such as cataracts, macular degeneration, glaucoma, diabetic retinopathy or from a stroke. The lens changes with age as the proteins in the cells begin to denature. This results in a loss of elasticity, which makes focusing on objects ('accommodation') at different distances difficult.

# Symptoms, causes and treatment of cataracts, glaucoma and macula degeneration

**Cataract**: a cataract is a clouding of the normally clear lens of the eye. It is like the mirror in the bathroom becoming fogged up with steam. A cataract can occur in one or both eyes. **Symptoms** can include: blurring of vision, glare or light sensitivity, poor night vision, decreased vision, loss of colour perception and needing a brighter light to read. An optometrist or ophthalmologist can diagnose a cataract with an eye examination.

**Causes of cataracts:** the prevalence of cataracts rises from about 2.5% for people in their forties to 99% of people in their 90s. Nearly half of people in their 90s have had cataract surgery. Other causes of cataracts include smoking, sunlight exposure, diabetes, and some blood pressure lowering medications.

**Cataract treatment:** cataracts are treated by surgically removing the cloudy lens of the eye. After the cloudy lens is removed, it is replaced with an artificial lens to restore the focusing power of the eye. The surgery is normally a 1-day procedure and does not require an overnight stay in hospital. After cataract surgery 85% of people have vision restored well enough to drive a car.

**Glaucoma:** in glaucoma the optic nerve is damaged. In the worst cases, the damage can lead to blindness. This is preventable if glaucoma is detected and treated early. There are two main types of glaucoma.

- **Open angle glaucoma** is the most common form. It occurs because drainage channels for the aqueous fluid become blocked over time. This causes the eye pressure to rise very slowly but painlessly. This is a 'silent disease' which, if it goes untreated, damages the optic nerve and reduces the field of vision. Eventually, only a small area of central vision remains (tunnel vision) before sight is lost completely. Glaucoma affects 1% of people over 40 and 5% over 65. Eye tests every two years are important. Glaucoma tests involve viewing your optic nerve by shining a light from a special electric torch into your eye and measuring the pressure in your eye using a special instrument. The main **treatment** for this type of glaucoma aims to reduce the pressure by using eye drops which act by reducing the amount of fluid produced in the eye or by opening up the drainage channels. If this does not help, your specialist may suggest either laser treatment or an operation to improve the drainage of fluids from your eye.

- **Closed angle glaucoma** (also called acute glaucoma) is much less common. The pressure in the eye rises rapidly because the periphery of the iris and the cornea come into contact so that aqueous fluid is not able to reach the tiny drainage channels in the angle between them. The eye becomes red and extremely painful. This may be accompanied by a headache, blurred vision and vomiting. Closed angle glaucoma needs immediate treatment involving medicine to reduce the pressure in the eye, followed by laser treatment or surgery. If treated quickly there can be almost complete and permanent recovery of vision.

**Age-related macular degeneration (AMD)** An estimated 500 000 people in the UK suffer from AMD. 40% of these are over the age of 75. It affects the central part of the vision (macula) and the side, or peripheral, vision is always left intact.

- **Dry AMD** is the more common form of AMD and affects almost 80% of those with the condition. It is caused by degeneration of the macula of the retina due to hardening of the arteries that provide oxygen and nutrients to the retina. As a result there is a gradual reduction in central vision. The onset of this condition tends to be slow. It usually affects both eyes simultaneously. Dry AMD tends to affect the ability to read and to see fine detail very difficult. Quality of life includes wearing powerful spectacles, using a magnifying glass and/or bright lights and using large print books.

**Wet AMD** is less common, but tends to have a more severe and rapid effect on the central area of vision. In this condition, blood vessels grow in an abnormal fashion into the macular area. These blood vessels may leak or bleed, causing a rapid and significant reduction in central vision. If the condition is picked up early, some people with wet AMD may be suitable for laser treatment.

**Figure 1** Sight loss caused by **a** cataracts and **b** macular degeneration

## Questions

1  The lens may become cloudy. This is known as a
   ................. . It can be treated
   ................. by removing the lens and replacing it with an
   ................. one.
   ................. is the name for a group of eye diseases in which the optic nerve is damaged. The central part of the retina is destroyed in ................. so
   ................. vision disappears.

2  Suggest four activities that would be beneficial to older people in slowing down the effects of ageing on the brain.

## Symptoms and possible causes of Alzheimer's disease

Alzheimer's disease is a **degenerative disease** of the nervous system that appears to affect particularly the temporal and frontal lobes of the brain. Initially, when Alzheimer's disease begins to destroy brain cells, no outward symptoms are evident. After a while, small memory lapses occur which grow more serious.

The duration of each stage and which symptoms appear when varies from person to person. Because the stages overlap, it is difficult to assign a person to a particular stage. However, the progression is always toward a worsening of symptoms and increasing dependence on caregivers. The end result of Alzheimer's is death, whether caused by the inability of the brain to maintain the body, or by another disease or injury along the way.

*Possible causes:* scientists do not yet fully understand what causes Alzheimer's disease. It is likely to be **multifactorial** (have many factors contributing to its onset).

- *Age* is the most important known risk factor. The number of people with the disease doubles every 5 years beyond age 65.
- A family history of the disease is a risk factor.
- Research has shown that the same *factors that increase the risk of heart disease* (smoking, high blood pressure or high cholesterol levels) may also increase the risk of Alzheimer's.
- Severe head or whiplash injuries appear to increase the risk of developing Alzheimer's.

| Stage | Symptoms |
|---|---|
| Early-stage Alzheimer's (mild) | Memory loss or other cognitive deficits are noticeable, yet the person can compensate for them and continue to function independently. They may forget familiar names or forget words to express what they want to say. They may forget the location of everyday objects |
| Mid-stage Alzheimer's (moderate) | Mental ability declines, the personality changes and physical problems develop. The person becomes more and more dependent on caregivers. Most victims are no longer aware that they don't remember things. Sufferers cannot communicate coherently |
| Late-stage Alzheimer's (severe) | Complete deterioration of the personality and loss of control over bodily functions. They depend totally on others for even the most basic activities of daily living. They are confused and disorientated as to date, time or place. In very late stages paranoia and delusions may occur. They will require constant supervision. Language becomes severely disorganised, and then is lost altogether. They may eventually lose the ability to swallow food and fluid, and this can ultimately lead to death |

**Table 1** Alzheimer's disease

- 15% of people with Alzheimer's disease have a family history of **Down's syndrome**. Recent research has concentrated on the gene responsible for the production of **beta-amyloid protein**. This protein is found in the tangled fibre masses that develop in the brain of people with Alzheimer's disease, and also in those of older people with Down's syndrome.
- The risk of developing the disease is higher if a parent, sister or brother has it. Scientists have identified some **genes** that are known to directly cause Alzheimer's. Inheriting these is almost certain to lead to early onset of the disease, usually before age 65, and sometimes as early as in the 30s or 40s. But these genes have been found in only a few hundred families worldwide. Alzheimers caused by these genes is known as *familial Alzheimer's disease.*
- A number of *environmental factors* have been put forward as possible contributory causes of Alzheimer's disease in some people. Among these is ingesting aluminium salts. There is circumstantial evidence linking aluminium with Alzheimer's disease but it has not yet been proved to be a cause. As evidence for other causes continues to grow, a direct link with aluminium seems increasingly unlikely.
- Some studies suggest that *remaining mentally active* throughout your life, especially as you get older, reduces the risk of Alzheimer's. But the evidence is currently inconclusive.

### Pathology

When tissues from the brains of Alzheimer's patients are viewed under the microscope, the major features are neurofibrillary tangles and senile plaques. Neurofibrillary tangles are mainly composed of **Tau protein**.

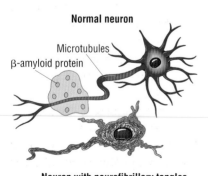

**Normal neuron**

Microtubules
β-amyloid protein

**Neuron with neurofibrillary tangles**

**Figure 1** Neurofibrillary tangles

Tangles are accumulations of twisted protein fragments that clog up the inside of neurones. Plaques encroach from the outside – they have a core of a protein called β-amyloid, which is infiltrated by clusters of degenerating axons and dendrites. The axons and dendrites are slowly engulfed.

A recent theory suggests that Alzheimer's disease is caused by inflammatory processes associated with ageing, and not by amyloid plaque deposits. It claims that the cause of the disease is the formation of toxic proteins within brain nerve cells. These proteins are not β-amyloid itself, but a derivative of β-amyloid (β-amyloid-derived diffusible ligands or ADDLs). While β-amyloid fibrils kill a wide range of nerve cells, the ADDLs affect only those types of nerve cells that are implicated in Alzheimer's disease. It has been suggested that the toxic protein may be an organophosphorous molecule. Organophosphourous molecules are known to attach to special proteins (Tau proteins) involved in assembling the microtubules found in axons, causing them to tangle.

### Examiner tip

Check your knowledge of peptides. What is the difference between a polypeptide and a protein? Review transcription and translation.

## Issues in the care of patients with dementia

Most Alzheimer's patients are cared for by family caregivers who are placed in a situation of escalating personal demands. Things that were taken for granted before, such as home safety and socialising will now require some planning. There is a need to communicate in new ways and make changes to the home environment.

Caring for an elderly person with dementia entails emotional, physical, social and financial burden. Caregivers of dementia patients can experience various emotional problems during the course of the illness. The dementia patient eventually may not even recognise the person who is caring for them. This is frequently the greatest concern of those carers who live with the dementia patient. It may be necessary to have a carer in attendance for 24 hours a day as the dementia progresses and the patient needs to be fed, taken to the toilet, washed and dressed. Eventually it may be necessary for the person to move into a residential care home. This can lead to financial problems. There is also a shortage of care homes in the UK.

## The social consequences of an ageing population

The 'baby boom' generation born just after World War II is approaching retirement. This unusually large number of ageing people, together with continued increases in longevity, is mainly responsible for the increasing age profile of populations in Western Europe. Lifestyle changes such as delayed childbearing and greater numbers of women going on to further education have also had a long-term impact on fertility rates. The number of children being born is below replacement levels (the levels needed to maintain the current population size) in most countries, and this trend is unlikely to reverse in the future.

At present, children aged under 16 outnumber people of state pension age, but this situation is changing. By 2025 those eligible to claim a pension will outnumber children by almost two million. This could affect the quality and size of the workforce, and hence the economy's earning power. It could affect the economy's productivity and could have significant effects on government tax revenue and expenditure. A particularly important issue is the effect on savings. Standard models of saving suggest that individuals save a higher proportion of their income in middle age in anticipation of their retirement, and 'dissave' (run down savings) when retired.

Financial provision for old age is becoming the responsibility of the individual. We cannot depend on the next generation to pay for the pensions of this one. People need to choose between reducing their current spending levels in order to save more while at work and/or re-evaluate the age at which they expect to retire. The impact of increased longevity will also result in an increase in the number of very old, frail and dependent people.

If each successive generation can expect a longer life, can it expect a longer healthier life? What are the prospects for disability-free life expectancy – that proportion of life beyond retirement when one is active and reasonably impairment-free? Currently, at the age of 65 one has another 12 years left, of which around 7.5 should be disability-free. The aim is obviously to add not only years to life but also life to years.

## Questions

1 Alzheimer's disease is a form of ................. in which cells in parts of the cerebral cortex die. An accumulation of ................. fibres and/or plaques form around the nerves in the cerebral cortex.

2 State three problems encountered when caring for a dementia patient who is a relative.

3 Name four possible causes of Alzheimer's disease.

4 What are the likely symptoms of early-stage Alzheimer's disease?

5 Why is the UK said to be 'an ageing population'?

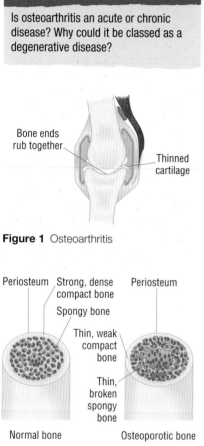

Bone ends rub together

Thinned cartilage

**Figure 1** Osteoarthritis

Periosteum — Strong, dense compact bone

Periosteum

Spongy bone

Thin, weak compact bone

Thin, broken spongy bone

Normal bone          Osteoporotic bone

**Figure 2** Osteoporotic bone

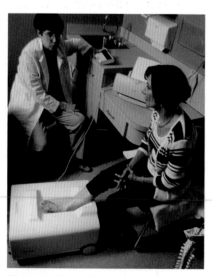

**Figure 3** Testing bone density

## Effects of ageing on the skeletal system

Bone mass begins to reduce between the ages of 30 and 40. Osteoblast activity (bone production) begins to decline, while osteoclast activity (bone reabsorption) continues at a normal level. Because of changes in hormone levels, blood cell production and bone marrow content bones become more fragile. They break easily and do not repair well. Vertebrae may collapse, distorting the vertebral joints and putting more pressure on spinal nerves. Intervertebral discs shrink. Compressed discs and the loss of bone mass lead to a decrease in body height, and the thoracic spine curves.

Arthritis means inflammation of the joints. **Osteoarthritis** (OA) is the most common form of arthritis in the UK. OA becomes more common with increasing age. Most cases develop in people over 50. By the age of 65, at least 50% of people have some degree of OA. It is mild in many cases, but about 1 in 10 people over 65 have a major disability due to OA, mainly in the hip or knee. Pain, stiffness, and limitations to movement of the joint are typical. The stiffness is usually worse first thing in the morning but tends to 'loosen up' after half an hour or so.

With ageing, the water content of the cartilage increases and the glycoprotein make-up of cartilage degenerates. Repetitive use of the joints over the years irritates and inflames the cartilage, causing joint pain and swelling. Eventually, cartilage begins to degenerate by flaking and tiny crevasses form. In advanced cases, there is a total loss of the cartilage cushion between the bones of the joints. This loss causes friction between the bones, leading to pain and loss of joint mobility. Inflammation of the cartilage can also stimulate new bone outgrowths (spurs) to form around the joints. Hips, knees, finger joints, thumb joints and lower spine are the most commonly affected joints.

**Why do some people develop osteoarthritis?** There is some evidence that the way joints are used earlier in life may have an effect. People who play a lot of sport seem to be more likely to suffer from OA of the knee joint. There is possibly a genetic link because the disease can occasionally be found in multiple members of the same family. Knee and hip OA are more likely to develop, or be more severe, in obese people. There may also be an ethnic link as hip OA is more common in white Europeans than in people of African or Asian origin.

**Treatment: paracetamol** is usually used to treat OA. It is more effective in easing pain if taken regularly, rather than 'now and again' when pain flares up.

**Anti-inflammatory painkillers:** these medicines are not used as often as paracetamol as there is a risk of serious side-effects, particularly with regular use in older people. Some people take an anti-inflammatory painkiller for short spells, perhaps for a week or two when symptoms flare-up. They then return to paracetamol.

**Hip and knee replacements** are very successful and can restore mobility.

**Osteoporosis** literally means 'porous bones'. Bones affected by osteoporosis are histologically normal in composition but less dense than normal bone – there is just less bone. Bones are more likely to break as a result of a minor bump or fall, or even without an injury. Hip and wrist fractures are the most common breakages, but they can occur in any bone. Osteoporosis can result in small fractures in the bones of your spine, causing a loss of height and a curved back (sometimes known as 'dowager's hump').

When you're young, your body makes new bone faster than it breaks down old bone, and your bone mass increases. By the age of 70, some women have lost 30% of their bone material. In the UK, about half of women and about 1 in 5 men over the age of 50 will fracture a bone, many as a result of osteoporosis.

## How can osteoporosis be prevented?

Prevention of osteoporosis begins from childhood. Osteoporosis is prevented by reaching the peak bone mass (maximum bone density and strength) during the childhood and teenage years and by continuing to build more bone as one gets older, particularly after the age of 30. (The higher your peak bone mass, the more bone you have 'in the bank' and the less likely you are to develop osteoporosis as you age.)

- get enough calcium and vitamin D by drinking milk or eating milk-based products
- exercise, particularly weight bearing exercise
- do not smoke, and avoid excessive intake of alcohol
- home safety to prevent falls and fractures
- to maintain bone mass, post-menopausal women may need adequate hormone replacement therapy to replace lost oestrogen
- be aware that long-term use of some medications, including excessive use of aluminium-containing antacids can lead to a loss of bone density.

## The effects of ageing on the cardiovascular and respiratory systems

There is a gradual reduction in efficiency of the **respiratory system** with age. The lungs accumulate damage due to air pollution, smoking, and respiratory infections. The ageing lung becomes less elastic which makes it less effective. The rib cage does not move as freely because of arthritic changes. This, in combination with the changes in elasticity, causes a reduction in chest movement which limits respiratory volume. The result is a decrease in the vital capacity, a rise in residual volume, a fall in functional reserve capacity, a fall in the volume of air expired in 1 second, and a fall in the peak expiratory flow rate.

For most people gas exchange is still perfectly adequate, but problems arise when infection sets in. Several respiratory disorders are more common in the elderly. These include emphysema, tuberculosis, bronchitis and pneumonia. The risk of contracting these increases with age but is partly due to the diminishing effectiveness of the immune system. Approximately one-quarter of older people suffer from pneumonia and other respiratory problems as a complication of influenza. Vaccination has been available since the late 1960s. It is offered annually to patients aged 65 and all those aged 6 months and over in clinical risk groups.

The **cardiovascular system** is one of the body systems most affected by age. The artery walls gradually become less elastic and may show signs of calcification – arteriosclerosis or 'hardening' of the arteries. Fatty materials are also deposited in the walls of the arteries causing atherosclerosis. This causes blood pressure to rise as the lumen of the artery is narrower and resistance to bloodflow increases.

Exercise cardiac output reduces with age and, because the heart is not able to pump blood as efficiently, circulation is slowed. In addition, the heart cannot respond as quickly, or as forcefully, to the increased workload of the exercised heart. Exertion, sudden movements, or changes in position may cause a decrease in cardiac output, resulting in dizziness and loss of balance. The coronary arteries narrow, restricting the vital blood supply to the myocardium. High blood pressure (hypertension) causes the left ventricle to work harder. It may enlarge and outgrow its blood supply and thus becomes weaker. Heart muscle loses elasticity and becomes more rigid and less able to accommodate the surge of pressure during systole. Heart valves become thickened by fibrosis and more rigid (leading to murmurs). The number of pacemaker cells decreases so the sino-atrial node beats more slowly and becomes less able to alter its rate.

Cardiovascular disease is a leading cause of death in this country in men over 45 and women over 65. Regular aerobic exercise, such as swimming can improve fitness and diminish the effects of these changes.

### Use of bone density tests for detection of osteoarthritis

A DEXA scan is a fast and accurate test, a normal X-ray for detecting bone density because it is more sensitive.

There are two different types of DEXA scanning devices, central and peripheral. Central DEXA devices are large machines that can measure bone density in the centre of your skeleton, such as your hip and spine. Peripheral DEXA devices are smaller, portable machines that are used to measure bone density on the periphery of your skeleton, such as your wrist, heel or finger. Central DEXA devices are more sensitive than peripheral devices. Bone mass often varies between parts of the body, so it is more accurate to measure the spine or hip than a heel or wrist. However, peripheral devices may help predict the risk of fracture in your spine or hip.

## Questions

1 Osteoarthritis is caused by the breakdown and eventual loss of the .................. of one or more joints whereas osteoporosis is the result of bones that are .................. than normal bones. Osteoporosis may be prevented by reaching the peak .............. during the childhood and teenage years.

2 How is bone density measured and what is the importance of this scan?

3 Arthritic change in the rib cage, coupled with loss of .................. in the alveoli result in a reduction in chest movement which limits .................. and a .................. in vital capacity of the lungs.

4 How does increasing age affect the left ventricle and the mitral valve?

# Summary and practice questions

1 Menopause is the ..................... of the menstrual cycle. It occurs when the ..................... stop working. Many women experience ....................., a feeling of suddenly increasing warmth. ..................... can help to overcome this symptom. The hormone ...... ..................... inhibits the reabsorption of bone. After menopause bone loss increases, resulting in ..................... which can result in an increase in the number of fractures. In males the level of ...... ..................... reduces.

2 State two alternative methods of treating the menopause.

3 As we age the rate at which impulses are conducted along neurones ..................... . This results from the ..................... sheath becoming ..................... . The effects of ageing on the peripheral nervous system are particularly noticeable in our hearing and seeing. Cataracts are a ..................... of the ..................... of the eye. A successful day patient operation can be carried out to remove the ..................... and replace it with an ..................... one. ..................... ..................... is the name for a group of eye diseases in which the optic nerve is damaged. There are two forms of age related macula degeneration in which cells from the ..................... of the ..................... are destroyed resulting in a loss of .....................vision making it difficult to read.

4 State three effects of ageing on the brain.

5 Alzheimer's affects 1 person in 5 over the age of 80. It costs the NHS more to treat Alzheimer's disease than any other single disorder. What are the main symptoms of the three stages of Azheimer's disease and what sort of problems are likely to be encountered when caring for an Alzheimer's patient?

6 If people have been immobile for a while because of illness or trauma such as falls and surgery and conditions such as arthritis, they will lose strength and power. Falls are the biggest cause of accidental death in the UK, beating both road accidents and fires. On average in the UK an older person dies just over every five hours as the result of a fall at home. Falls can devastate health and quality of life. As we get older we fall more frequently – GP and hospital records show that one in three over 65's fall every year, but that figure rises to one in two for over 85 year olds.

..................... is an age related disorder, characterised by low ..................... and increased bone fragility putting the patient at risk for ..................... . The underlying cause is attributed to various factors, insufficient vitamin ...... and ..................... intake, postmenopausal hormonal condition, nutritional disorders, lack of weight bearing ..................... and consumption of drugs such as cortisone, among others. ..................... is a type of arthritis that is caused by the breakdown and eventual loss of the ..................... of one or more joints. It commonly affects the hands, spine, and large weight-bearing joints, such as the ..................... and ..................... it becomes more common as we get older and affects about eight million people in the UK.

7 What does ageing reduce in the respiratory system?

8 Cardiovascular and circulatory problems, such as stroke, blood pressure problems and heart attacks, are the commonest causes of death and disability for older people. Nearly one-third of unexplained falls are due to a sudden drop in blood pressure or an abnormal slowing of the heart rate. Outline the effect of ageing on the heart.

9 List the possible social consequences of an ageing population.

**10** The table shows the percentage of women of different age groups reporting to a large hospital with fractures of the femur.

| Age group/years | Percentage of women in age group reporting with fractures of the femur |
|---|---|
| 20–29 | 0 |
| 30–39 | 0 |
| 40–49 | 0.4 |
| 50–59 | 1.1 |
| 60–69 | 2.6 |
| 70–79 | 7.4 |

Plot a graph to show the trend shown by the figures and describe this trend.

Suggest an explanation for the trend shown by the figures in this table.

**11** What are the advantages and the disadvantages of delaying the menopause?

**12 (a)** Why do you think that a cataract develops from the middle of the lens towards the outside edge?

**(b)** Describe how a cataract is removed.

**(c)** Why do people who have had a cataract operation often still need to wear glasses?

**13** Studies have shown that about 5% of the neurones in the part of the brain called the hippocampus disappear with each decade after the age of 50.

For every 100 neurones present in the hippocampus at age 50, calculate how many will be present by the age of 70. Show your working.

**14 (a)** In the eye of an older person, the ciliary muscles do not contract with as much force as in the eye of a younger person. Explain why this might make reading a book difficult for an older person.

**(b)** Why is it difficult to read if a person has macular degeneration?

**15** Ageing is one factor which increases the risk of developing Alzheimer's disease.

**(a)** Name two other risks.

**(b)** Outline the effect of ageing on the sense organs and the peripheral nervous system.

## Question 1

In the genetic disease phenylketonuria, the enzyme phenylalanine hydroxylase (PAH) does not function. This enzyme converts the amino acid phenylalanine into another amino acid, tyrosine. Figure 1.1 is a diagram outlining the pathway.

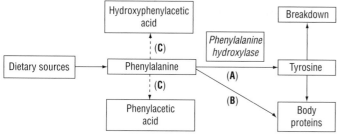

**Figure 1.1**

The gene for PAH is carried on chromosome 12.

**(a)** With reference to the information in Figure 1.1, explain how a mutation in the gene for PAH could lead to the enzyme being unable to carry out its function.     [5 + 1 QWC]

**(b)** PKU is inherited as an autosomal recessive condition. Using a genetic diagram, explain how two parents with a normal phenotype could have a child with PKU.     [4]

**(c)** People who are heterozygous for PKU are thought to be less susceptible to toxins produced by some moulds. These moulds grow on foods in damp, wet climates such as that found in NW Europe.

Suggest why the allele for PKU is found more commonly in NW European populations.     [2]

## Question 2

This question is based on the following passage about the human genome project.

The human genome project has resulted in the sequencing of DNA on each of the human chromosomes and the location of genes on chromosomes has been identified. On chromosome 11, there is a cluster of genes called the LOC genes. These genes code for receptor molecules found in the nose and are responsible for our sense of smell. LOC120009 is 31 110 nucleotides long. It has 11 exons but most of the gene consists of introns. In addition to the genes which have been identified, the chromosome also contains intergenic regions. These are lengths of DNA which lack any genes and they exist on every chromosome. These regions contain repeating sequences of nucleotides. The number of repeats varies from one person to the next and these variations are used in DNA profiling or 'genetic fingerprinting'. Only small samples of DNA are required for profiling. This is due to the use of the polymerase chain reaction or PCR. Following PCR, enough DNA is available to cut with restriction enzymes. Separating out the fragments by electrophoresis gives a pattern of bands which is used to distinguish DNA from different individuals.

**(a)** Explain the meaning of the following terms:

    **(i)** nucleotide

    **(ii)** exon

    **(iii)** intron.     [6]

**(b)** What term is used to describe the repeating sequences of nucleotides in the intergenic region?     [1]

**(c)** Explain how the polymerase chain reaction is used in preparing small samples of DNA for analysis by electrophoresis.     [5 + 1 QWC for use of technical terms]

**(d)** The United Kingdom Human Genetics Commission (HGC) was asked to consider the advantages and disadvantages of taking DNA profiles of every new born baby born in the UK. Suggest one advantage and one disadvantage of taking DNA profiles of all new born babies.     [4]

# Question 3

Figure 3.1 shows a diagram of a rod cell from the retina.

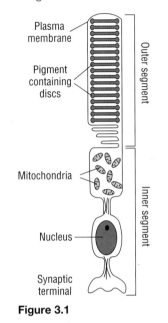

**Figure 3.1**

**(a)** Name the pigment present in rod cells. [1]

**(b)** Give two structural differences between rod cells and cone cells. [2]

**(c)** Identify the structure labelled X and explain their function in rod cells. [4]

**(d)** Explain how light energy falling on a rod cell results in the formation of an action potential.
[5 + 1 QWC for logical sequence of steps]

**(e)** A Snellen chart is used to assess visual acuity. People with good visual acuity are often described as having 20/20 vision. Explain how the Snellen chart would be used to carry out a visual acuity test. [4]

# Question 4

The number of cases of Alzheimer's disease is forecast to increase.

**(a)** Describe the symptoms of Alzheimer's disease. [3]

**(b)** Outline the changes in the brain which are characteristic of Alzheimer's disease. [4]

Globally, more than 23.4 million people are currently estimated to have Alzheimer's. 60% of people with Alzheimer are living in developing countries. In developed countries, the rate of Alzheimer's is predicted to double by 2040. In China, the rate is forecast to increase by 300% over the same period.

**(c)** Suggest what precautions you would need to take to ensure that data, such as that given above, was valid and reliable. [3]

> **Examiner tip**
>
> Questions such as 2d, 3e and 4c are testing some of the How Science Works objectives.

**(d)** Comment on the 300% rise in the rate of Alzheimer's predicted to occur in China. [2]

# Answers

### 1.1.1.1
1 **(a)** So that no pathogens/bacteria are introduced into the body.
   **(b)** To make sure there are no bacteria/pathogens on the skin, as these could enter the wound and cause an infection.
   **(c)** Veins are wider and nearer the surface and have a thinner wall, so it is easier to insert a needle. Also, the blood in a vein is under less pressure, so there will be less bleeding afterwards.
2 **(a)** So that the cells can be distinguished from each other when viewed under a microscope.
   **(b)** To enable the blood cells, and the structures within them, to show up more easily.
   **(c)** So that the blood cells will stick to the slide.

### 1.1.1.2
1 It only contains diluting fluid and has not been able to mix with the blood.
2 A single square may not be representative of the whole sample.
3 There are very few white blood cells compared to red blood cells. They will not be easily visible if red blood cells are present.
4 White blood cells are transparent, but when the stain is added they can be seen more easily, and the structures inside the cells (such as the nuclei) will show up more easily.
5 14 cells in one triple-lined square. One triple-lined square has a volume of $0.1 \times 0.1 \times 0.2$ mm$^3$ = 0.04 mm$^3$; $1/0.04 = 250$; $250 \times 14 = 3500$. Therefore, there are 3500 cells in 1 mm$^3$ of the dilution. The original sample was diluted 200 times. Therefore, there were $200 \times 3500 = 700\,000$ cells in 1 mm$^3$ of the original blood sample.

### 1.1.1.3
1 A: erythrocyte; B: leucocyte; C: platelet.
2 A: approximately 8 μm; B: approximately 7 μm.

### 1.1.1.4
1 Ribosomes/rough endoplasmic reticulum.
2 Golgi apparatus.
3 So they can join on to other cells or associate with other molecules outside the cell.
4 At the cell surface, they control the substances that enter and leave the cell. Inside the cell, they surround many organelles and help to form compartments in which certain reactions take place, or in which some molecules are collected together.

### 1.1.1.5
1 Chloroplast; cell wall; vacuole.
2 Root cell; xylem; phloem sieve tube; etc.

### 1.1.2.1
1

| Bond | Primary structure | Secondary structure | Tertiary structure |
|---|---|---|---|
| Peptide bond | ✓ | ✗ | ✗ |
| Hydrogen bond | ✗ | ✓ | ✓ |
| Sulfur bond | ✗ | ✗ | ✓ |

2 Answers will vary.

### 1.1.2.2
1 It becomes denatured and changes shape. Heat breaks the hydrogen bonds holding the protein in its tertiary structure, so the protein changes shape. The molecules tangle together forming a white, insoluble mass.
2 The polypeptide chains have tangled together, so they cannot re-form into their original shape.
3 Insufficient oxygen is delivered to the retina of the eye. Cells in the retina become damaged as they do not receive enough oxygen to respire actively.

### 1.1.2.3
1 In the secondary/tertiary structure of a protein; between bases in DNA molecules.

2

| Component | Vein leaving the small intestine | Artery leaving the heart |
|---|---|---|
| Oxygen | Lower | Higher |
| Glucose | Higher | Lower |
| Carbon dioxide | Higher | Lower |

3

| Component | Blood plasma | Serum | Tissue fluid | Lymph |
|---|---|---|---|---|
| Fibrinogen | Present | Absent | Absent | Absent |
| Albumin | Present | Present | Absent | Absent |

## 1.1.2.4

1 Similarities: both involve movement from an area of high concentration to an area of lower concentration/are passive/do not require ATP. Difference: osmosis refers to water only, but diffusion refers to many different molecules and ions.

2 It has different areas of positive and negative charges.

3 (a) The cell will take in water by osmosis and lyse or burst.
(b) The cell will lose water by osmosis and shrivel up.

4 The water will move into the cells, moving into the area of lowest concentration first.

## 1.1.2.5

1 Similarity: both require a protein carrier molecule. Differences: active transport requires energy/uses ATP, but facilitated diffusion does not; active transport is movement against a concentration gradient, but facilitated diffusion occurs down a concentration gradient.

2 Sodium ions enter the cell by active transport because they enter the cell against a concentration gradient. Chloride ions enter the cell by diffusion, because they have a similar concentration inside and outside the cell.

## 1.1.2.6

1 The high glucose level in the blood lowers the water potential of the blood plasma. This means that water will leave the red blood cell by osmosis and the red blood cell will shrivel up/shrink.

2 It is insoluble so does not affect osmosis. It is compact and branched, so glucose can be released when needed.

3 When blood glucose levels are low, there is less glucose for respiration so less ATP is made.

## 1.1.2.7

1 Joining two amino acids together, or joining two monosaccharides.

2 The lipids are complexed with proteins, which are soluble and so able to be carried in blood plasma.

## 1.1.3.1

1 So that pathogens from the first-aider do not contaminate the wound of the injured person, and so that any pathogens that may be in the blood of the injured person (e.g. HIV or hepatitis) are not transferred to the first-aider.

2 It may be sealing a blood vessel and limiting the loss of blood.

3 It must be firm to allow a blood clot to form, but if it is too tight it may cut off the blood supply to the limb.

4 This lowers blood flow to the limb, as the heart has to pump the blood against gravity.

5 If you remove the first pad, you will damage or remove the blood clot that is forming.

6 Hydrolysis is splitting a large molecule into two smaller molecules by adding a molecule of water to the bond that is broken. Examples are splitting a polypeptide into amino acids, a polysaccharide into fatty acids and glycerol, or a polysaccharide into monosaccharides.

7 Only fibrinogen will fit into the enzyme's active site.

## 1.1.3.2

1 An affected person will have blood that is very slow to clot, so they will get large bruises if they are injured, and will bleed profusely from a small cut.

## 1.1.4.1

1 The bonds holding it in its tertiary structure have broken, and the molecule becomes tangled up with other molecules. This stops it reforming into its original shape.

2 Albumin, fibrinogen, prothrombin, etc.

## 1.1.4.2

1 If blood cells are frozen, ice crystals will form and rupture the membranes of the cells.

2 (a) They need to supply the growing fetus with plenty of oxygen and glucose for respiration. This supply comes from the mother's blood. If the mother gives blood, she may reduce the supply of oxygen and glucose to her fetus.
(b) There is a very small risk of picking up an infection such as hepatitis or HIV if the tattooing was not carried out properly.
(c) A prostitute is more likely to have picked up an infection such as HIV or hepatitis.

## 1.2.1.1

1 It has to pump blood all round the body. The right ventricle only has to pump blood to the lungs.

2 One cardiac cycle lasts 0.6 seconds, therefore the heart rate = 60/0.6 = 100 beats per minute.

3 It would follow the same pattern as the line for the left ventricle, but it would be lower as the right ventricle does not develop as much pressure as the left ventricle.

## 1.2.1.2

1 This means that the ventricles can contract from the bottom upwards, ensuring that they empty completely.

2 This is because there is a delay at the AVN, allowing the atria to empty completely before the wave of contraction passes to the ventricles.

3 (a) One cardiac cycle is approx 0.9 seconds, therefore heart rate = 60/0.9 = 66.7/68 beats per minute.
  (b) One cardiac cycle is approx 0.4 seconds, therefore heart rate = 60/0.4 = 150 beats per minute.
  (c) One cardiac cycle is approx 1.2 seconds, therefore heart rate = 60/1.2 = 50 beats per minute.

## 1.2.1.3
1 5700/74 = 77 beats per minute.
2 18 900/195 = 96.9/97 $cm^3$.
3 The heart muscle has become stretched during exercise, as there is a greater volume of blood returning to the right atrium.
4 It means that an athlete's heart rate does not increase as much as a non-athlete's for the same amount of exercise. This means that an athlete can exercise more strenuously and for longer than a non-athlete.

## 1.2.2.1
1 The blood in an artery is under greater pressure, and flows faster than the blood in a vein.
2 (a) The artery has a thicker wall and smaller lumen than the vein.
  (b) The wall is thinner than the artery and has distorted during preparation of the section.
  (c) Artery: approximately 1000 μm; vein: approximately 500 μm.

## 1.2.2.2
1 The blood returns to the heart after it has been oxygenated in the lungs. This means it has higher pressure as it circulates round the body.
2 Blood plasma contains large proteins.
3 Lymph contains less oxygen and glucose than tissue fluid, and contains more waste products such as carbon dioxide.

## 1.2.2.3
1 Otherwise you may be measuring a short-term rise in blood pressure resulting from exercise or stress.
2 High blood pressure means that more tissue fluid is formed from the capillaries, but this cannot all return into the lymph vessels. Tissue fluid is especially likely to accumulate in the legs and feet because of the effects of gravity.
3 This is because there is not enough blood circulating round the body to supply all the cells with the glucose and oxygen that they need for respiration.
4 The same volume of blood is passing through a narrower vessel, so the blood pressure increases.

## 1.2.3.1
1 In the capillaries the squamous epithelial cells form a thin permeable wall. In the alveoli, they form a thin diffusion pathway for gas exchange.
2 Proteins made in the ribosomes on the endoplasmic reticulum, then carbohydrate is added in the Golgi apparatus. The mucus would then be packaged into vesicles and secreted from the cell.
3 The damaged cilia would not remove the mucus in the respiratory tract back to the throat. Instead, the mucus would gradually descend into the lungs and block the airways. This would lead to a cough as the person is trying to clear the airways. Pathogens in the mucus would be carried into the lungs where they cause infections.

## 1.2.3.2
1 Shortage of breath; skin/lips go blue; rapid/shallow breathing or baby stops breathing; grunting while breathing.
2 The oxygen would dissolve in the fluid lining the alveolus. It would then diffuse across the membrane of the alveolus cell, into the cytoplasm and then across the membrane on the other side of the cell. It would then diffuse across the endothelium cell in the wall of the capillary, then across the membrane of a red blood cell. Here it would combine with a haemoglobin molecule.

## 1.2.3.3
1 (a) 0.3 $dm^3$.
  (b) Approximately 0.2.
2 (a) Falls in peak flow rate correspond to exposure to house dust mites.
  (b) The bronchi and bronchioles are constricted, so air cannot flow along them as quickly.

## 1.2.3.4
1 To avoid catching an infection from the casualty, and to avoid spreading infection to the casualty.
2 They may have a head, neck or spinal injury – twisting or moving them may make it worse.
3 If there is a pulse, the heart is beating. If there is no pulse, CPR should be carried out.
4 There is still a lot of oxygen in expired air, so the person is getting oxygen. Also, the extra carbon dioxide stimulates the breathing mechanism.

## 2.1.1.1
1 TAAGCT.
2 30%.

**3** A molecule of water is removed when the phosphate group of one nucleotide binds to the deoxyribose molecule in another molecule. The resulting bond is called a phosphodiester bond.

**4** This shows that the amount of adenine in DNA is roughly equal to the amount of thymine, and the amount of cytosine is roughly equivalent to the amount of guanine. This supports Watson and Crick's model because it shows that A always pairs with T, and C always pairs with G.

## 2.1.1.2

**1** 4.

**2 (a)** A chromosome may consist of two chromatids. A chromatid is a replicate chromosome, and cannot be called a chromosome in its own right until it separates from its sister chromatid.

**(b)** A centromere holds two chromatids together. A centriole sets up the spindle fibres in a cell.

**3** The fibres are too thin to be seen using a light microscope.

## 2.1.1.3

**1 (a)** 810/960 × 100 = 84.4%.

**(b)** Prophase, because there are more cells in prophase than any other stage.

**2 (a)** 6.

**(b)** 12.

**(c)** 6.

**3** They do not have phosphatidylserine on the outside of their cell membranes.

## 2.1.1.4

**1** It can compress the surrounding brain tissue – the brain cannot expand because it is confined by the skull.

**2** If cancer is detected early, it is unlikely to have spread, or metastasised. It is much harder to destroy all the cancer cells if there are metastases.

**3** There has been more time for a series of mutations to have accumulated, and the person has been exposed to carcinogens for a longer time.

## 2.1.1.5

**1** After this stage, the cells start to differentiate and are no longer totipotent.

**2** They are synthesising large amounts of the protein haemoglobin.

**3** It loses its nucleus, so it no longer contains DNA which codes for the proteins it needs. It also loses most of its organelles.

## 2.1.2.1

**1** A suitable answer will consider having a large number of people in the study; the people would have a health check before the study begins, to see whether they have cancer already; a way of finding out what these people eat in a week, perhaps recorded on a proforma; and a way of analysing the results.

**2** There are other factors that are not controlled, e.g. whether these people smoke; have genetic factors that predispose them to bowel cancer; how much they exercise; and how old they are.

**3** These are chemicals that cause cancer. It is hard to find out whether a chemical causes cancer because these changes may take a long time. They can be tested on cell cultures, which are not the same as testing them on a whole organism, or on animals, which may not respond in the same way as a human.

## 2.1.2.2

**1** X-rays can cause mutations, so exposure to X-rays could increase the chance of causing cancer in the radiographer.

**2** Cancer cells divide rapidly, so they are more metabolically active than other cells.

**3** It avoids unnecessary exposure to X-rays.

## 2.1.2.3

**1** It damages DNA, and DNA is more likely to be damaged when it is dividing. Cancer cells divide rapidly so there is a lot of DNA replication occurring.

**2** Herceptin is the right shape to bind to one kind of protein receptor in cancer cells. This means it will not work for different kinds of cancer with a different shape of protein receptor.

**3** There is always a small risk when any vaccine is given. Women will not catch HPV unless they become sexually active, and women are most at risk if they are promiscuous. If a woman remains celibate for life she will not get HPV, and if she only has one sexual partner she is at very low risk. Therefore, some people think that vaccinating young girls is encouraging promiscuity.

## 2.1.2.4

**1** This is done to see whether the drug is effective and whether there are any side effects.

**2** The physiology of a human is different from that of an animal, so it is possible that the drug may not be as effective on humans as on animals, or that there may be side effects in a human that did not occur in animal tests.

**3** They may not be effective; they may produce unwanted side effects; or they may not work as well as pre-existing drugs.

**4** Advantages: if there is an unwanted side effect, or they do not work, there are fewer ethical issues as the patients were going to die of cancer anyway. Disadvantages: these people have very advanced cancer, so a drug that is effective in the early stages will not be identified.

**5** If the person does not know which drug they are getting, they will not be biased. In a double-blind study, the doctor does not know either, and this will avoid bias in the reports made by the doctor.

## 2.2.1.1

**1** 23.

**2 (a)** 46.
  **(b)** 23.

**3 (a)** In metaphase 1 of meiosis, the chromosomes are arranged in homologous pairs/bivalents, whereas in mitosis the chromosomes line up individually.
  **(b)** Metaphase 2 of meiosis is very similar to metaphase of mitosis, except that in metaphase of mitosis the diploid number of chromosomes is present, but in metaphase 1, the haploid number of chromosomes is present.

## 2.2.1.2

**1** Anaphase I: the spindle fibres pull the chromosomes to opposite poles of the cell. One chromosome from each homologous pair is pulled to each pole.

**2** This is because each gamete is different because of independent assortment and crossing over. Random fertilisation means that each zygote is unique.

**3 (a)** This is interphase, when DNA replication takes place.
  **(b)** This is when the diploid cell becomes two haploid cells at the end of meiosis 1.
  **(c)** This is when the two haploid cells formed in meiosis 1 become four haploid gametes at the end of meiosis 2 (when the chromatids split in anaphase 2).

## 2.2.1.3

**1** Preconceptual care is the care a mother should take of herself before she becomes pregnant, while postconceptual care is the care she should take of herself and her fetus during pregnancy.

**2** If a woman is anaemic, she does not have enough haemoglobin in her red blood cells. This means that there will be less oxygen transported round her body to the developing fetus, and to the mother's growing tissues.

## 2.2.1.4

**1** Calcium ions are needed as cofactors for thromboplastin to convert prothrombin to thrombin.

**2** This means that the placenta may detach early in labour as the uterus contracts, leaving the baby short of oxygen. It may also be delivered before the baby, again leaving the baby short of oxygen.

**3 (a)** Suitable graph drawn.
  **(b)** Between 7 and 7.5 weeks.
  **(c)** Extrapolation of the graph should give a suitable estimate.

**4 (a)** To ensure that the baby is growing properly and receiving sufficient nutrients.
  **(b)** Approximately 20 weeks.
  **(c)** Genetic factors; differences in proportion/shape of head.

**5** If she is eating a healthy diet with plenty of green leafy vegetables and wholemeal bread, she will not need an extra supply of iron.

## 2.2.1.5

**1**

| | Advantages | Disadvantages |
|---|---|---|
| Amniocentesis | There is a slightly lower risk of miscarriage than with CVS | Fewer cells obtained so it takes longer to get a result. If a termination is decided upon, the fetus will be 18–20 weeks |
| Chorionic villus sampling | More cells are obtained so results available more quickly | Slight risk of deformity in the fetus. Slightly higher risk of miscarriage than with amniocentesis |

**2** The baby's chromosomes are normal. This baby is a girl.

## 2.2.2.1

**1** The baby wriggles and there are errors as a result. This makes sure that the result is reliable.

**2** Infant scales are more accurate, and will record small increases in weight.

**3** This is to make sure that the baby is growing in size and proportion and not just body mass.

**4** This could be because of genetic factors, or because it was born prematurely (or was a twin birth).

## 2.2.2.2

1 This shows how quickly the child has grown over a specific period. For example, if a child has been very ill, relative growth rate may show a period of faster growth following recovery from the illness.

2 **(a)** Both boys and girls grow much faster in the first year of life. This rate of growth slows down gradually until they are five years old. Both boys and girls increase their rate of growth at puberty, although girls reach puberty sooner than boys. The growth spurt for boys is greater than the growth spurt for girls. After puberty, the rate of growth for both sexes slows down. Girls stop growing by 17–18 years, and boys stop growing at about 20 years.

**(b)** This is because boys reach puberty later than girls, but grow larger than girls.

## 2.3.1.1

1 A prokaryotic cell has a cell wall made of peptidoglycan. Eukaryotic plant cells have a cell wall of cellulose. Prokaryotic cells do not have a nucleus with a nuclear membrane but eukaryotic cells do. Prokaryotic DNA is circular and not complexed with protein, but eukaryotic DNA is linear and complexed with protein. Prokaryotic ribosomes are smaller than eukaryotic ribosomes. Prokaryotic cells do not have membrane-bound organelles, such as mitochondria and endoplasmic reticulum, but eukaryotic cells do.

2 It is a droplet infection, so people are more likely to catch it if they are living in overcrowded, badly-ventilated housing as they are more likely to breathe in infected droplets there.

3 More travel to countries where TB is prevalent, and immigration into the UK of people who carry TB. Also, there are more cases of people with health conditions such as HIV and diabetes that make them more susceptible to TB.

## 2.3.1.2

1 They do not show any signs of life, e.g. reproduction, respiration etc. and can only replicate inside a living cell. They do not have a cell structure or any metabolism of their own.

2 There is no firm answer to this. If an HIV positive mother breast feeds her baby, there is a chance she will pass on HIV in the breast milk. However, breast feeding is the best way to feed a baby, and offers the baby protection against infectious diseases. If the mother and baby are living in conditions where the baby is at risk of infections, and there is no reliable alternative sterile baby milk available, the mother may decide to breast feed.

## 2.3.1.3

1 Antibiotics target different stages in metabolism, but viruses do not have any metabolism of their own.

2 If there are any bacteria left, they are likely to be those most resistant to the antibiotic. These will multiply, and may pass on their resistance to other bacteria. If she continues the course of antibiotics, this will ensure that all the bacteria are destroyed.

## 2.3.1.4

1 There are many different answers here. The answer should include a way of growing bacteria, e.g. on an agar plate. The plant extract could be placed on the agar plate and the agar plate incubated. If the compound has antimicrobial activity, there will be a region around the extract where the bacteria are unable to grow.

2 There are various answers including; it may have toxic side effects; it may not be as good as existing drugs; it may be broken down in the body so it is not effective as a medicine.

3 The answer should consider the sample size; whether the study should be blind or double-blind; a control group, either with placebo or with an existing cancer drug; how to measure the effectiveness of the drug; timescale of study.

4 This makes sure that the seeds are as varied genetically as possible.

5 Points may include the importance of biodiversity; the possible uses of plants; the possibility of reintroducing species in the future; and the problems of deforestation and habitat destruction.

## 2.3.2.1

1 The skin is an effective barrier against infection, but if the skin is broken, pathogens can enter the tissues directly.

2 This is because of inflammation, the result of vasodilation and migration of phagocytic white blood cells such as neutrophils and monocytes to the area.

3 A different strain of flu will have different antigens, so the antibodies to one strain will not 'fit' on to the antigens of a different strain.

## 2.3.2.2

1 The protein will be synthesised at the ribosomes, on the rough endoplasmic reticulum. This protein will pass to the Golgi apparatus where it is modified into a glycoprotein. It will then be packaged into vesicles for secretion from the cell.

2 The cold virus keeps changing its antigens, so antibodies against one strain will not be effective against the antigens of a different strain.

### 2.3.2.3

1  If a very young child catches measles, they are more at risk of suffering complications or even dying. Therefore, the vaccine is recommended as it will give them sufficient immunity.

2  Factors affecting this will be how easily the pathogen spreads from one person to another, and how effective the vaccination is in producing immunity in people who are vaccinated.

### 2.3.2.4

1  Group AB has both antigens on its red blood cells, and no antibodies in the plasma, so it can accept blood of any blood group.

2  A different strain of HIV will have different antigens, so one kind of antibody will not 'fit' on to the antigens of a different strain.

3  A person who has been exposed to TB will have memory cells that respond to the TB antigen. This means they will mount a powerful secondary response when the Mantoux test is administered.

4  (a) This will be the same as Figure 2, except that TB antigens will be present in the well instead of HIV antigens.

   (b) This will be the same as Figure 2, except that the dish will be coated with HIV antigens (variable regions pointing upwards so that HIV antigens can bind). The secondary antibody will be able to bind to HIV antigens.

### 2.3.3.1

1  This means there are few new cases of the disease, but people who have the disease are surviving for a long time.

2  (a) This gives Health Authorities an idea of how effective vaccination programmes are, as it is recommended that children are vaccinated against mumps.

   (b) An increased incidence of food poisoning could mean that there is a food shop or restaurant that is the source of the problem, and it needs to be investigated.

   (c) People who have been in contact with a person who has TB may need to be screened, and they may need to be offered vaccination.

3  People may not always go to their GP to report an illness, especially if it is mild.

### 2.4.1.1

1  (a) Atheroma is the fatty plaque that builds up in artery walls. Angina is a condition that results from the narrowing of the coronary arteries caused by atheroma.

   (b) Myocardial infarction occurs when the heart muscle cannot receive enough oxygen, as a result of blockage of a coronary artery. Cardiac arrest is when the heart stops beating. This may result from a myocardial infarction, but there are other possible causes as well.

2  (a) This is to make sure there is plenty of oxygen in the person's lungs.

   (b) This is to stimulate the heart to start beating again.

3  These are both diseases that are more likely to occur as a result of ageing.

### 2.4.1.2

1  It could disrupt the electrical activity in the heart and stop the heart beating.

2  A donor vein would contain antigens that are different from the person's own antigens, so the new vein would be rejected by the immune system.

3  This is because the patient does not have to stay in a hospital, so the cost of running the hospital and caring for a person in residential care are avoided.

### 2.4.1.3

1  This depends on factors such as how many people smoke; the kind of diet that people eat; genetic factors; and how much exercise people have.

2  Environmental factors: passive smoking; behavioural factors: exercise, smoking; social factors: diet; genetic factors: gender, genes causing predisposition to CHD.

### 2.4.2.1

1  The decrease in figures following the smoking ban suggests that there is a link. Scientists would need to investigate more closely to be sure no other factors were involved.

2  Because at the time England had no smoking ban in place so the figures would be a reasonable comparison between environments with and without smoke.

3  Lung cancer and COPD take many years to develop so a reduction will only be visible after many years of a smoking ban.

### 2.4.2.2

1  Steroids work long-term to reduce inflammation in the airways, so these should prevent asthma. However, beta-agonists are fast-acting and relax the muscles in the airways, so these are effective if a person has an asthma attack.

**2** Animals introduce the child to more different antigens, so they will have more kinds of memory cells present that can produce antibodies against a wider range of antigens.

**3** Children living in cities are more exposed to air pollution.

## 2.4.3.1

**1** High glucose levels in the blood causes a reduction in blood water potential. This means water will pass from the tissues into the blood by osmosis. As a result, the person will feel thirsty.

**2** As water is drawn into the blood by osmosis, this increases the volume of the blood, but the capacity of the blood vessels does not increase.

**3** Salt also lowers the water potential of the blood. This will further increase the person's blood pressure.

**4** This will lower their blood pressure and improve their general health. It also stimulates the rate of respiration and reduces blood glucose levels.

**5** The glucose test meter contains an enzyme which is specific to glucose. No other molecule will fit into the active site of the enzyme.

**6** One possible answer would be to have two groups of Asian children, one group who eat a traditional Asian diet and another group who eat a more Westernised diet. The children could be tested after a certain period of time to see how many people in each group have developed diabetes.

**7** This would allow consumers to understand which foods contain the most sugar. It would also help people to identify foods that will contribute to obesity.

**8** Junk food tends to be high in energy content, contributing to obesity. It is often high in sugar and low in complex carbohydrates.

## 2.4.3.2

**1** This is so the test is not affected by food or drink that the person has recently consumed.

**2** This is to see how quickly the blood glucose level falls. If the person is secreting insulin and responding to its effects, the blood glucose level will fall rapidly.

**3** Prevalence refers to the number of people in a population who have the disease.

**4** Exact answers will vary according to the readings estimated from the graph. In China, prevalence has risen from about 19 million to 28 million cases, an increase of $9/19 \times 100 = 47.4\%$; in Brazil, the prevalence has risen from about six million to seven million, an increase of $1/6 \times 100 = 16.7\%$. Possible reasons include the rapid changes in China, where people are becoming wealthier

and eating more Westernised food, including 'junk' food. People in Brazil may be more likely to eat a healthy diet including fresh fruit and vegetables.

## 4.1.1.1

**1** Like DNA and RNA, ATP contains the organic purine base adenine and a 5-carbon ribose sugar. ATP is also similar in that it is composed of three parts, with adenine and ribose acting as a backbone. In DNA and RNA the backbone is made up of a ribose sugar and a phosphate group.

**2** When the ATP is broken down to ADP and Pi it releases a small amount of energy and waste heat. Many thousands of molecules are broken down continually so this release of heat is sufficient to maintain core body temperature.

**3** Glucose would release too much energy and heat in one go if it was to be hydrolysed completely and this could not be channelled into cellular reactions. It would be an inefficient way to use the available energy. ATP when hydrolysed releases just enough energy to drive one chemical reaction.

## 4.1.1.2

**1** It involves a series of enzyme-driven steps that each need to function in a specific order to allow successful breakdown of glucose.

**2** Splitting of sugar.

**3** This allows the cell to produce a small amount of ATP in the absence of oxygen – important for example during times of oxygen stress such as during anaerobic respiration.

**4**

| | Starting molecules | Products of reactions | Site of reactions | Oxygen required |
|---|---|---|---|---|
| Glycolysis | Glucose, ATP, NAD | Pyruvate, reduced NAD, ATP | Cell cytoplasm | No |
| Link reaction | Pyruvate and coenzyme A | Acetyl coenzyme A, $CO_2$, reduced NAD | Mitochondrial matrix | Yes |

## 4.1.1.3

**1** 3.

**2** Dehydrogenases catalyse the removal of electrons and protons from the substrate and use these to reduce coenzymes such as NAD and FAD.

**3** This occurs in the matrix of the mitochondria where the enzymes are found – not found in the cytoplasm.

**4** Allosteric inhibition is where molecules attach to a part of the enzyme away from the active site and prevent the substrate from binding. Competitive inhibition is where the inhibitor molecule looks similar to the substrate and blocks the active site temporarily.

## 4.1.1.4

**1** Its inner membranes cover a large area and have many stalked particles on them to produce ATP.

**2** Its role is to mop up electrons and protons and ensure that membrane-bound coenzymes can be continually reduced and oxidised.

**3** Protons and electrons are brought into the fixed coenzymes on the cristae membrane via reduced NAD and reduced FAD. The electrons pass along a series of carrier molecules, pumping protons into the intermembrane space. These protons leak back through protein channels connected to ATP synthase stalks generating ATP. Electrons and protons are mopped up by oxygen molecules, creating water as a waste product.

**4** Protons have a charge and therefore cannot pass through the non-polar lipid membrane – they can only pass through polar protein channels.

**5** 32.

## 4.1.1.5

**1** Links, Krebs and the electron transport chain.

**2** If key enzymes in the electron transport chain are blocked then there is no production of ATP because protons cannot be pumped, creating a concentration gradient and so no protons can flow through the stalked particles.

**3** These coenzymes have to be recycled and regenerated.

**4** Any four from: movement; active transport; maintenance; repair; cell division; production of substances; maintaining body temperature; bioluminescence.

**5** They would be tired, slow to react and respond to stimuli, and possibly suffer from muscular cramps.

## 4.1.1.6

**1** Pyruvate has the chemical formula $CH_3COCOOH$, whereas lactate has the formula $CH_3CHOHCOOH$.

**2** Oxygen stress such as at the start or after prolonged strenuous exercise.

**3** Protein cannot be stored and excess proteins (or during the absence of other substrates) can be converted to fatty acids by the removal of their amino groups and will enter the Krebs cycle where they go on to produce ATP in a similar way to glucose.

**4**

| Method of respiration | Site of reaction(s) | Reactants | Products | Relative contribution to cells ATP total |
|---|---|---|---|---|
| Aerobic | Cytoplasm and mitochondria | Glucose, oxygen, NAD, FAD, ADP and Pi | ATP, $CO_2$ | 36 |
| Anaerobic | Cytoplasm | Glucose, NAD, ADP and Pi | ATP, lactate | 2 |

## 4.1.1.7

**1** 18/27 = 0.67.

**2** Will use different substrates during different times or situations, e.g. lipids during colder periods as they generate more heat and raise body temperature.

**3** Use the apparatus shown in the diagram but immerse the tube containing the organism in a water bath set at different temperatures. It is important to leave the organism to acclimatise to the temperature for several minutes before taking any readings.

**4** Eventually when the lactate is oxidised, extra oxygen will be needed to remove this substrate as well as carrying out normal respiration.

## 4.1.2.1

**1** Increases breathing rate, tidal volume, rate of diaphragm contractions in the short-term. In the long-term, increased muscle size, capillary bed in muscles, concentration of myoglobin, glycogen store, respiratory enzyme total, VO2 max, number of red blood cells, heart muscle, stroke volume, resting heart rate, blood pressure, recovery from exercise, maximum breathing rate and vital capacity.

**2** Nitric oxide dilates the blood vessels and will allow more blood to go to the heart muscle, stopping the pain from angina.

**3** These are the fibres most used in sprinting and short bursts of exercise, and the upper leg muscles are important here. The muscle would appear very dark red in colour as they would contain large stores of myoglobin to allow prolonged aerobic respiration.

**4** Rate is monitored by receptors in the medulla oblongata and stretch receptors around the intercostal muscles which will detect increases in breathing rate. There is also monitoring from chemoreceptors which are monitoring the blood pH and $CO_2$ level.

## 4.1.2.2

1 Glycogen is a complex polysaccharide composed of thousands of alpha glucose molecules joined together with glycosidic bonds in 1–4 linkage. It is a highly branched molecule with side branches forming from 1–6 carbon links.
2 Blood groups may differ, causing clots and also there is the risk of infection with such diseases as hepatitis C and HIV.
3 Steroids are lipids, which are non-polar, and so can easily diffuse through the lipid bilayer of membranes.
4 This allows their body to naturally replenish the blood taken.

## 4.1.2.3

1

| mRNA | tRNA | rRNA |
| --- | --- | --- |
| Involved in copying the DNA molecule – transcription | Involved in bringing specific amino acids to link up to the codons presented on the mRNA – translation | This is the material that ribosomes are composed of and act as the sites for protein synthesis |

2 Haemoglobin is a 3D protein with a quaternary structure. It is made up of four polypeptide chains bound tightly together with haem groups, which are not made of protein. Enzymes are 3D proteins with a tertiary structure and are composed of one polypeptide chain only.
3 DNA molecule unzips at the site of a particular gene; mRNA nucleotides copy complimentary base pairs on the leading strand of the DNA; mRNA codes join together using the enzyme RNA polymerase, forming a long single strand; this strand leaves the nucleus through a pore and attaches to a ribosome; tRNA molecules attach to specific amino acids and it has an anti-codon that links up with complimentary codons on the mRNA; this continues and adjacent amino acids form peptide bonds and link together forming a polypeptide chain; this continues until the entire protein molecule is coded for.
4 Peptide bonds between amino acids and ionic, disulfide bridges, hydrogen bonds, hydrophobic interactions between R groups.

## 4.1.2.4

1 Haemoglobin is a 3D protein with a quaternary structure. It is made up of four polypeptide chains bound tightly together with haem groups, which are not made of

protein. Myoglobin contains only one haem group instead of four in haemoglobin. It also only contains one polypeptide chain and has a tertiary protein structure.
2 It can carry more oxygen than myoglobin and is more efficient at doing so. Myoglobin is also restricted to muscle cells only.
3 The curve follows the loading of oxygen onto haemoglobin in the lungs to the unloading in respiring tissues. Because of the differing affinities for oxygen at different partial pressures, the dissociation curve is not linear. For example, it is hard to lose all of the oxygen from the haemoglobin molecule and also hard to load it up with the first molecule of oxygen in the lungs.
4 The initial first burst of exercise will always be anaerobic because the heart and breathing rate take a while to reach the level that the body is demanding. As a result, for an initial brief period some cells will be starved of oxygen and so build up a level of lactate.

## 4.1.2.5

1 These enzymes bring together nucleotides, making long chains of either DNA or RNA, and also check that the correct complimentary base pairing has occurred.
2 Oxygen is needed for the Links, Krebs and the electron transport chain. Less oxygen will create anaerobic conditions which only produce 2 ATP molecules per molecule of glucose.
3 Deletion and addition of bases producing frame shift mutations.
4 Presence of a virus, presence of a toxin, activity of a specific gene or series of genes such as during the creation of fingers in the development of the fetus.

## 4.1.2.6

1 Cardiac muscle must be involuntary and not under any conscious control, whereas skeletal muscle must be voluntary.
2 This is made up of alternating light and dark bands, filaments of the protein actin and myosin. The sarcomere is the key structure of a muscle fibre and is characterised by having clearly identified z lines, A bands, H zones and I bands.
3 This structure stores and releases calcium ions when the neuromuscular junctions are stimulated by nervous action. This release causes a change in the behaviour of actin and myosin and allows contraction of muscle units.

## 4.2.1.2

1 Sperm cells contain primarily just a nucleus and the means to propel that nucleus into the ovum and through

the female reproductive tract. The ovum contains all of the necessary nutrients to support the early divisions of the blastocyst. It also contains all of the mitochondria that will be used by the developing fetus.

2 This process is controlled by GnRH, LH and FSH and the subsequent involvement of the male and female sex hormones. Both LH and FSH stimulate the production of gametes and stimulate cells to produce and release the sex hormones oestrogen and testosterone.

3 Negative feedback mechanisms are self-regulating systems and in the case of the menstrual cycle involve one hormone such as oestrogen inhibiting the release and action of LH and FSH. When oestrogen levels drop, these other hormones can play a role and cause ovulation to occur.

## 4.2.1.3

1 As soon as one sperm enters the cortical reaction takes place, changing the surface proteins of the ovum and preventing any further entry by other sperm.

2 It allows the blastocyst to rest and develop into an embryo. Projections from the embryo and the uterine wall go on to form the placenta.

3 The blastocyst contains nuclear material and subsequent proteins different to the mother. It is recognised as foreign and the female reacts as if this was an invading pathogen and launches a defence mechanism.

4 Answers will vary but should be a good summary of the content of the spread that can be used for revision.

## 4.2.1.4

1 Oral contraceptives tend to contain small quantities of both progesterone and oestrogen. These act to inhibit the release of LH and FSH and so prevent follicle development and ovulation.

2 Advantages: an almost 100% chance of not releasing gametes and it is a permanent solution to contraception; disadvantages: it cannot be reversed and that it offers no protection from STIs.

3 Certain religious groups consider preventing conception to be destroying the potential for life and the use of contraceptives such as condoms are seen as unnatural methods preventing life from beginning.

4

| Contraceptive device | Condom | IUD | Implanon |
|---|---|---|---|
| Advantage of use | Cheap and easy to use; protects from STIs | Can be left in uterus for long periods of time; almost 100% effective at preventing pregnancy | Slowly releases female contraceptive hormones over a long period into the blood so no need to remember to take tablets |
| Disadvantage of use | Can split if not used properly | Need to be fitted; if pregnancy does occur this can endanger the life of both fetus and mother | Does not protect from STIs |

## 4.2.2.1

1 Hormones can be taken that stimulate the release of GnRH. This will in turn stimulate greater production of FSH and LH, allowing more follicles to mature and increasing the chance of pregnancy – super ovulation.

2 FSH acts to promote the final development and maturation of the follicle containing the secondary oocyte and LH rises in concentration, leading to ovulation.

3 Antibodies will cause sperm to clump together and agglutinate (also be destroyed by phagocytic white blood cells) so preventing them from reaching the Fallopian tube or from fertilising the secondary oocyte.

4 The virus is very small and can easily pass from sperm into secondary oocyte and there is a risk that the fetus could be born with HIV and that during gestation and labour the mother could become infected.

5 It might be better to remove the sperm produced and use artificial insemination techniques. IVF could be used and embryos could also be screened for the presence of cystic fibrosis so that the parents could prepare themselves for the birth or make the decision as to whether terminate the pregnancy.

## 4.2.2.2

1 Frozen embryos cannot be stored indefinitely and so any unused ones will be destroyed – are we destroying life or potential life? Stored embryos can also be used by other

couples, but this raises concerns about the ownership of frozen embryos and there have been disputes between parents of frozen embryos as to who decides their fate.

2 This is to prevent any degradation of the embryo which might result in possible genetic defects and mutations.

3 Monoclonal antibodies are produced by the process of genetic engineering so that they exactly recognise the hormone hCG. Because the variable region and the binding site are specific for each antibody, like the lock and key on an enzyme, they will only recognise hCG molecules.

4 If one of the twins dies early on it can often allow the other fetus to develop normally. The nutrients from the dead embryo will be reused by the mother and the surviving fetus.

### 4.2.3.1

1 Both produce ATP in similar ways using ATP synthase enzymes and by using proton gradients. Chloroplasts and mitochondria have inner membranes covering a large surface area which act as reaction sites and each uses coenzymes, dehydrogenases. They are similar sized organelles and contain DNA enabling them to replicate and produce their own proteins.

2 Photosynthesis produces the carbohydrates, lipids and proteins that make up the structures in plants which are the basis of all food chains. All the food we eat has received its nutrients and energy from the products of photosynthesis.

3 Autotrophic feeders obtain all their nutrients from inorganic sources and use these to produce their own food using energy from the Sun. Heterotrophic feeders obtain all of their food and energy from consuming other organisms.

4

| | Light-dependent stage | Light-independent stage |
|---|---|---|
| Site of reaction | Thylakoid membranes on chloroplast | Stroma |
| Reaction molecules involved | $H_2O$, NADP, ADP, Pi, photosystems 1 and 2 | RuBP, rubisco |
| Reactants | Light and water | $CO_2$, ATP and reduced NADP |
| Products | ATP, reduced NADP and $O_2$ | TP, ADP, NADP, RuBP |

### 4.2.3.2

1 Amino acids can be polymerised to produce proteins needed for all structures in the plant such as membranes. Lipids can be used as energy stores in the plant such as in seeds and will also join with proteins to form structures such as membranes.

2 Plants would not be able to make sufficient protein. The result would be poor growth and the plants would be stunted in appearance.

3 It becomes too inefficient to feed and organisms at the extreme of a food chain, ie. the top consumer, would need to spend vast amounts of time and energy consuming food that would offer very little in return. The energy payback would be slight.

### 4.2.3.3

1 Extensive agriculture takes place over last areas of generally low grade land such as upland areas. There are typically low numbers of animals involved, producing low yields and requiring little input or care from the farmer. The soils are naturally fertilised by the animals. Intensive agriculture involves high stock levels and lots of fertilisers, antibiotics or other chemicals to boost yields. Return is high for the input and the majority of farms act in this way, adopting a factory approach towards farming.

2 This is the approach to farming that involves replacing from the land what is taken and attempting to make little impact on the surrounding ecosystem as a result of the farming.

3 Hedgerows protect wildlife and promote biodiversity as well as acting as wildlife corridors. They prevent loss of soil by protecting fields from adverse weather conditions and they can house beneficial pest control agents and pollinators. Hedgerows can interrupt and block farm machinery, they can reduce the farm size and also can harbour insect pests and weed species. They also prevent the farmer from ploughing or harvesting up to the margin of the field.

4 If antibiotics are overused then they can stimulate resistance in bacteria that can affect humans. Antibiotics used in farming are the same as used in human populations. The result is that we have no defence against certain infectious agents.

### 4.2.3.4

1 Succession is the natural change of an ecosystem through a series of changes to a system such as a woodland. A deflected succession is when that climax situation is not reached, such as through the process of grazing.

2 Acid heathland; peat bog; moorland; golf courses.

3 Neither, because both will change the ecosystem through grazing or trampling or the release of nutrient-rich faeces, and so a climatic climax will not be reached.

4 Soil changes through the process of succession along with the changes seen in the flora and fauna. As ecosystems develop over time, so will the soil profile deepen, change colour and contain increasing quantities of nutrients, humus and organic matter.

## 4.2.4.1

1 DD factors might include lack of food, disease, lack of space, infertility, increase in contraception, war. DI factors could include climate changes, fires, tsunamis, drought leading to famine, earthquakes and volcanic eruptions.

2 He suggested that the poorer members in the population would always continue to have larger families because they understood that they would inevitably lose some to disease or starvation. As a result they had large families and this kept the overall population growing exponentially. Only a large scale problem such as an epidemic disease could curtail the population growth. He rather suggested that the problems of the human race were the doing of the poor and his views of how to deal with this rise in population – that is, leave them to the peril of nature seems rather cruel and callous.

3 No – there will be certain pressures that exert ultimate control and stop the population rising. There is a limit to the resources and land masses that can hold such an ever-increasing population without degrading all other ecosystems and species.

4 Birth control can be altered through better contraceptive devices and education. This can also be done through governmental policies – such as one child per family in China. Greater food production, living standards and health care also means that there is less need to have large families as there is only a very slim chance that your children will die in childhood. A high life expectancy means less need to have a high birth rate.

## 4.2.4.2

1 It is impossible to correctly count fish populations and estimates can be widely inaccurate and also subject to bias. You need to take into account the fecundity of the species, other species that interact with that particular species, ecological and environmental conditions and any other pressures upon these particular fish, such as susceptibility to diseases and pollution. This makes it very difficult to estimate population sizes, predicted catch sizes and therefore makes the idea of sustainable fishing nonsense.

2 Rising human populations use up natural resources and encroach on the land and ecosystems of other species. A rising population affects distant ecosytems. Pollution such as $CO_2$, methane and sulfur dioxide can enter the atmosphere, travel vast distances and pollute un-related ecosystems or species.

3 They can be used for ecotourism, medicines and future life saving drugs, agriculture such as the production of nuts and oils and selectively logged under strict guidelines so as only to remove certain species and only a certain number of trees in a given area.

## 4.3.4.3

1 When glucose, lipids or amino acids are used to generate energy through respiration in living species, $CO_2$ is given off as a waste gas which can return to the atmosphere. Dead organisms or waste will be broken down by decomposers which will use the carbon and release some of it back as $CO_2$.

2 The fluctuations represent when the trees in the northern hemispheres have leaves that are photosynthesising and when these trees have lost these leaves and are carrying out much less photosynthesis. As a result there will be a global atmospheric increase in $CO_2$ when the leaves have fallen and a decrease when the leaves are photosynthesising.

3 Intensive farming produces a lot of extra waste manure. Decomposing produces $CO_2$ and also lots of additional methane, $CH_4$, which is a powerful greenhouse gas. Intensive farming also uses lots of energy and fossil fuels for things like heating animal enclosures, which will contribute to global warming.

4 It is only a correlation between the rise in $CO_2$ and the rise in global temperatures. Some research shows that the trend for warming actually starts before a rise in $CO_2$ and some research even suggests that we are in a period of global cooling. It is also important to stress that there is often bias in some of the scientific reporting. It is important to assess the producers of the research and who is funding the research.

## 4.2.4.4

1 The rise is thought by many scientists to be linked to a rise in global temperatures and subsequent climate change.

2 Meat is produced primarily by intensive agriculture which uses high energy inputs to create the final product and produces considerable $CO_2$ and $CH_4$ waste gases.

3 Considerably lower.

4 We are not trying to cut back the amount of carbon we release but the amount of carbon dioxide gas released.

We also cannot really have a carbon footprint as all organic molecules contain carbon and everything we do will turn carbon from one form to another. The idea is really to do with living sustainably and trying to limit our use of fossil fuels, which are a huge contributor to the greenhouse effect.

## 5.1.1.1

1 Allele; locus; recessive.
2 An alteration of the DNA sequence or a change in the number or structure of chromosomes.
3 The enzyme phenylaline hydroxylase changes shape and cannot convert phenylalanine to tyrosine. The build-up of phenylalanine can cause brain damage in young children.
4 Parents: Ff × Ff

|   | F | f |
|---|---|---|
| F | Ff | Ff |
| f | Ff | ff |

Ff = child with cystic fibrosis. Therefore a 1 in 4 or 25% chance.

## 5.1.1.2

1 Both alleles are expressed in the phenotype.
2 $Hb^AHb^S$ and $Hb^AHb^S$ as those with the genotype of $Hb^SHb^S$ are unlikely to survive long enough to reproduce.
3 $Hb^S$ confers resistance to malaria. There is no malaria in the UK, therefore no selective advantage for heterozygote in the UK. Frequency of $Hb^AHb^S$ will consequently decrease as $Hb^SHb^S$ likely to die before reproducing.
4 Parents are heterozygotes so father will be $Hb^AHb^O$ and mother $Hb^BHb^O$.
Parents: $Hb^AHb^O \times Hb^BHb^O$

|   | $Hb^A$ | $Hb^O$ |
|---|---|---|
| $Hb^B$ | $Hb^AHb^B$ blood group AB | $Hb^BHb^O$ blood group B |
| $Hb^O$ | $Hb^A Hb^O$ blood group A | $Hb^OHb^O$ blood group O |

A 1 in 4 chance of being blood group O.

## 5.1.1.3

1 X; Y; X; autosomes.
2 Females have two X chromosomes while males have only one, so males lack a 'back up' copy to correct for the defective gene.
3 (a) No as both males and females have the disease.
   (b) Recessive as in generation 2 no one is an albino but in generation 3 it reappears, so allele must have been present in generation 2.

(c) Heterozygous as allele reappears in generation 3.
(d) If both parents are heterozygous (carriers) for albinism, then each child has a 1 in 2 chance of being an albino (genetic diagram to show cross).

## 5.1.1.4

1 Crossover frequency.
2 Chromosome.
3 Down's syndrome.
4 Answers will vary.

## 5.1.2.1

1 Restriction; blunt; sticky; ligase.
2 Small circular strand of DNA found in some bacteria.
3 In order to identify which bacteria have taken up the recombinant plasmid DNA.
4 Because it carries the recombinant DNA into a bacterium.

## 5.1.2.2

1 If the protein is a glycoprotein then sugars have to be added to the protein and bacteria cannot do this.
2 It is expressed in milk secreted by the animal.
3 Exons; gel electrophoresis; anode; furthest.
4 Polymerase chain reaction.
5 Forensic science; paternity testing; matching organ donors.

## 5.1.2.3

1 Using genes to treat or prevent disease.
2 Somatic cell – all body cells with the exception of egg or sperm cells. Germ cells – egg or sperm cells.
3 To map and sequence every gene on every human chromosome.

## 5.1.3.1

1 (a) No – it is not sex-linked so could have been a paternal grandparent.
   (b) Yes, unless it is a new mutation.
2 Advice – by drawing out a pedigree chart she could point out the girl's chances of developing breast cancer. She could talk about the tests suggested and the positive results if the cancer is caught at an early stage.
3 In the first consultation the counsellor would investigate the family history and use the information given to draw a pedigree chart. This would be used to discuss the probabilities. The counsellor would discuss the available options, including treatments for cystic fibrosis that are now available. The role of the counsellor is to give the couple the facts so that they are able to make an informed decision.

## 5.1.4.1

1  2.
2  Compatibility is the ability to accept transplanted tissue.
3  Major histocompatability complex; 6; human leucocyte antigens.
4  Cadavers – establishing brain death, permission; living donors – need a good tissue match; xenotransplantation (genetically engineered animals) – immunological rejection, transfer of disease; sale of organs – monetary pressure, poorly executed operations to remove the wanted organ.

## 5.1.4.2

1  Cells that have the potential to develop into many different cell types.
2  Inside blastocysts.
3  Tissue transplant and possibly organ transplants sometime in the future, testing drugs, designer babies.
4  Therapeutic.

## 5.2.1.1

1  Cornea; lens.
2  Rods: elongated structure; outer segment; stack of discs containing the visual pigment; contains one type of visual purple – rhodopsin. Cones: shorter and broader and more tapered; no discs – continuous folded system instead; contains three types of iodopsin (pigment).
3  Straight on – focus on fovea which only contains cones (colour vision) whereas to the side of the fovea are rods which are able to work in low light intensity.
4  Structure nearest to the light stimulus are the ganglia and bipolar cells. The light receptors are furthest away from light stimulus.

## 5.2.1.2

1  –40mV; sodium–potassium; sodium; potassium; outer; inner.
2  Rhodopsin; all-*trans*-retinal; opsin; potential; bipolar; action; ganglion; optic.
3  Pupil size in RAPD, burst blood vessels – hypertension. Diabetic retinopathy: specific to those with diabetes, damages blood vessels in the retina. Glaucoma: results from an increase in fluid pressure inside the eye that leads to progressive optic nerve damage and loss of vision. People with diabetes are nearly twice as likely to develop glaucoma as other adults.
4  Cannot distinguish between certain colours, e.g. red/green. Very unlikely to be totally colour blind.

## 5.2.2.1

1  Cerebellum.
2  Protect and cushion the brain.
3  Medulla oblongata.
4  Cerebrum; sense organs; limbic; memory.

## 5.2.2.2

1  **(a)** Dorsal root ganglion.
   **(b)** In the spinal cord (grey matter).
2  A long process that carries the impulse away from the cell body.

3

| Sensory neurone | Motor neurone |
| --- | --- |
| Long dendron carrying impulses to cell body and axon carrying impulses away | Long axon and many short dendrites |
| Carries impulses from receptor to CNS | Carries impulses from CNS to effector |

4  These cells wrap themselves many times around the axons and dendrons – many layers of plasma membranes called myelin. Function: act as an insulator and to speed up the transmission of a nerve impulse.
5  Rapid response; withdrawal of part of body about to be damaged.

## 5.2.2.3

1  Sodium/potassium; three; potassium.
2  Action potential; sodium; reverses; depolarised.
3  To ensure that a neurone responds to a certain specific level of stimulation.
4  Due to the refractory period; during this period the sodium and potassium channels cannot be opened by a stimulus so action potential can only travel forwards to where the channels can be opened.
5  Myelin acts as an electrical insulator so the action potential has to 'jump' from one node of Ranvier to the next.

## 5.2.2.4

1  Synaptic cleft.
2  Action potential arrives at axon terminal; calcium ion channels open and calcium ions diffuse into the synaptic knob; synaptic vesicles containing ACh move to presynaptic membrane; vesicles fuse with this membrane and release ACh into the synaptic cleft; Ach diffuses across the cleft and attaches to receptors on the postsynaptic membrane; this causes the sodium channels to open; sodium ions enter the postsynaptic membrane and start a new action potential.

**3** Vesicles containing ACh are only in the presynaptic side of the synapse and the protein receptors are only on the postsynaptic side.

**4** ACh cannot be formed in time – acetylcholinesterase hydrolyses ACh into choline and ethanoic acid. ATP released by the mitochondria is used to recombine choline and ethanoic acid into acetylcholine.

## 5.2.2.5

**1** CT.

**2** MRI.

**3** fMRI.

**4** Nerve; electrodes; electric; distance; time; slower.

## 5.2.2.6

**1** Short-term memory.

**2** Look at photos of family; play memory games; videos of home; family; constant reminders of day; date; events; etc.

**3** Answers will vary.

## 5.2.3.1

**1** Therapeutic; caffeine; alcohol; illegal.

**2** Habitual.

**3** Dopamine.

**4** Diamorphine.

## 5.2.3.2

**1** Multiple sclerosis; hypoglycaemia.

**2** No.

**3** Physical: when drug is taken to avoid physical discomfort or withdrawal; psychological: need for stimulation for pleasure or to escape reality.

**4** Alcoholics Anonymous.

## 5.3.1.1

**1** Set point; a change; effector; set point.

**2** Maintenance of a stable internal environment; temperature control; control of concentration of glucose.

**3** Can turn the system off so it returns to set point.

**4** Once a change is detected it is then corrected in order to keep the internal environment at a set point.

## 5.3.1.2

**1** Vasodoliation.

**2** More food is eaten in winter as more energy is needed to maintain constant body temperature. Respiration releases energy in the form of heat.

**3** Sweating stops; vasoconstriction; shivering; erector muscles contract = goose pimples.

**4** Quicker response.

## 5.3.1.3

**1** Eat more high energy foods as energy needed to maintain a constant body temperature whatever the outside temperature.

**2** Not as active, may be frugal with heating, may not eat a suitable diet.

**3** Keep indoors during mid-day hours, wear loose clothing and a hat, drink plenty of liquid.

## 5.3.2.1

**1** Exocrine: secretes through a duct; endocrine: secretes into the blood.

**2** In the pancreas.

**3** Glycogen: a polysaccharide (animal starch); glucagon: a hormone (polypeptide).

**4** By the action of insulin, glucose goes to glycogen; more glucose taken up by the liver, fat and muscle cells; increased use of glucose in respiration.

**5** Glycogen is insoluble, glucose is soluble; glycogen is inert so will not affect the water potential of the cell.

## 5.3.2.2

**1** Insulin dependent.

**2** Age; obesity; high blood pressure; family history.

**3** Insulin injections.

**4** Not an instant intake of glucose – takes time for the starch to be digested into glucose.

## 5.3.3.1

**1** Metabolism; lungs; kidneys; respiration; amine; amino acid; urea; medulla; cortex; renal artery.

**2** Build-up of toxic substances in the body: urea and carbon dioxide.

## 5.3.3.2

**1** Renal capsule; proximal convoluted tubule; loop of Henle; distal convoluted tubule.

**2** Proximal convoluted tubule.

**3** Build-up of pressure, therefore ultrafiltration increases.

**4** $180 \, dm^3$.

## 5.3.3.3

**1** Hypothalamus.

**2** True.

**3** **(a)** Increase.
   **(b)** Decrease.

**4** **(a)** Increased from 100 to 320 in first hour, then fell back until it was lower than at the start.
   **(b)** Osmoregulation.

(c) 60–320, 120–150, i.e. less than half the volume because ADH secretion has increased so wall of distal convoluted tubule and collecting duct becomes permeable to water, so water passes into the blood and therefore less in the urine.

## 5.3.3.4

1 Diabetes mellitus; bacterial infection; uncontrolled high blood pressure.
2 Sodium would build-up in the blood as when kidneys fail sodium is not removed. Build-up of sodium would alter water potential of blood plasma. Excessive salt intake leads to excessive thirst and retention of body fluid (oedema), eventually to cardiac arrest.
3 Stimulates the body to produce more red blood cells in bone marrow.
4 Look for levels of creatinine – creatinine levels increase as kidney disease progresses.

## 5.3.4.2

1 Increase rate of diffusion.
2 Stop it clotting in machine.
3 Could cause excessive bleeding in patient if enters bloodstream.
4 It would no longer exchange materials as dialysate would contain the excess toxic materials such as salts, so gradually diffusion would stop (dialysate same concentration as fluid in peritoneal cavity).

## 5.3.4.3

1 Answers will vary.
2 Answers will vary.
3 Answers will vary.

## 5.4.1.1

1 Menopause; oestrogen; ovary; oocytes; oestrogen; osteoporosis.
2 Chance of pregnancy starts to decline at about 30 years of age. From about 40 years old the decline is rapid. At 50 there is only a 0.00001 chance of pregnancy in a month.
3 Reduces the number of sperm, the motility of sperm and the percentage of normal sperm in an ejaculation.
4 Reduce hot flushes, mood swings.

## 5.4.1.2

1 Graph.
2 Cyclic.
3 Phytoestrogens; antioxidants; herbal remedies.
4 Increase in endometrial cancer, ovarian and breast cancer, increase in heart attack, blood clots.

## 5.4.2.1

1 Cataract; surgically; artificial; glaucoma; two; macular degeneration; central.
2 Used all right-handed subjects with no history of disease or history of alcohol or substance abuses.

## 5.4.2.2

1 Dementia; tangle; mental.
2 Talking to an early stage patient who is in denial; caring for a loved one who no longer recognises the carer; relating to a dying patient; financial worries.
3 Age; family history; smoking; having Down's syndrome; severe head injuries.
4 Memory loss. May forget names of people or places, words, the location of everyday objects.
5 Low birth rate, more people over 65. By 2025, people eligible to claim a pension will outnumber children by almost two million.

## 5.4.2.3

1 Cartilage; less dense; bone mass.
2 DEXA scan: 'dual energy X-ray absorptiometry'. It is a test that measures the density of bones, central and peripheral. A DEXA scan uses low energy X-rays. A machine sends X-rays from two different sources through the bone being tested. Bone blocks a certain amount of the X-rays. The denser the bone is, the fewer X-rays get through to the detector. By using two different X-ray sources rather than one it greatly improves the accuracy in measuring the bone density. The amount of X-rays that comes through the bone from each of the two X-ray sources is measured by a detector. This information is sent to a computer which calculates a score of the average density of the bone. A low score indicates that the bone is less dense than it should be, some material of the bone has been lost, and is more prone to fracture. Peripheral scans can be done as a clinic in local GP surgery so will sample a large number of people. Can predict risk of a fracture in spine or hip due to low bone density.
3 Elasticity; respiratory volume; reduces.
4 Left ventricle: may enlarge, outgrow its blood supply and therefore the cardiac muscles will not be able to contract so forcefully; mitral valve: becomes thickened and more rigid – likely to leak (heart murmur).

# Summary answers

## Unit 1 Module 1

### Q1

(a)

(b) (i) Peptide bond.

   (ii) Condensation.

(c) Secondary structure held in place by hydrogen bonds; tertiary structure involves other bonds/named bonds, e.g. ionic bonds as well.

### Q2

(a) neutrophil

(b) Answer to be supplied when final photograph seen.

(c) Four from: spread a drop of blood thinly across a slide; allow film to dry; fix with alcohol; add stain/named stain, e.g. Leishman's stain; detail, e.g. flooding slide; wash stain off.

### Q3

(a)

| Macromolecule | Example | Sub-unit molecule(s) | Chemical element(s) |
|---|---|---|---|
| Polysaccharide | Glycogen | Glucose | Carbon, hydrogen, oxygen |
| Protein | Haemoglobin | Amino acids | Carbon, hydrogen, oxygen, nitrogen, sulfur |
| Lipids | Triglycerides | Glycerol and fatty acids | Carbon, hydrogen, oxygen |

(b) Three from: insoluble; so does not affect osmosis; branched; lots of 'ends' to break glucose units off/easily broken down to release glucose; compact/can store a lot in small space.

### Q4

(a) A = nucleus; B = chloroplast; C = mitochondrion.

(b) Modifies proteins; packages them for secretion/into vesicles.

(c) Lymphocyte does not have cell wall; lymphocyte does not have chloroplast; lymphocyte does not have large vacuole.

## Unit 1 Module 2

### Q1

(a) (i) Ciliated epithelium

   (ii) A = cilia; B = basement membrane; C = lysosomes/vesicles/mucus.
   Answer to be supplied when final drawing size is known

(c) Three from: pathogens/dirt get stuck in mucus; cilia move mucus towards throat; pathogens destroyed by acid in stomach; prevents pathogens being carried into lungs.

### Q2

(a) A = alveolus; B = erythrocytes/red blood cells.

(b) Two from: stops alveoli sticking to each other; reduce surface tension; allow lungs to inflate and deflate easily.

(c) Four from: sterilise mouthpiece; fill chamber with air/oxygen; take in the deepest breath possible; force all the air out of your lungs; repeat several times and find average.

### Q3

(a) A = aorta; B = atrio-ventricular valve/tricuspid valve; C = right ventricle; D = left ventricle.

(b) Arrows showing blood coming in through vena cava, into right atrium, into right ventricle and then out through the pulmonary artery.

(c) Left ventricle has to pump blood round the body; right ventricle only pumps blood to the lungs.

(d) Three from: oxygenated and deoxygenated blood would mix; blood entering aorta would not be fully oxygenated; less oxygen for respiration; person would feel tired/exercise would be difficult.

## Unit 2 Module 1

### Q1

B, F, D, A, C, E.

### Q2

Three from: mutation of proto-oncogenes; named carcinogens e.g. uv light, ionising radiation; weakened immune system; some kinds of viruses.

### Q3

(a) (i) T = thymine/organic base; E = pentose sugar/5C sugar/deoxyribose; F = phosphate.

   (ii) Hydrogen bond.

(b) 600; One deoxyribose per base.

(c) DNA 'unzips'/hydrogen bonds between bases break; each strand of DNA becomes a template; new nucleotides line up alongside complementary bases; joined together by DNA polymerase.

## Q4

(a) (i)   Interphase/S phase/synthesis phase.
   (ii)  Mitosis/metaphase/end of prophase.
   (iii) Interphase.

(b) Interphase/S phase; this is when DNA replication occurs.

(c) Two from: mitosis stops at metaphase/metaphase can't happen; cell cannot complete mitosis; inhibits/stops cell division.

## Q5

(a) (i)   BCA.
   (ii)  A = anaphase; B = prophase; C = metaphase.

(b) (i)   Chromatid is a replicate of a chromosome, so one chromosome may consist of two chromatids; centromere holds two chromatids together, but a centriole sets up the spindle fibres.

## Q6

(a) Thermography; X-ray/mammography; MRI scan; PET scan.

(b) At this stage it is unlikely to have spread/metastasised.

(c) Six from: lumpectomy is when tumour alone is removed; mastectomy is when whole breast is removed; mastectomy more likely if tumour is advanced/has spread; chemotherapy is using drugs to kill cancer cells; tamoxifen interferes with effects of the hormone oestrogen; binds to receptors on cell membranes stopping oestrogen from binding; chemotherapy may be used before surgery to shrink tumour; may be used after surgery to kill remaining cancer cells.

## Unit 2 Module 2

### Q1

(a) Component of haemoglobin.

(b) (i)   Prevention of neural tube defects/spina bifida in fetus.
   (ii)  Any four from: to find out if mother is immune to rubella; rubella in pregnancy can harm fetus; details of harm, e.g. eye/brain/ears may be damaged; advisable to find out if you are immune to rubella before becoming pregnant; done by blood test.

### Q2

(a) Any six valid points: e.g. protein needed for growth of fetus/maternal tissues; specific example, e.g. haemoglobin; energy from complex carbohydrates, e.g. starch; folic acid to prevent neural tube defects in fetus; calcium for bone growth; iron for haemoglobin.

(b) Any five from: alcohol passes through placenta; moderate drinking leads to lower birthweight in baby; heavy drinking causes fetal alcohol syndrome; nervous system does not develop properly; reduced brain growth; child may be born with learning diufficulties; AVP, e.g. poor muscle tone, heart defects.

### Q3

(a) The ultrasound pulse is produced by the probe. The probe sends a pulse of ultrasound waves in the direction that the operator wants to investigate. The ultrasound waves bounce back from the organ, baby, tumour or other structure as a series of echoes. The echoes are used to build up a picture of your internal organs for analysis.

(b) $350 - 313 = 37$; $37/313 \times 100 = 11.8\%$

### Q4

(a) (i)   Baby born with large brain; able to learn complex skills, e.g. language, tool use.
   (ii)  Any three from: results in longer childhood/delayed puberty; human spends longer learning complex skills before becoming a parent themselves.

(b) Any three from: the absolute growth rate is the change in mass or height divided by the time period; the relative growth rate is the change in weight or mass, divided by the weight or mass at the beginning of the time period; absolute growth rate shows growth rate over a long period of time; relative growth rate shows the efficiency of growth over shorter time.

### Q5

(a) (i)   352416
   (ii)  A = nuclear envelope; B = centromere; C = bivalent; D = centriole.

(b) Any six from: homologous chromosomes pair; chromatids cross over/form chiasmata; exchange pieces of DNA; independent assortment/random segregation; of chromosomes in anaphase 1; of chromatids in anaphase 2; results in new combinations of alleles.

## Unit 2 Module 3

### Q1

(a) Any two from: infectious diseases are major killers worldwide; no effective treatments previously available; idea of antibiotics being very effective.

(b) Any five from: course not completed; antibiotics incorrectly prescribed/use of broad spectrum antibiotics/no prior selection of specific antibiotic; natural selection/have selective advantage; the antibiotic is the

selective agent; pass on resistance genes to other bacteria; idea of divide/resistant bacteria survive; can transfer resistance genes to other bacterial species; further mutations may enhance survival ability; bacteria may acquire several resistance genes/become resistant to more than one antibiotic.

(c) Any four from: natural selection/have selective advantage; the antibiotic is the selective agent; pass on resistance genes to other bacteria; idea of divide/ resistant bacteria survive; can transfer resistance genes to other bacterial species; further mutations may enhance survival ability; bacteria may acquire several resistance genes/become resistant to more than one antibiotic.

(d) Any three from: avoid prescribing antibiotics unless really necessary; wash hands with alcohol rub between patients; isolate patients who are infected with resistant bacteria; screen patients on admission to see if they carry resistant bacteria; specific example, e.g. use new disposable plastic apron for each patient.

(e) Elderly patients have weaker immune systems; may have other health problems that make them less able to cope with infection.

## Q2

(a) Any three from: used to check if TB is present or has been present/TB antibodies; no reaction if person has no past or present contact with TB; vaccine given; if TB present/has been present site will swell/redden; no vaccine given; further tests.

## Q3

(a) Any two from: response made by lymphocytes/specific immune system; B/T cells; to presence of specific antigen/pathogen/non-self cells; example of immune response, e.g. antibody production.

(b) (i) Any two from: secondary response faster than primary one; rate of production of antibodies faster in secondary response; secondary response produces more antibodies than primary one; concentration of antibodies stays higher for longer; use of comparative figures.

(ii) Any four from: vaccine provokes primary immune response; vaccine contains antigens/weakened version of pathogen/named pathogen; B/T lymphocytes form memory cells; antibodies produced; if same antigen encountered again, memory cells produce secondary response; much faster than primary response; person unlikely to feel unwell/no symptoms.

## Q4

(a) Any three from: HIV is the pathogen/virus; causes AIDS; destroys immune system; AIDS is not a specific disease/ is a syndrome; opportunistic; multiple infections/named example.

(b) Any three from: blood transfusions; semen to blood/ vaginal fluids/unprotected sex; across placenta from mother; breast milk; sharing contaminated needles.

(c) Antibiotics only effective against bacteria/HIV is a virus not a bacterium.

(d) (i) Possible to be HIV positive and have no symptoms/ dormant; UK has fluctuating population/emigration/ immigration; no routine testing for HIV; those at risk will not necessarily be tested.

(ii) Invasion of privacy/no freedom of choice; people may suffer prejudice/described; life insurance companies may refuse to insure HIV positive people; employment considerations for doctors/dentists with HIV; people may prefer not to know.

## Q5

(a) (i) Any two from: no nucleus/nuclear membrane; naked DNA; correct reference to small size; no membrane bound organelles/no enveloped organelles/no named organelle; circular DNA/ plasmid; capsule.

(ii) Any two from: droplet infection/inhaling water droplets from infected person; sneezing/coughing/ spitting/talking/water particles; drinking milk from infected cattle.

(b) Any two from: HIV weakens/compromises the immune system; attacks T helper cells; more likely to become infected when in contact with TB; greater incidence of TB so increasing the rate of subsequent infection.

(c) Any seven from:
*B lymphocytes:* humoral immunity/response; antigens bind to antigen receptors on lymphocytes; B lymphocyte activated to form clone/large number of genetically identical cells produced; most turn into plasma cells; some into memory cells; plasma cells secrete specific antibodies; antibodies attach to pathogen and lead to its destruction; plasma cells die once pathogen is destroyed; memory cells remain to give future protection from same pathogen/quicker response to subsequent infection from same pathogen.
*T lymphocytes:* cell mediated immunity/response; T lymphocyte receptors recognise pathogen antigens; T lymphocyte binds to antigen on infected cell; T lymphocyte activated and divides to form a clone; Clones enter circulation and attach to other infected

cells; T lymphocytes destroy pathogens inside cells; by releasing lytic enzymes/chemical messengers/cytokines which attract phagocytes to engulf cells; some T lymphocytes then turn into memory cells to fight future infection from same pathogen.

## Unit 2 Module 4

### Q1

(a) (i) To allow comparison/samples different sizes.

(ii) Any two from: general upward trend; peaking in 1994; downward trend (after 1994); figures to support; higher rate of increase in 0–4 than 15+; more cases/higher frequency in 0–4.
Any three from:
**Reasons for increase:** immune system of 0–4 years old more susceptible than 15+; better diagnosis; increase in pollutants/named.
**Reasons for decrease:** better medication; better understanding of the causes; hypoallergenic products/named; decrease in pollutants/government legislation.

(b) Any two from: animal hairs/dander; exercise; pollen; dust/dust mites; lightning/thunder storms; cold weather; pollution; smoking; chest infections.

(c) Any three from: mimic effects of adrenaline/noradrenaline; relax smooth muscle in trachea/bronchioles; opens up airways; reduce mucus blockage (by coughing) reduce leaking capillaries.

### Q2

(a) Any three from: chronic: long term; slow/gradual onset; any example, e.g. Alzheimer's, TB, MS. Acute: short term; sudden onset with rapid changes; any example, e.g. heart attack, flu.

(b) Any six from: tumour/cancer; specific location; persistent cough; coughing up blood; emphysema; reduced surface area; burst/damaged alveoli; loss of recoil; difficulty in breathing/barrel chest/breathlessness; (chronic) bronchitis; wheezing/coughing; inflamed lining; more mucus production; bacterial growth; COPD.

### Q3

(a) Any two from: choking/blocked airway; bronchiolitis; seizures; meningitis; asthma; drug overdose; premature babies; cardiac arrest; drowning; brain stem injury.

(b) Any two from: exhaled air contains oxygen; 16%; this oxygen used to oxygenate unconscious person's blood; increase in carbon dioxide stimulates breathing/respiratory centre.

### Q4

(a) Any two from: build up of atheroma/fatty deposits/cholesterol; in coronary artery; restricting blood flow; to heart muscle.

(b) (i) For comparison/countries are different sizes.

(ii) Any two from: smoking; lack of exercise; diabetes; genetic inheritance; alcohol.

### Q5

(a) Any two from: test strip placed in meter; contains glucose oxidase; sterile lancet/drop of blood/swab with alcohol; gluconolactone produced; produces electric current; picked up by electrode on test strip.

(b) Any two from: high glucose in blood lowers blood water potential; water enters blood by osmosis; need to drink to provide this water.

(c) Any three from: avoid sugar; eat polysaccharides/starch; broken down more slowly; high fibre in diet to slow absorption rate; exercise to increase respiration rate.

## Unit 4 Module 1

### Q1

Active transport; protein synthesis; phosphorylation of glucose (in glycolysis).

### Q2

Phosphorylation; adenine; ribose; phosphate; phosphate; chloroplast; mitochondrion; glycolysis.

### Q3

Energy transfer molecules, a source of high-energy electrons used to synthesise ATP by oxidative phosphorylation.

### Q4

Hexose; 6; pyruvate; ATP; reduced NAD; pyruvate; oxygen; acetyl coenzyme A; decarboxylation; acetyl coenzyme A; oxaloacetate; carbon dioxide; reduced NAD; reduced FAD; ATP.

### Q5

Pyruvate acts as a hydrogen acceptor and picks up the hydrogen from reduced NAD. The NAD is regenerated and pyruvate is reduced to lactate.

### Q6

Hydrogen atoms are passed from reduced NAD and reduced FAD to hydrogen acceptors in the inner mitochondrial membrane. The hydrogen is split into protons and electrons. The electrons are transferred from carrier to

carrier in the inner membrane of the mitochondria. The energy released is used to pump protons into the intermembranous space. These are then released through channels in the ATP synthase stalked particles and the energy released is captured as ATP. The final acceptor for the hydrogen ions and the electrons is molecular oxygen and water is formed. Your diagram should show how these steps relate to the mitochondrion with the structures mentioned above clearly labelled.

## Q7

| Substrate | Where this is broken down in the respiratory pathway and what is its energy contribution? |
|---|---|
| Fatty acids | Via acetyl coenzyme A in the mitochondrial matrix. Energy value is approximately 37 kJ g$^{-1}$ |
| Amino acids | First deaminated and then enter glycolysis. Energy value is 17 kJ g$^{-1}$ |

## Q8

Ratio of carbon dioxide produced to oxygen used.

## Q9

Dehydrogenation uses dehydrogenase enzymes to remove an atom of hydrogen. Decarboxylation is carried out by decarboxylase enzymes and results in the production of carbon dioxide.

## Q10

Aerobic respiration is more efficient because more of the energy from glucose is captured as ATP. This is because in aerobic respiration, oxygen is available to act as the terminal hydrogen and electron acceptor and so ATP can be made by oxidative phosphorylation from the reduced coenzymes produced in the Link reaction and the Krebs cycle.

## Q11

Carbo loading is a short way of referring to carbohydrate loading and means increasing the glycogen store in muscle cells.

## Q12

Polypeptides; nucleotides; polymerase; nucleotides; phosphate; adenine; guanine; thymine; cytosine.

## Q13

In **transcription**, RNA nucleotides pair up with the sense strand of a section of DNA coding for a polypeptide and a molecule of mRNA is synthesised. In **translation**, the mRNA leaves the nucleus and binds to a ribosome. Transfer RNA

molecules bring their specific amino acids to the ribosome and the amino acids are joined in the order of the codons on the mRNA.

## Q14

mRNA has a different **function** to tRNA – mRNA carries a complementary copy of the genetic code for a polypeptide to the ribosome while tRNA molecules bind to specific amino acids and transport them to the ribosome. mRNA also has a different **structure** as it is single-stranded where parts of the tRNA molecule are double-stranded and it is shaped like a 'hairpin'.

## Q15

Short-term: heart rate, stroke volume and cardiac output rise; ventilation rate increases as tidal volume and breathing rate increases; blood pressure increases. Long-term: resting heart rate falls as cardiac muscle increases in size; resting blood pressure is lower; muscle cells adapt with increased numbers of mitochondria, bigger myoglobin stores and glycogen stores.

## Q16

Protein; quaternary; 4; high; muscle; carbon dioxide; pH.

## Q17

The shift to the right of the oxygen dissociation curve when levels of carbon dioxide are high.

## Q18

ATP is broken down by the myosin head – an ATPase. This releases the myosin from the actin, allowing the 'power stroke' to be repeated and the myofilaments to slide over each other to produce muscle contraction.

## Q19

| Muscle component | Description at rest | Description during contraction |
|---|---|---|
| z lines | Maximum distance apart | Move closer together |
| I band | Paler band | Smaller |
| A band | H band seen | A band stays same size, H band disappears |

## Unit 4 Module 2

## Q1

Meiosis is a reductive division; the number of chromosomes is reduced from diploid to haploid to prevent a doubling of chromosome number at fertilisation.

## Q2

Both contain haploid number of chromosomes and are a product of meiosis. In males, all four products of meiosis become gametes and the gametes are adapted to be motile. In females, only one of the products of meiosis becomes a gamete. The oocyte is not motile and has a large cytoplasm content.

## Q3

A is the acrosome, a modified lysosome, which houses the enzymes necessary to achieve penetration by the male nucleus at fertilisation.

## Q4

Disadvantage: Norplant had side effects such as uterine bleeding (it has been largely replaced by Implanon). Main advantages of any implant method includes not having to remember to take any contraception precautions and it maintains a more constant blood level.

## Q5

Spermatogenesis – sperm mother cell divides by MITOSIS. One cell becomes the primary spermatocyte. Meiosis occurs, forming firstly two haploid secondary spermatocytes and subsequently four haploid spermatids. The spermatids embed into sertoli cells and each one becomes a mature sperm cell. The process is continuous. In oogenesis, meiosis is not continuous. It starts before birth with the formation of primary follicles. Prior to ovulation, the primary oocyte undergoes the first meiotic division but only one cell progresses. The other set of chromosomes forms the polar body. The second stage of meiosis does not occur until fertilisation and again a polar body is formed. In oogenesis, only one product of meiosis becomes a secondary oocyte.

## Q6

LH triggers secretion of oestrogen. A sudden rise in LH triggers ovulation. FSH initiates development of the primary follicle. It also stimulates production of oestrogen. Oestrogen promotes repair of endometrium, acts as feedback inhibitor of FSH and, from about day 11, acts as positive feedback on LH and FSH causing the surge which triggers ovulation.

## Q7

For LH and FSH in females, see 6. In males, LH and FSH also act as gonadotrophins. LH stimulates the production of testosterone and FSH and testosterone control spermatogenesis.

## Q8

Mitosis; blastocyst; embryo; human chorionic gonadotrophin; FSH and LH; oxytocin; prolactin.

## Q9

Surgery to unblock the Fallopian tube or IVF; surgery (if a blockage is suspected) or IVF; IVF using selected sperm which may be injected into the oocyte; intrauterine insemination – inject semen into the uterus.

## Q10

See above – this means sperm will not be destroyed in the vagina and have a greater chance of reaching the Fallopian tube to fertilise the secondary oocyte.

## Q11

hCG produced by the blastocyst (the 'ball of cells') maintains the endometrium. Protease enzymes from the trophoblast digest part of the endometrium and the blastocyst inserts into the endometrium.

## Q12

Multiple pregnancy may result from multiple ovulation and fertilisation OR from a single fertilisation event after which there is a cleavage ('splitting') to form identical twins.

## Q13

GnRH stimulates the release of LH and FSH. This can, in turn, trigger ovulation.

## Q14

Not likely to result in a live birth so not 'cost effective' (see NICE guidelines).

## Q15

Monoclonal antibodies are specific to hCG. They are attached to the urine testing kit. If hCG is present, it binds to the variable region of the antibody triggering a chain of events which results in a colour change. Their absolute specificity and the ability to respond to even low levels of hCG makes them sensitive enough to detect the 'pregnancy' very soon after fertilisation.

## Q16

Low birth rates can result from an increase in income in a population or from education or other social considerations. A low birth rate could be due to biological factors such as a fall in viable sperm count in males. It can be as a result of policy decisions. Death rates can increase as a result of disease – this could be infectious or non-infectious such as a rise in CHD due to obesity. You should group your ideas into social, economic and biological reasons.

## Q17

See 16. As societies become more advanced, population size tends to stabilise as families tend to have fewer children – for example, by leaving child-bearing until later in life. More children survive and people may live longer, but fewer children are actually born. The answer could be different if 'population' is considered in global terms – are advances in technology available to all countries?

## Q18

Ecosystem: all the living organisms and the non-living components of the area where they live and the way all these elements interact. Biodiversity: the number of different types of organisms present in an area. Deforestation: the removal of woodland from an area in such a way that it does not regenerate. Habitat: the place where an organism lives.

## Q19

Addition of extra carbon dioxide to the atmosphere by use of fossil fuels PLUS removal of forest on a scale which cuts the global rate of photosynthesis OR changing marine conditions such that marine photosynthetic rates are reduced – i.e. less carbon dioxide is being removed. You might be asked for examples of activities which result in either of the above.

## Q20

Aesthetic reasons: pleasant to look at/walk in/holiday in. Medicinal reasons: organisms as a source of potential medicines. Agricultural reasons: could be sources of genetic 'traits' which could be introduced into domestic species to improve production or resistance to disease.

## Q21

Light is captured by chlorophyll. Energy is used to synthesise ATP by photophosphorylation. Reduced NADP is also produced and these two products from the light-dependent reaction are used in the light-independent reaction to convert GP to TP.

## Q22

Soil; nitrogen fixing; amino acids; nitrosomonas; nitrobacter.

## Q23

Succession: a directional change in a community where the changing community alters the environment, resulting in further changes to the community. Deflected succession: intervention in the normal sequence of series, by grazing or mowing for example, to produce a plagioclimax. Ecosystem:

all the living organisms of all species and all the non-living or abiotic components of their environment which interact in a defined area. Carbon footprint: the mass of carbon dioxide released into the atmosphere by an individual, a community or an activity.

## Unit 5 Module 1

## Q1

Gene; genotype; the same; homozygous.

## Q2

DNA sequence; number; structure.

## Q3

Triplet repeat; dominant; recessive.

## Q4

Co-dominance.

## Q5

Group A.

## Q6

| Parents' phenotype | group O | group AB |
|---|---|---|
| Parents' genotype | $I^O I^O$ | $I^A I^B$ |

| | $I^O$ | $I^O$ |
|---|---|---|
| $I^A$ | $I^A I^O$ | $I^A I^O$ |
| $I^B$ | $I^B I^O$ | $I^B I^O$ |

50% chance of being blood group A and 50% of being blood group B.

## Q7

Gene causing that particular disease is on one of the sex chromosomes.

## Q8

Pedigree chart.

## Q9

Linked; crossover frequency.

## Q10

21.

## Q11

Karyotype.

## Q12

H – normal blood allele; h – haemophilia allele
Queen Victoria – $X^H X^h$; Prince Albert – $X^H Y$; 50%.

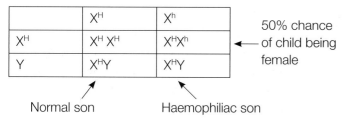

|   | $X^H$ | $X^h$ |
|---|---|---|
| $X^H$ | $X^H X^H$ | $X^H X^h$ |
| Y | $X^H Y$ | $X^H Y$ |

50% chance of child being female

Normal son     Haemophiliac son

## Q13

Extra chromosome 21 (trisomy 21).

## Q14

Sex linked – probably on non-homologous part of the Y chromosome.

## Q15

Checking the antigens in cell membranes; checking the match of the haplotypes; using cadavers – acceptance of definition of death; should transplants be carried out except as matter of life and death? (face transplants); sale of organs; removal of organs from unwilling donors (e. g. prisoners).

# Unit 5 Module 2

## Q1

Motor; sensory; sodium; pump; resting; sodium; potassium.

## Q2

Presynaptic synapse membrane becomes more permeable to calcium ions; inflow of calcium ions; synaptic vesicles move and then fuse with the surface membrane of the presynaptic knob; transmitter substance released; exocytosis; ACh diffuses across the cleft; binds with a receptor on postsynaptic membrane; causes a change in permeability of the post synaptic membrane and an action potential in postsynaptic neurone.

## Q3

To ensure a neurone responds to a specific minimum level of stimulation.

## Q4

Conducts impulses more quickly.

## Q5

(i)   Controls coordination of movement.
(ii)  Controls breathing, heart rate and blood pressure.
(iii) Controls secretions of the pituitary gland.

## Q6

Blood–brain barrier protects the brain from the many chemicals flowing within the blood; when functioning normally, the blood–brain barrier prevents foreign invaders (such as viruses) and immune cells from passing out of the bloodstream and into the central nervous system; it also allows oxygen and essential nutrients to enter the central nervous system; in multiple sclerosis, immune cells are allowed to enter the central nervous system, implying that the blood–brain barrier is damaged or compromised in some way. These immune cells, called T lymphocytes, attack the myelin in the brain and spinal cord, causing the lesions which lead to MS symptoms.

## Q7

Receptor in the retina which conveys information to the optic nerve; the impulse travels along sensory neurones to the brain (CNS); impulse crosses synapse to the inter neurone and then to the motor neurone to the effector - the constrictor muscle of the iris.

## Q8

Sclera; refracted; cone; rhodopsin; 11 trans- retinal; sodium; bipolar; bipolar; generator.

## Q9

Its victim/food would stop breathing and therefore moving; it possibly acts by stopping the secretion of neurotransmitters so no impulse passes to the muscles of respiratory system (diaphragm and intercostals) from the brain.

## Q10

(i)  No photoreceptor cells at Y; no rods or cones at Y.
(ii) X has many more/only cones than Z which each synapse to a bipolar cell.

## Q12

D; B; C

# Unit 5 Module 3

## Q1

Maintenance of a constant internal environment.

## Q2

Receptor; effector; constrict; vasoconstriction; dilate; vasodilation; radiation; convection; evaporates; thyroxine.

## Q3

Hypothermia means having a body temperature which is below 35°C. Hyperthermia means having a body temperature significantly higher than 37°C.

## Q4
Endocrine; pancreas; glucagons; receptors; beta cells; glycogen; increases; two; one; middle age.

## Q5
BMI over 27; a sedentary lifestyle; a high waist to hip ratio (apple shaped).

## Q6
Excretion is the removal of waste products from cell metabolism, such as urea or carbon dioxide. Elimination is, for example, the egestion of faeces.

## Q7
Amino; nephrons; selective reabsorbtion; proximal convoluted tubule; loop of Henle; distal convoluted tubule; renal capsule; wider; raises.

## Q8
Proximal convoluted tubule.

## Q9
Osmosis.

## Q10
Antidiuretic hormone; posterior pituitary; osmoreceptors; hypothalamus; blood plasma; collecting duct; more; water.

## Q11
Erythropoietin; calcitriol (and the enzyme renin).

## Q12
Peritoneal dialysis: good – allows patient to walk freely; bad – greater risk of infection.
Haemodialysis: good – more efficient at removing waste and excess solutes; bad – takes hours on a dialysis machine.
Kidney transplant: good – if successful the person can lead a normal life; bad – may have to take immunosuppressant drugs.

## Q13
(a) At osmoreceptors in the hypothalamus of the brain.
(b) In blood plasma.
(c) Because the response cancels out the stimulus. The water potential will be fluctuating around a set point as a rise or fall is detected and the appropriate response corrects this.
(d) The volume of urine would increase.

## Q14
Nervous control is rapid and generally produces short-term responses. The impulse results in a response from a limited range of effectors, sometimes only one. Hormonal control can produce more long-term responses and several effectors may be involved. Think of the target organs for insulin or adrenaline.

## Q15
A drop in temperature is detected by the hypothalamus which releases a thyrotrophin-releasing hormone. This in turn stimulates the release of thyroid stimulating hormone from the anterior pituitary gland. This in turn stimulates the release of thyroxine from the thyroid gland.

## Q16
(a) Proteins are too large to cross the basement membrane of the capillaries in the glomerulus.
(b) Because the glucose in the filtrate is reabsorbed in the proximal convoluted tubule.
(c) Because the absorption of water from the filtrate in the proximal convoluted tubule, loop of Henle and collecting ducts concentrates the urine, giving an increased concentration of urea.

## Unit 5 Module 4
## Q1
Stopping; ovaries; hot flushes; HRT; osteoporosis; testosterone.

## Q2
Phytoestrogens; antioxidants; herbal medicines.

## Q3
Slow/decrease; myelin; thinner; clouding; lens; lens; artificial; macula/central area; retina; central.

## Q4
Decrease in brain weight and volume; cortex thins; white matter decrease; enlargement of ventricular system.

## Q5
Early: loss of memory.
Mid: mental abilities decline, personality changes, more dependent on carers.
Late: complete deterioration of personality and loss of control of body functions, possibly paranoia and delusions.
Problems faced by carers: escalating demands, home safety, physically and mentally demanding, emotionally demanding as patient may not recognise family member who is doing the caring.

## Q6
Osteoporosis; bone density; fractures; D; calcium; exercise; osteoarthritis; cartilage; hips and knees.

## Q7

Vital capacity.

## Q8

Cardiac muscle loses elasticity; heart valves become thickened and rigid; left ventricle may enlarge; number of pacemaker cells decrease so sino-atrial node beats more slowly and heart is less able to alter its rate.

## Q9

Less people of working age so shortage in work force; less people earning and contributing to pension fund but more people drawing on fund; elderly population may be greater drain on health services.

## Q10

Graph; trend: increasing percentage of fractures with increasing age, very slow increase until 50–59 age group (1.1%), next age group to 69 increases by 1.5% over the 1.1% and then there is a steep increase – 7.4% increase in the next 10 years; increase in osteoporosis, decrease in cardiovascular function which could lead to dizziness and falls, decreasing vision (cataracts, glaucoma and macular degeneration) so obstacles such as rugs and uneven pavements not seen.

## Q11

Advantages: may decrease risk of osteoporosis, may decrease risk of heart disease.
Disadvantages: older cells less likely to divide accurately, risk of miscarriage higher, risk of foetal abnormality e.g. Down's syndrome, may increase risk of breast cancer, if baby born then physical demand on older mother.

## Q12

(a) Further from capillaries so less oxygen; centre of lens is further away from the surface so greater distance for removal of toxic waste; may be more protein in the centre and it is the protein that changes to make the lens cloud.

(b) Daycase surgery: pupils dilated using eyedrops, local anaesthetic put in eye, small incision in surface of eye, lens liquefied by electric current, high frequency sound waves (phacoemulsification) are sometimes used to break up the lens before extraction, lens removed but the majority of the lens capsule is left intact, artificial lens inserted.

(c) Artificial lens cannot accommodate.

## Q13

100 neurones at 50; for every 10 years 5% lost i.e. five neurones, therefore by age 70 10 neurones in every 100 will be lost.

## Q14

Lens will have difficulty in accommodating so focusing on the pages of a book will be difficult; lens does not become sufficiently convex/rays of light refracted/bent less; focusing on near objects requires maximum refraction; rays of light focused beyond retina.

## Q15

(a) Genetic/family history; severe head injury; smoking; lack of mental stimulation.

(b) Slower nerve impulse conduction; reduction in hearing and vision; glaucoma; macula degeneration; cataract; hair cells in cochlea degenerate; ear drum loses its elasticity.

# Examination answers

## Unit 1

### Q1

(a) Large nucleus; stained purple; small amount of cytoplasm. [3]

(b) A – golgi apparatus; C – smooth endoplasmic reticulum; D – mitochondrion; E – lysosome; F – ribosomes; H – nuclear envelope. [6]

(c) (i) Large quantity of RER (candidates must make it clear that it is the quantity and not just the presence of RER), to synthesise proteins and direct them to the Golgi apparatus. Large numbers of mitochondria to provide the energy (ATP) from aerobic respiration. [3]
(ii) Golgi apparatus.

(d) Glycoproteins packed into vesicles, vesicles transported to cell surface membrane and fuse with cell surface membrane. Contents released. This is exocytosis.

### Q2

(a) Globular; four; prosthetic; haem; peptide bonds; bone marrow. [6]

(b) Soluble FIBRINOGEN in the blood plasma is converted to FIBRIN by the enzyme THROMBIN. Thrombin is produced from PROTHROMBIN in the presence of (CALCIUM IONS, THROMBOPLASTIN, CLOTTING FACTORS). Thromboplastin is released in response to DAMAGE. [4]

(c) ENZYME – THROMBIN; SUBSTRATE – FIBRINOGEN; PRODUCT – FIBRIN (There are other possibilities within the clotting cascade). [3]

## Q3

(a) (i) Sphygmomanometer.
(ii) Artery.
(iii) Arrow towards the hand. [3]
(b) (i) No blood flow, pressure in cuff is higher than systolic pressure. [2]
(ii) Blood flow is intermittent, only flowing when the blood pressure is high, during systole, a figure quote from the graph (*y*-axis – giving units). [2]
(iii) Longer period of blood flow; lumen open for longer. [2]
(iv) 120/80 mmHg. [2]

## Q4

(a) Sterile mouthpiece, medical grade oxygen, other answers are possible. [2]
(b) 30, breaths per minute. [2]
(c) Air stays in **bronchi/bronchioles/trachea**, this is **dead space**, only 25% of air reaches the alveoli, where **gas exchange** takes place. Four marks for content and one for using the terms underlined. [5]

## Unit 2

## Q1

(a) S, G2, mitosis. [3]
(b) Programmed cell death. [1]
(c) (i) A mass of cells, which are undifferentiated. [2]
(ii) A benign tumour does not spread whereas a malignant tumour undergoes metastasis and secondary tumours can form in other parts of the body. [3]
(d) (i) UV radiation; tobacco smoke – there are other causes. [2]
(ii) DNA damage could be to proto-oncogenes which are mutated to become oncogenes. The cell carries on through the cycle, no apoptosis occurs and cell division by mitosis becomes out of control. [4]

## Q2

(a) (i) The distance between the two sides of the head. [1]
(ii) Measure the SCALE BAR in cm so you have a 'conversion' factor – a magnification. THEN measure a crown–rump length on the picture and multiply by this factor. Remember to write down each step as you go along. [3]

(b) Energy requirement increases to support growth of the FOETUS, UTERUS and PLACENTA. BUT increase is only needed in the final trimester (three months) as activity also slows down. Protein increases to support growth (of the fetus, uterus and placenta). Calcium – DOES NOT RISE IN PREGNANCY will need to rise during LACTATION to provide calcium for bone formation IN THE BABY. Iron does not rise as MENSTRUATION stops. Folic acid supplements are given to prevent SPINA BIFIDA/NEURAL TUBE DEFECTS. (You will notice the highlighted words. In some questions, a mark is awarded for the correct use of technical terms. If you have used SOME of these terms (usually four or five) you receive an additional mark. The question paper will make it clear when this will happen. [6 + 1]
(c) Fetal Alcohol Syndrome. You should be able to give at least three further features of this syndrome to gain the marks. These could include low birth weight, small head, flat face with snub nose, smooth area between nose and mouth. Thin upper lip, small, widely spaced eyes, learning difficulties, hyperactivity. [4]

## Q3

(a) Population stayed constant at 3.8 million bacteria – the second mark is for quoting the data accurately from the graph.
(b) The QWC mark is for making sure you really compare the two cells. You should be using words such as 'whereas' or 'on the other hand'.
The neutrophil has a DNA in linear chromosomes within a nucleus whereas the bacterial cell has no nucleus, the DNA is circular and it is not attached to proteins as a chromosome. Bacterial cells also have additional DNA as plasmids. Neutrophils have other membrane bound organelles such as mitochondria but bacteria have no membrane bound organelles. Bacteria also have smaller ribosomes. They have a cell wall and cell surface membrane unlike the neutrophil which just has a cell surface membrane. (There are other differences). [7]
(c) They stop the bacteria reproducing and allow time for the immune system to respond, for example, through phagocytic cells engulfing the bacteria or by the production of antibodies.

## Q4

(a) A long term study – from 1948 until now – using large numbers of people (quote the figures given).
(b) (i) Smoking, lack of exercise, obesity, high levels of salt in the diet – there are more possible answers.

**(ii)** CHD has a genetic component for example genes could lead to high blood cholesterol – it is a multifactorial disease.

**(c)** Description of catheter and detail of where it is inserted could be given. The balloon is inside a STENT. The balloon is inflated when the stent is in position in the coronary artery. The balloon is then withdrawn leaving the lumen wider. Time in hospital is shorter and recovery time is faster – you could give figures. This keeps cost down and also people can return to work or normal life faster. Bypass surgery requires much longer periods of time in hospital and recovering.

## Unit 4

### Q1

**(a) (i)** X = crista(ae); Y = matrix. [2]

**(ii)** Measure the line in mm. Multiply by 1000 to convert to micrometres. Divide by the magnification. Give the answer to the nearest micrometer. [2]

**(b)** Carbon dioxide; ATP; Reduced NAD/FAD (reduced co-enzymes). [3]

**(c)** Electron carriers within the inner membrane; outer membrane impermeable to protons; intermembranous space allows build up of protons/H ions; stalked particles on inner membrane; channel for hydrogen ions to flow back into matrix; and site for ATP synthesis. Folded inner membrane increases surface area for ECT and ATP synthase. [5]

**(d)** More folds/cristae; more (Krebs cycle) enzymes; more stalked particles. [2]

### Q2

**(a) (i)** Where more than one polypeptide chain joins to make the functional protein, for example, the four globin chains.

**(ii)** Where the protein is attached to a non-protein group, for example, the haem group in haemoglobin.

**(b)** The gene is transcribed, a complementary mRNA copy of the gene is synthesised. The mRNA then leaves the nucleus and binds to a ribosome. Transfer RNA molecules bring amino acids to the ribosome. Each transfer RNA is specific to one amino acid. The anticodons on the tRNA position the correct amino acids in the correct sequence by complementary base pairing. Each three bases (on DNA/RNA) codes for one amino acid.

**(c) (i)** Curve should be same shape but to the LEFT of the normal adult curve.

**(ii)** Oxyhaemoglobin does not release oxygen to muscle cells from fetal haemoglobin as easily/until much lower partial pressures of oxygen, which could lead to more anaerobic respiration and a build up of lactic acid in muscles, leading to cramp/fatigue.

### Q3

**(a)**

| One | Four |
|---|---|
| Two | None |
| Follicle/germinal epithelium of ovary | Germinal epithelium of testis |
| Follicle cell | Sertoli cell |
| Puberty until menopause | Puberty until death |

[5]

**(b)** Achieved by meiosis; independent orientation followed by independent assortment at metaphase and anaphase during first and second divisions. This reassorts the chromosomes. Crossing over can also happen/exchange of genetic material between homologous chromosomes. [4]

**(c)** The glycoprotein and plasma proteins are removed from the outer surface of the sperm 'head'. The sperm head is now more sensitive to signals from the oocyte and the sperm swims more strongly. This is achieved by proteases.

**(d) (i)** UUACGGUUCGGA

**(ii)** By hydrogen bonds, two between U and A and three between C and G. [3]

**(iii)** As a contraceptive. [1]

### Q4

**(a)** TP – triose phosphate. [1]

**(b)** ATP and REDUCED NADP. [2]

**(c)** Decomposers or putrifying bacteria convert the nitrogen compounds in dead organic matter to ammonium. *Nitrifying* bacteria convert this to nitrate. Nitrosomonas converts ammonium ions to nitrite and *Nitrobacter* converts nitrite to nitrate. [4]

**(d)** *Look to make a balanced argument. Advantages* – Increased quantities of grain produced without using more land. This provides food for increasing populations. Less waste material/straw to dispose of. *Disadvantages* – Large areas of land lead to hedgerow removal, high nitrate levels in the soil prevent some plant species competing – *both of these* result in loss of biodiversity. Short plants require more chemicals such as herbicides to out-compete weeds. [4]

# Unit 5

## Q1

**(a)** A change in a DNA triplet, such as a base substitution, could lead to a different amino acid or primary structure. The chain could then have a different tertiary structure. The shape of the active site would no longer complement the phenylalanine so no enzyme substrate complex forms and no tyrosine is produced.

**(b)** Choose symbols correctly for ALLELES
(e.g. P = normal; p = PKU).
Parents both Pp with gametes P and p,
Offspring genotypes PP, Pp, Pp and pp
Correctly match pp genotype with PKU phenotype.

**(c)** Heterozygotes more likely to survive and some of their children would inherit this allele. The toxins provide a selection pressure, heterozygotes have a selective advantage.

## Q2

**(a)** (Nucleotide) a pentose sugar, joined to an inorganic phosphate group, and a nitrogenous base. (exon) the coding regions of the gene, which will appear in the final mRNA transcript, (intron) non coding regions, which do not appear in final mRNA transcript. [6]

**(b)** Minisatellites.

**(c)** PCR amplifies (sections of ) the DNA or makes multiple copies. It uses primers which bind to complementary regions or anneal, when the DNA is heated and then cooled. A DNA polymerase (TAQ polymerase) and activated nucleotides are then added and the strands extend from the primers. A thermal cycler repeats this over and over again. *Examiners would be looking for terms such as primer, TAQ polymerase, annealing, thermal cycler.*

**(d)** ADVANTAGES – more effective targeting of drugs, as response can vary with genotype; advance planning of medical treatments, such as hypercholesterolaemia; used as part of identity profile, in criminal cases/missing persons.
DISADVANTAGE – too expensive; NHS resources limited/could deflect money from other areas/link between genotype and health still being establishes. Ethical issues, could be used to discriminate/when would you inform the children of a disease risk/how secure would the data be.

## Q3

**(a)** Rhodopsin.

**(b)** CONES HAVE iodopsin rather than rhodopsin. Pigment is in a continuous, folded surface membrane and not in membrane discs.

**(c)** Mitochondria, produce ATP by aerobic respiration. ATP is needed to reform rhodopsin.

**(d)** Light absorbed by rhodopsin converts rhodopsin into opsin and retinal. Retinal membrane protein changes shape and membrane becomes impermeable to sodium ions. The membrane becomes hyperpolarised and inhibitory neurotransmitter is no longer released. The bi-polar neurone depolarises triggering an action potential. This triggers neurotransmitter release into synapse and an action potential in the ganglion cell and on through sensory neurones in optic nerve to the brain.

**(e)** Cover one eye, read chart from top row down. Subject seated 6 metres/20 ft from chart. If they can read to the bottom row they have 20/20 vision.

## Q4

**(a)** Symptoms include, increasing memory loss; particularly short-term memory; difficulty in concentrating; anxiety; personality change; deterioration of speech; loss of muscle control; incontinence. [3]

**(b)** Formation of beta amyloid plaques between neurones and tangles of fibres within neurones made from tau proteins. [4]

**(c)** A reliable test for Alzheimer's, that was common to all countries being included, a reliable way of counting the population in each country/recording the age of the population. A model of population growth based on the age profile and current numbers. [3]

**(d)** China has large numbers of middle aged people NOW *(remember – this is a 40 year projection so answers involving birth rate are not correct!)*, better standard of living/more affluent so will probably live longer, implications for health care – it will cost more in the future to provide care/no national health care provision/heavy burden on state and/or families – *The command word is 'comment' so you could explain or consider a consequence.* [2]